AF566946

ANNALS OF THE NEW YORK ACADEMY OF SCIENCES

Volume 458

The New York Academy of Sciences
2 East 63rd Street
New York, New York 10021

CONGENITAL ADRENAL HYPERPLASIA

ANNALS OF THE NEW YORK ACADEMY OF SCIENCES
Volume 458

CONGENITAL ADRENAL HYPERPLASIA

Edited by Maria I. New

The New York Academy of Sciences
New York, New York
1985

Cover: *The cover shows the pathogenesis of virilization and salt-wasting in 21-hydroxylase deficiency (see page 166).*

Library of Congress Cataloging-in-Publication Data

Main entry under title:

Congenital adrenal hyperplasia.

(Annals of the New York Academy of Sciences, ISSN 0077-8923; v. 458) (Nov. 1985)
Papers presented at the Symposium on Advances in Congenital Adrenal Hyperplasia, held on July 7–9, 1984, in New York City, as part of the satellite meeting of the VIIth International Congress of Endocrinology, sponsored by the National Institute of Child Health and Human Development, the National Institutes of Health, and the Cornell University Medical College.
Includes bibliographies and index.
1. Adrenocortical syndrome—Congresses. 2. Adrenocortical syndrome—Immunological aspects—Congresses. 3. HLA histocompatibility antigens—Diagnostic use—Congresses. 4. Prenatal diagnosis—Congresses. I. New, Maria I., 1928– . II. Symposium on Advances in Congenital Adrenal Hyperplasia (1984: New York, N.Y.). III. International Congress of Endocrinology (7th: 1984: Quebéc, Quebéc). IV. National Institute of Child Health and Human Development (U.S.). V. Cornell University. Medical College. VI. Series.
[DNLM: 1. Adrenal Hyperplasia, Congenital—congresses. W1 AN626YL v.458 / WK 700 C749 1984]
Q11.N5 Vol. 458 [RJ420.A3] 500 s 85-24663
ISBN 0-89766-311-x [618.92′45]
ISBN 0-89766-312-8 (pbk.)

SP
Printed in the United States of America
ISBN 0-89766-311-x (cloth)
ISBN 0-89766-312-8 (paper)

ANNALS OF THE NEW YORK ACADEMY OF SCIENCES

Volume 458

November 30, 1985

CONGENITAL ADRENAL HYPERPLASIA[a]

Editor

MARIA I. NEW

Organizing Committee

M. I. NEW, B. DUPONT, G. P. CHROUSOS, and D. LORIAUX

CONTENTS

Introduction. *By* ALFRED M. BONGIOVANNI ix

Clinical and Endocrinological Aspects of 21-Hydroxylase Deficiency. *By* MARIA I. NEW 1

Extended MHC Haplotypes in Salt-Losing 21-Hydroxylase Deficiency. *By* CHESTER A. ALPER, ELLEN FLEISCHNICK, ZUHEIR AWDEH, DONALD RAUM, JOHN F. CRIGLER, JR., PARK S. GERALD, and EDMOND J. YUNIS 28

The HLA Associations in Congenital Adrenal Hyperplasia due to 21-Hydroxylase Deficiency in a Yugoslav Population. *By* A. KAŠTELAN, LJ. BRKLJAČIĆ-ŠURKALOVIĆ, and M. DUMIĆ 36

HLA Associations in 21-Hydroxylase Deficiency (Congenital and Late-Onset Adrenal Hyperplasia) in France. *By* P. COUILLIN, R. RAPPAPORT, F. KUTTENN, J. HORS, J. FEINGOLD, J. BOUÉ, and A. BOUÉ 41

The Association Between Congenital Adrenal Hyperplasia and HLA in Southern Italy. *By* SERAFINO ZAPPACOSTA, MICHELE MAIO, MARIO DE FELICE, and ROSSELLA VALENTINO 46

HLA Associations in Late-Onset 21-Hydroxylase Deficiency in Israel. *By* Z. LARON, A. ROITMAN, A. PERTZELAN, H. KAUFMAN, and R. ZAMIR 52

HLA-B14 and Nonclassical 21-Hydroxylase Deficiency in a Heterogeneous New York Population. *By* LENORE S. LEVINE 65

Different Gene Defects in the Salt-Wasting (SW), Simple Virilizing (SV), and Nonclassical (NC) Types of Congenital Adrenal Hyperplasia (CAH). *By* D. KNORR, E. D. ALBERT, F. BIDLINGMAIER, W. HÖLLER, and S. SCHOLZ.. 71

[a]The papers in this volume were presented at the Symposium on Advances in Congenital Adrenal Hyperplasia, which was held on July 7–9, 1984 in New York City as part of the Satellite Meeting of the VIIth International Congress of Endocrinology. The symposium was sponsored by the National Institute of Child Health and Human Development, the National Institutes of Health (Bethesda, Maryland 20205), and the Cornell University Medical College (New York, New York 10021). Additional support was provided by The Lindemann Foundation, Hoffmann-LaRoche Inc., Ross Laboratories, and Wyeth Laboratories.

The Coexistence of IgA Deficiency and 21-Hydroxylase Deficiency Marked by Specific MHC Supratypes. *By* T. J. COBAIN, M. S. STUCKEY, J. MCCLUSKEY, A. N. WILTON, A. GEDEON, M. J. GARLEPP, F. T. CHRISTIANSEN, and R. L. DAWKINS 76

Neonatal Screening Program for Congenital Adrenal Hyperplasia in a Homogeneous Caucasian Population. *By* E. CACCIARI, A. BALSAMO, A. CASSIO, S. PIAZZI, F. BERNARDI, S. SALARDI, A. CICOGNANI, P. PIRAZZOLI, F. ZAPPULLA, M. CAPELLI, M. PAOLINI, and C. I. CORDARO 85

Newborn Screening for Congenital Adrenal Hyperplasia with Special Reference to Screening in Alaska. *By* SONGYA PANG, DAVID A. SPENCE, and MARIA I. NEW 90

Neonatal Screening for Congenital Adrenal Hyperplasia in Japan. *By* H. NARUSE, E. SUZUKI, M. IRIE, A. TSUJI, N. TAKASUGI, M. FUKUSHI, N. MATSUURA, and K. SHIMOZAWA 103

Pitfalls of Prenatal Diagnosis of 21-Hydroxylase Deficiency Congenital Adrenal Hyperplasia. *By* SONGYA PANG, MARILYN S. POLLACK, MAY LOO, ORVILLE GREEN, ROBERT NUSSBAUM, GEORGE CLAYTON, BO DUPONT, and MARIA I. NEW 111

Pitfalls in Prenatal Diagnosis of 21-Hydroxylase Deficiency by Amniotic Fluid Steroid Analysis?: A Six Years Experience in 102 Pregnancies at Risk. *By* MAGUELONE G. FOREST 130

The Use of Gamma Interferon to Increase HLA Antigen Expression on Cultured Amniotic Cells Used for the Prenatal Diagnosis of 21-Hydroxylase Deficiency. *By* DAVID H. MAURER and MARILYN S. POLLACK 148

Prenatal Therapy in Congenital Adrenal Hyperplasia: Attempted Prevention of Abnormal External Genital Masculinization by Pharmacologic Suppression of the Fetal Adrenal Gland *in Utero*. *By* GEORGE P. CHROUSOS, MARK I. EVANS, D. LYNN LORIAUX, JAMES MCCLUSKEY, JOHN C. FLETCHER, and JOSEPH D. SCHULMAN 156

Modern Medical Therapy of Congenital Adrenal Hyperplasia: A Decade of Experience. *By* JEREMY S. D. WINTER and ROBERT M. COUCH 165

LHRH Analog Treatment of Central Precocious Puberty Complicating Congenital Adrenal Hyperplasia. *By* ORA HIRSCH PESCOVITZ, FERNANDO CASSORLA, FLORENCE COMITE, D. LYNN LORIAUX, and GORDON B. CUTLER, JR. 174

Effect of Hydrocortisone Dose Schedule on Adrenal Steroid Secretion in Congenital Adrenal Hyperplasia. *By* JORG WINTERER, GEORGE P. CHROUSOS, D. LYNN LORIAUX, and GORDON B. CUTLER, JR. 182

Monitoring Treatment in Congenital Adrenal Hyperplasia: Use of Serial Measurements of 17-OH-Progesterone in Plasma, Capillary Blood, and Saliva. *By* I. A. HUGHES, J. DYAS, J. ROBINSON, R. F. WALKER, and D. R. FAHMY 193

Cytochrome P-450: Physiology of Steroidogenesis. *By* PETER F. HALL 203

Biochemical Properties of Cytochrome P-450 in Relation to Steroid Oxygenation. *By* MINOR J. COON and KUNIYO INOUYE 216

Evidence that Variation among Untreated Rabbits in Hepatic Progesterone 21-Hydroxylase Activity Is Indicative of Enzyme Heterogeneity Rather than a Transient Inductive Effect. *By* U. MULLER-EBERHARD, L. GHIZZONI, H. H. LIEM, M. NEW, M. FINLAYSON, and E. F. JOHNSON ... 225

Deoxycorticosterone Biosynthesis in Kidney Tissue of Experimental Animals: Characterization of Steroid 21-Hydroxylase Activity in Guinea Pig Kidney. *By* M. LINETTE CASEY, PAUL C. MACDONALD, and CRAIG A. WINKEL ... 232

An Approach to the Molecular Biology of Congenital Adrenal Hyperplasia. *By* BON-CHU CHUNG, KARLA J. MATTESON, JOHN E. MORIN, SYNTHIA H. MELLON, and WALTER L. MILLER ... 238

Regulation of Cytochrome P-$450_{17\alpha}$ Activity and Synthesis in Bovine Adrenocortical Cells. *By* MAURICIO X. ZUBER, EVAN R. SIMPSON, and MICHAEL R. WATERMAN ... 252

The Human Major Histocompatibility Complex: Genes and Proteins. *By* JACK L. STROMINGER ... 262

Structure Organization and Expression of the Major Histocompatibility Class III Genes. *By* HARVEY R. COLTEN ... 269

Molecular Cloning of Steroid 21-Hydroxylase. *By* PERRIN C. WHITE, MARIA I. NEW, and BO DUPONT ... 277

Index of Contributors ... 289

Introduction

ALFRED M. BONGIOVANNI

School of Medicine
University of Pennsylvania
Philadelphia, Pennsylvania 19104

The study of congenital adrenal hyperplasia in man has engaged the attention of many investigators in different disciplines. It is interesting that the first unmistakable description of this condition came from the observations of what today we would call a basic scientist. This was Luigi De Crecchio, Professor of Anatomy at the University of Naples.[1] De Crecchio's classical and lengthy description appeared in 1865. I quote in part from my personal loose translation from the Italian language: "I wish to describe this case with the purpose of teaching the young people who attend my lectures. I propose in this narrative that it is sometimes extremely difficult and even impossible to determine sex during life. In one of the anatomical theaters of the hospital for incurables, there arrived towards the end of January a cadaver which in life was the body of a certain Joseph Marzo. . . . The general physiognomy was decidedly male in all respects. There were no feminine curves to the body. There was a heavy beard. There was some delicacy of structure with muscles that were not very well developed. . . . The distribution of the pubic hair was typical of the male. Perhaps the lower extremities were somewhat delicate, resembling the female, and were covered with hair. . . . The penis was curved posteriorly and measured six cms. Or with stretching, ten cms. The corona was three cms long and eight cms in circumference. There was an ample prepuce. There was a first grade hypospadias. . . . There were two folds of skin coming from the top of the penis and encircling it on either side. These were somewhat loose and perhaps resembled labia majora."

The author then describes the internal organs and reports that there was a vagina, uterus, tubes, and ovaries, all of which were normal in appearance. However, the author does describe some male structures. He found a prostate that was divided into three distinct portions. The fact that exposure to large amounts of androgens during fetal life in the genotypic female may cause some development of masculine structures has not often been noted in higher mammals, but from time to time similar observations have been made by others.

Over the past 30 years, congenital adrenal hyperplasia has been divided into several distinct forms based on deficiencies of discrete enzymatic functions of steroidal biogenesis. In the beginning, the different types were recognized by detailed and laborious analysis of body fluids for intermediary steroid metabolites. Within this arena, there worked many an illustrious organic steroid chemist and quite a few clinical investigators who learned the basic principles from the former. These early steps were necessary in order to apply the more recent technology of radioimmunoassay (RIA). Without the knowledge of the precise chemical marker of each form of the disease, the development of a suitable diagnostic RIA would not have been possible. Furthermore, in the course of these labors, certain forms of congenital adrenal hyperplasia were discovered that were not associated with virilization. Indeed, the enzymatic defects in these other forms interfered with the normal biosynthesis of testosterone so that the affected genotypic male was incompletely formed. Within these contexts, the experimental embryologist had his or her field day.

It is remarkable that the aforementioned Professor De Crecchio—remember that he was an anatomist—wrote, "it was of great importance to determine the habits, tendencies, passions, and general character of this individual: in the course of examining the cadaver it occurred to me that I could not do better than collect the story of his life, and I was determined to get as complete a story as possible, determined to get at the base of the facts and to avoid undue exaggeration which was rampant in the conversation of many of the people present at the time of the dissection." Many people were interviewed by Professor De Crecchio and it was established that Joseph Marzo conducted himself within the sexual arena exclusively as a male. The author states that Marzo contracted "the French disease" [*sic*] on two occasions. It was also ascertained that he had frequent vomiting and diarrhea, and that during such an episode he died. Might Joseph Marzo have been what today we call a salt-loser? Thus, the human behaviorists, such as the preeminent Dr. John Money, entered the field and in more recent years developed a better understanding of psychosexual development and orientation in man through the study of patients with congenital adrenal hyperplasia and their sexual incongruities in hormone secretions.

That there is vast heterogeneity has been often mentioned over the years and this has been further developed more recently by the group at Cornell.[2] Various degrees of enzymatic deficiency are apparent. When these are severe, there are clinical manifestations in very early life. In less severe forms, the clinical expression of the derangement is seen in later life, and in the older literature, this has sometimes been termed "late-onset" disease. Almost all descriptions of this latter condition are confined to affected females who display acquired unwanted hirsutism and/or disturbances of ovarian function. It is not clear why ascertainment in the older male, except in some family studies, has escaped attention. The evidence is that these disorders are transmitted by autosomal recessive genes. Therefore, both sexes should be affected equally. Finally, the most benign deficiencies have no clinical consequences and are detected only through refined dynamic tests with the measurement of intermediary steroid metabolites in the blood after appropriate adrenal stimulation.

Most recently, the geneticists have been having their own circus around the 21-hydroxylase deficiency. Following the rather surprising discovery that the gene for 21-hydroxylase lay within the HLA system on the 6th chromosone, there followed an explosion of new knowledge, of new means of classification, of better understanding of the genetic transmission of the disorder, and on and on.[3] Much of this conference will cover these details. In addition, other areas of the general topic will be discussed in the ensuing presentations. I trust that all of you will be amply rewarded by the matters that presently will be introduced around this fascinating category of metabolic disorders in man. As stated in 1657 by William Harvey,[4] nature may display her normal laws and secret mysteries through the study "of rarer forms of disease."

REFERENCES

1. DE CRECCHIO, L. 1865. Sopra un caso di apparenzi virili in una donna. Morgagni **7:** 154–188.
2. NEW, M. I., F. LORENZEN, A. J. LERNER *et al.* 1983. Genotyping steroid 21-hydroxylase deficiency: Hormonal reference data. J. Clin. Endocrinol. Metab. **57:** 320–326.
3. LEVINE, L. S., M. ZACHMANN, M. I. NEW *et al.* 1978. Genetic mapping of the 21-hydroxylase deficiency gene within the HLA linkage group. N. Engl. J. Med. **299:** 911–915.
4. HARVEY, W. Cited by A. E. GARROD. 1928. Lancet **I:** 1055–1057.

Clinical and Endocrinological Aspects of 21-Hydroxylase Deficiency[a]

MARIA I. NEW

Division of Pediatric Endocrinology
Department of Pediatrics
The New York Hospital
Cornell Medical Center
New York, New York 10021

INTRODUCTION

Congenital adrenal hyperplasia (CAH) is a family of inherited disorders of adrenal steroidogenesis. The disorders each have a characteristic pattern of hormonal abnormalities caused by an inherited deficiency of one of the several enzymes necessary for normal steroid synthesis. The most common form of CAH is caused by a deficiency of the 21-hydroxylase (21-OH) enzyme. A brief review of the mechanism and regulation of adrenal steroidogenesis is given below to aid in the understanding of the pathophysiology of 21-OH deficiency.

STEROIDOGENESIS AND ENZYMATIC CONVERSIONS OF ADRENAL STEROID HORMONES

Steroidogenesis

FIGURE 1 shows a simplified scheme of adrenal steroidogenesis; each hydroxylation step is indicated and the newly added hydroxyl group is circled. The three main classes of hormones synthesized by the adrenal cortex are mineralocorticoids (17-deoxy pathway), glucocorticoids (17-hydroxy pathway), and sex steroids. A detailed description of steroidogenesis is omitted here since it has been extensively reviewed in other texts.[1] For the purpose of understanding the clinical manifestations of CAH, reference to the scheme in FIGURE 1 will suffice.

Mechanism of Adrenal Steroid Regulation

The circulating level of plasma cortisol mediates the hypothalamic-pituitary-adrenal feedback system. Decreased cortisol secretion results in increased ACTH secretion. In those forms of CAH in which an enzyme deficiency causes impaired cortisol synthesis, there is excessive ACTH secretion and hyperplasia of the adrenal cortex (FIGURE 2).

Aldosterone secretion is primarily regulated via the renin-angiotensin system,

[a]Some of the work described in this review was supported by NIH grant no. HD–00072, and by a grant (RR–47) from the Division of Research Resources, General Clinical Research Centers Program, NIH.

which is responsive to the state of electrolyte balance and plasma volume (FIGURE 3). Aldosterone secretion is also stimulated directly by high serum K^+ concentration.

It has been proposed by New and Seaman[2] that the zona glomerulosa and the zona fasciculata behave as two separate glands with respect to regulation and secretion. According to this concept, steroidogenesis in the fasciculata is regulated by ACTH,

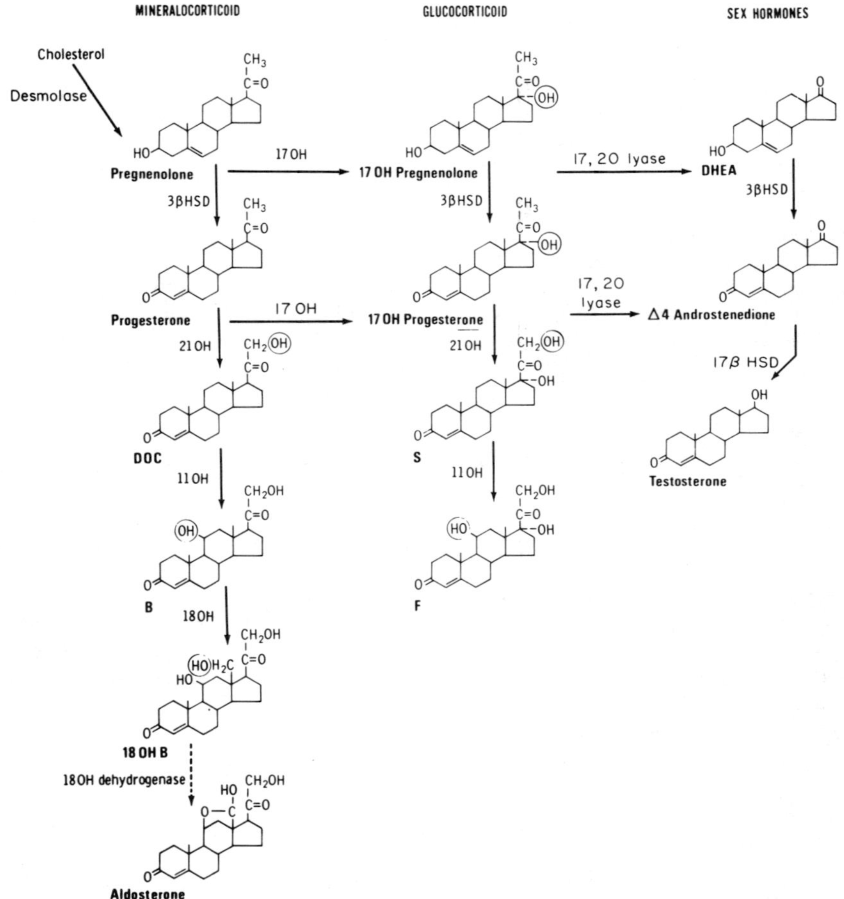

FIGURE 1. Simplified scheme for adrenal steroidogenesis. Each hydroxylation step is indicated and the newly added hydroxyl group is circled. (From New and Levine,[47] by permission of Plenum Publishing.)

while that in the glomerulosa is regulated by the renin-angiotensin system such that: (i) ACTH stimulates secretion of cortisol, corticosterone, and androgens by the zona fasciculata and zona reticularis, and (ii) angiotensin stimulates aldosterone secretion by the zona glomerulosa, with ACTH presumably exerting only a secondary influence on the glomerular secretion of aldosterone (FIGURE 4). Whereas the zona fasciculata

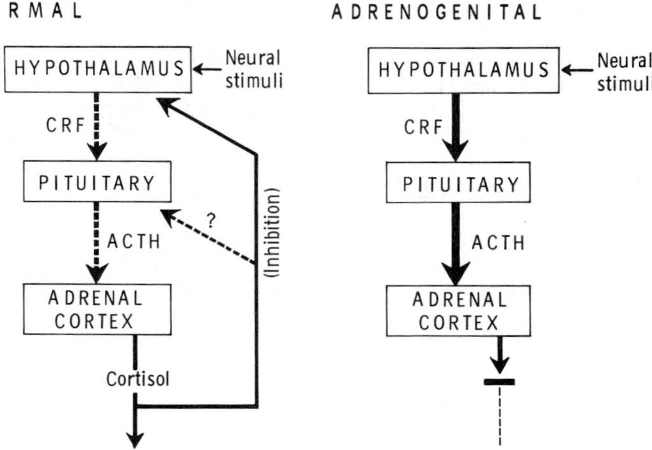

FIGURE 2. The regulation of cortisol secretion in normal subjects and in patients with congenital adrenal hyperplasia. (From New and Levine,[47] by permission of Plenum Publishing.)

lacks the enzyme necessary for the terminal step of aldosterone synthesis, the zona glomerulosa lacks the 17α-hydroxylase activity required for the production of 17-hydroxycorticoids and androgens.

FETAL SEXUAL DEVELOPMENT

In order to understand the pathophysiology of CAH, it is necessary to discuss normal sexual differentiation briefly.

According to the hypothesis developed by Jost,[3] normal differentiation of male

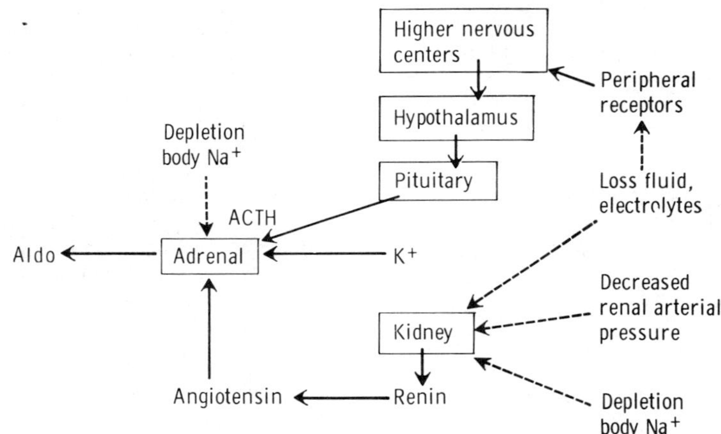

FIGURE 3. Regulation of aldosterone secretion. (From New and Peterson,[48] by permission.)

genitalia is dependent on two functions of the fetal testis:

(i) Secretion of the androgen testosterone, which stimulates the Wolffian ducts to develop into the male internal genitalia, and which is reduced in the target tissue to dihydrotestosterone (DHT). DHT causes differentiation of the male external genitalia. In the male, the normal source of androgen is the fetal testis, but androgen from the adrenal or exogenous sources can cause masculinization of the external genitalia.

(ii) Secretion of a nonsteroidal substance,[4] which inhibits Mullerian duct development such that normal males are born without a uterus. Since the fetal ovary secretes neither testosterone nor the inhibiting factor necessary to inhibit Mullerian structures, the normal female is born without male differentiation of external genitalia (i.e., with female external genitalia) and without Mullerian repression (i.e., with a uterus and Fallopian tubes).

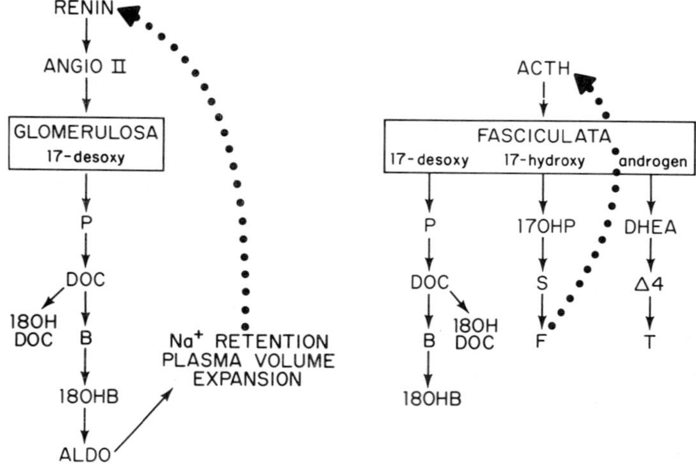

FIGURE 4. Regulation of adrenocortical steroidogenesis considering the fasciculata and glomerulosa as two separate glands. Dotted arrows indicate negative feedback. Symbols: P, progesterone; DOC, deoxycorticosterone; B, corticosterone; 17-OHP, 17-hydroxyprogesterone; S, 11-deoxycortisol; F, cortisol; Δ4, Δ^4-androstenedione; DHEA, dehydroepiandrosterone; T, testosterone; 18OHDOC, 18-hydroxydeoxycorticosterone; 18OHB, 18-hydroxycorticosterone; ALDO, aldosterone. (From New *et al.*,[49] by permission.)

Thus, in the above scheme, the ovary does not play a determining role in sex differentiation. Female fetuses exposed to high levels of androgen consequent to CAH, an androgen-producing tumor in the mother, or administration of androgens to the mother, manifest virilization of the external genitalia, but have normal internal female genitalia.

21-OH DEFICIENCY

Impairment of 21-hydroxylation is the most common enzymatic deficiency observed in CAH. Reference to FIGURE 1 indicates that an enzymatic deficiency of

21-OH results in decreased cortisol synthesis. The decreased cortisol synthesis induces increased ACTH secretion, which has been demonstrated in the blood.[5] Consequent to the increased ACTH secretion, there is overproduction of cortisol precursors and sex steroids (e.g., testosterone), which do not require 21-OH for their biosynthesis.

Patients with 21-OH deficiency excrete excessive urinary pregnanetriol, the metabolite of 17-hydroxyprogesterone (17-OHP). Urinary 17-ketosteroids, which result from the metabolism of dehydroepiandrosterone (DHEA), Δ^4-androstenedione (Δ^4), and testosterone, are also present in increased amounts. Simple and reliable radioimmunoassays for circulating serum levels of adrenal steroids have been developed to provide a more accurate laboratory method for the diagnosis of CAH than urinary steroid measurement alone.

Simple Virilizing 21-OH Deficiency

The most prominent feature of this form of 21-OH deficiency is virilization. Since adrenocortical function begins in the third month of gestation, the fetus is exposed at the critical time of sexual differentiation to the overproduced fetal adrenal androgens resulting from 21-OH deficiency. Thus, in the female fetus, the external genitalia are masculinized by excessive fetal adrenal androgens, resulting in female pseudohermaphroditism (FIGURE 5). In rare cases, the masculinization may be so profound that the urethra is penile.[6] Since the female fetus has no testis, Mullerian inhibiting factor is not produced and the female with CAH is born with a uterus and Fallopian tubes. The absence of the Wolffian system, despite increased androgen levels, suggests that the level of androgen necessary for Wolffian development is higher than that produced in CAH.

Because CAH due to 21-OH deficiency is the most common cause of ambiguous genitalia in the newborn, and because the female pseudohermaphrodite with this disorder has the capacity for an entirely normal female sex role, including fertility, it is very important to consider this disorder in any intersex problem present at birth.

Males with this disorder do not manifest genital abnormalities at birth. Without treatment, both males and females manifest progressive virilization resulting in early fusion of the epiphyses and eventual short stature.

Salt-Wasting 21-OH Deficiency

A salt-wasting form of CAH is also associated with 21-OH deficiency. The feature of virilism is the same as in simple virilizing CAH, but, in addition, there is profound aldosterone deficiency, which is present even when the patient is sodium-depleted. The salt-wasting form of 21-OH CAH may cause a life-threatening adrenal crisis within the first few weeks of life, and therefore, must be recognized soon after birth in order for treatment to be administered prior to crisis.

THE ZONA FASCICULATA AND ZONA GLOMERULOSA AS SEPARATE GLANDS

In concert with the theory that the zona fasciculata and zona glomerulosa function as two separate glands under different regulatory control, New and Seaman[2] have proposed that it is possible for an enzyme deficiency to occur in the fasciculata, but not in the glomerulosa. It has also been proposed that the independence of enzyme activity

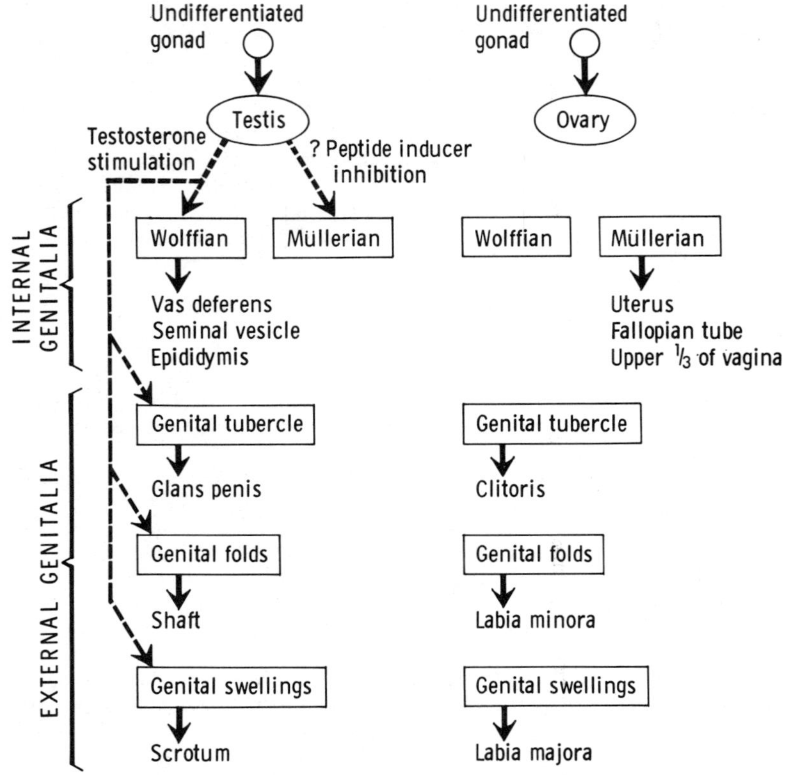

FIGURE 5. Fetal sex differentiation. (From New and Levine,[47] by permission of Plenum Publishing.)

of the zona fasciculata and the zona glomerulosa may account for the different clinical manifestations of the simple virilizing form and the salt-wasting form of 21-OH CAH.[7]

The hypotheses traditionally proposed to explain the difference between the salt-wasting and simple virilizing forms of CAH are based on either a "one-enzyme" or a "two-enzyme" defect. However, neither *in vivo* nor *in vitro* studies have been conclusive, and results compatible with both theories have been reported.[8]

The "one-enzyme" theory attributes the differences between salt-wasting and simple virilizing CAH to a different degree of enzymatic deficiency, the most severe deficiency leading to salt-wasting. The "two-enzyme" theory postulates that the simple virilizers have an enzymatic deficiency in the 17-hydroxy (glucocorticoid) pathway, whereas the salt-wasters have a deficiency in both the 17-hydroxy and 17-deoxy pathways of adrenal steroidogenesis. This theory is supported by findings of normal or even elevated aldosterone production in simple virilizers, in contrast to the impaired aldosterone production in salt-wasters.[8]

Studies from our laboratory[7,8] suggested a new hypothesis which conforms to the two gland model of the adrenal cortex to explain the difference in salt-wasting and simple virilizing CAH. This hypothesis states: (i) in both simple virilizers and salt-wasters, there is a fasciculata defect of 21-hydroxylation in both the 17-hydroxy and 17-deoxy pathways, and (ii) in the salt-waster, there is also a defect in 21-hydroxylation in the glomerulosa, while in the simple virilizer, the glomerulosa is spared this defect. This hypothesis is schematically presented in FIGURE 4, and the data supporting this hypothesis are presented in FIGURES 6 and 7.

The study of Kuhnle *et al.*[7] verifies this hypothesis by demonstrating a 21-OH deficiency in the 17-deoxy pathway of the fasciculata in both simple virilizers and salt-wasters. Upon ACTH stimulation, progesterone levels rose markedly, while deoxycorticosterone (DOC) and corticosterone responses were minimal. When the fasciculata was then suppressed with dexamethasone and the glomerulosa was stimulated with low sodium intake, both normal subjects and simple virilizers, but not salt-wasters, demonstrated a rise in serum and urinary aldosterone and 18-OH-corticosterone levels, indicative of normal glomerulosa 21-OH function. These results indicate the absence of a 21-OH deficiency in the zona glomerulosa of the simple virilizer, and show that in both normal subjects and simple virilizers, the zona glomerulosa can increase aldosterone secretion in response to renin stimulation independent of precursor hormones of the zona fasciculata (FIGURE 7).

These findings further suggest the possibility of separate genetic loci for the regulation of 21-hydroxylation in the zona fasciculata and the zona glomerulosa. Genetic studies of recombination, discussed below, are also suggestive of this possibility. The hormonal data also suggest that within the fasciculata, only one genetic locus

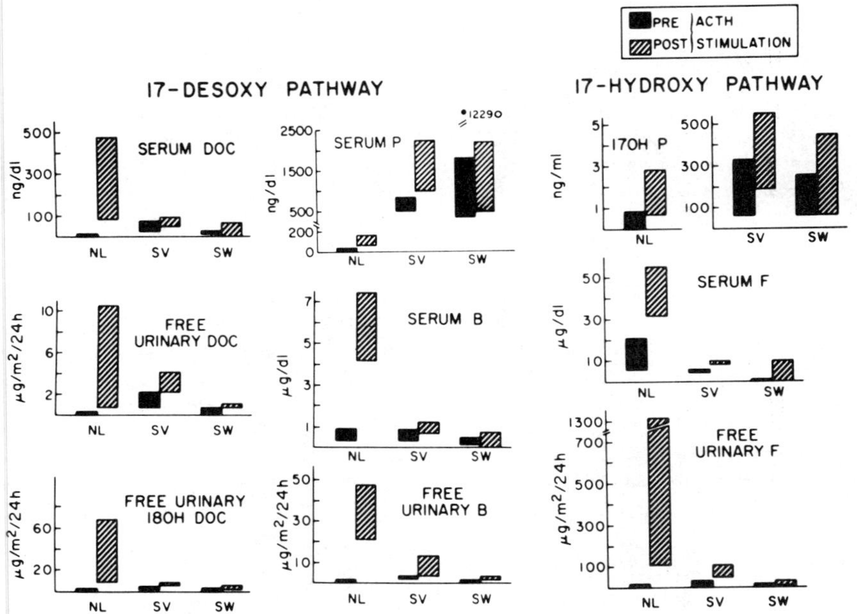

FIGURE 6. Hormonal response to ACTH stimulation of the zona fasciculata. NL, normal subject; SV, patient with the simple virilizing form of 21-hydroxylase deficiency; SW, patient with the salt-wasting form of 21-hydroxylase deficiency. (From Kuhnle *et al.*,[7] by permission of The Endocrine Society.)

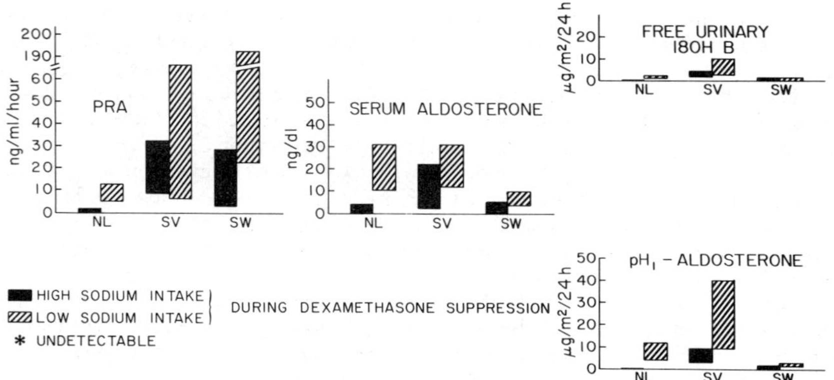

FIGURE 7. Hormonal response to sodium deprivation of the zona glomerulosa. See legend to FIGURE 6 for symbols. (From Kuhnle *et al.*,[7] by permission of The Endocrine Society.)

seems to regulate 21-OH enzyme activity in both the 17-hydroxy and 17-deoxy pathways.

GENETICS OF CONGENITAL ADRENAL HYPERPLASIA

Population Studies

Several surveys have established that the 21-OH deficiency is transmitted as an autosomal recessive trait,[6,9] with males and females equally at risk. With few exceptions, either the simple virilizing or salt-wasting form is found consistently within a family. The salt-wasting variety occurs in about 50–80 percent of patients with 21-OH deficiency.[10]

HLA Linkage

The genes for HLA (human leukocyte antigens), which are cell surface antigens important in transplantation, are located on chromosome 6. The HLA complex consists of at least four genetic loci which code for the antigens HLA-A, HLA-B, HLA-C, and HLA-D/DR. Multiple alleles have been demonstrated for each locus. In addition to the HLA loci, several other loci have been mapped on chromosome 6 in close linkage with HLA (FIGURE 8). Each individual inherits one chromosome 6 from his father and one from his mother. The HLA genes are codominantly expressed as illustrated below:

$$\frac{\text{A3, Bw47(w4), Cw6, DR7}}{\text{A28, Bw35(w6), Cw4, DR5}}$$

in which one haplotype (the set of A3, Bw47(w4), Cw6, DR7) is inherited from one parent, while the other haplotype (A28, Bw35(w6), Cw4, DR5) is inherited from the other parent.

Close genetic linkage between HLA and CAH due to 21-OH deficiency was first described in 1977.[11] Subsequently, studies of 34 unrelated families with a total of 48 patients were reported from New York and Zurich.[12] The findings of this study are illustrated in the two typical pedigrees shown in FIGURE 9. As can be observed in Family 7, the three affected sibs are HLA-identical. In Family 16, the unaffected sibs are all HLA-different from their affected sister. Each parent is an obligate heterozygote carrier and has transmitted one HLA haplotype carrying the gene for 21-OH deficiency to the patient. The brother and sister having only one haplotype linked to the gene for CAH are presumed heterozygotes. The sib sharing neither haplotype with the patient is presumed not to carry the gene for 21-OH deficiency. Thus, the HLA genotype is a marker for the CAH genotype.

International studies reported at the Eighth International Histocompatibility Workshop provided more detailed genetic mapping of the specific location of the 21-OH deficiency gene relative to the different loci that constitute the HLA complex. These studies have established that the 21-OH deficiency gene is located between the

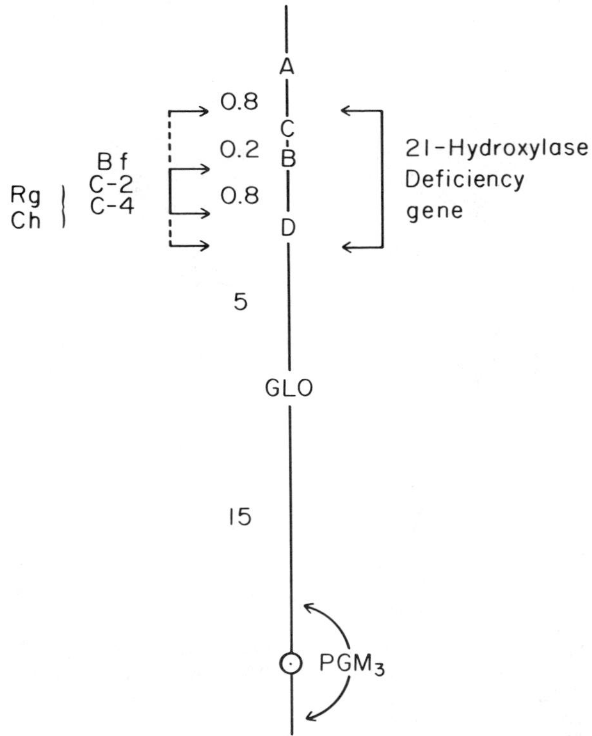

FIGURE 8. HLA linkage group on chromosome 6. The recombinant fractions for the known linkages between A:C, C:B, B:D, B:GLO, and GLO:PGM_3 are shown. The position of the genes for factor B (Bf), complement C-2, complement C-4, and Rodgers (Rg) and Chido (Ch) blood groups is also indicated. The 21-hydroxylase deficiency gene can be mapped between HLA-A and Glyoxalase I (GLO). The most likely position of the 21-hydroxylase deficiency gene is very close to HLA-B. (From Levine *et al.*,[12] by permission of *The New England Journal of Medicine.*)

A. FAMILY Zurich 7 (21-Hydroxylase Deficiency)

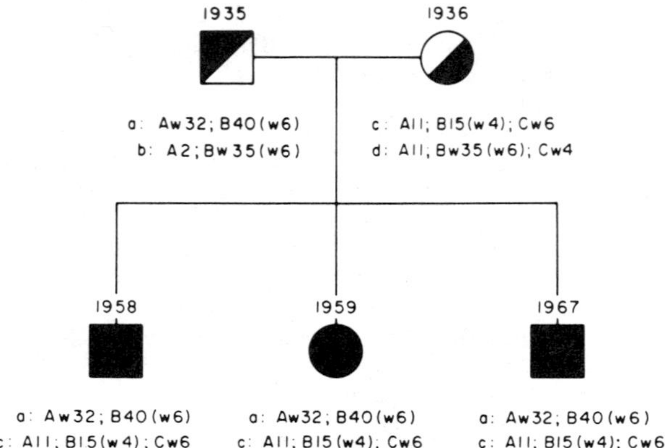

B. FAMILY N.Y. 16 (21-Hydroxylase Deficiency)

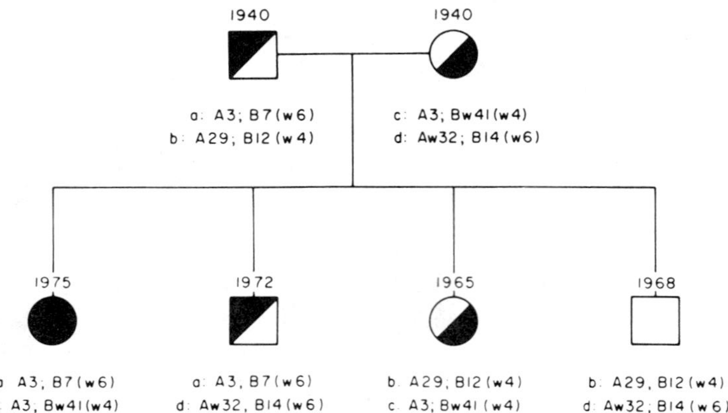

FIGURE 9. Pedigrees for two families with classical 21-hydroxylase deficiency. The HLA haplotypes for the HLA-A, HLA-B, and HLA-C alleles are given in each family. The paternal haplotypes are labeled a and b, and the maternal haplotypes c and d. The parents are obligate heterozygous carriers for the 21-hydroxylase deficiency gene (denoted by the half-black symbols). The affected children are denoted by black symbols. In A, three affected siblings are HLA genotypically identical. In B, one affected child is HLA genotypically different from the three unaffected siblings. One sibling who carries the parental a and d haplotypes is presumed to be a heterozygous carrier for 21-hydroxylase deficiency because he shares the a haplotype with the patient. Another sibling has the parental b and c haplotypes and shares the c haplotype with the patient, and should be a carrier of the classical 21-hydroxylase deficiency gene. The child with the b and d haplotypes should be normal for the gene. (From Levine *et al.*,[12] by permission of *The New England Journal of Medicine.*)

HLA-A locus and the centromere, but is further away from the centromere than the glyoxalase I (GLO) locus.[13] The gene is thus mapped to within 3–4 centiMorgans.

Statistical methods of genetic analysis have more formally demonstrated close genetic linkage between the 21-OH deficiency gene and HLA. The studies reported at the Eighth International Histocompatibility Workshop found a peak LOD Score of 15.65 at a recombination frequency of $\theta = 0.00$ for linkage between HLA and 21-OH deficiency CAH.[13] The LOD Score is a statistical index of the certainty with which one can arrive at a conclusion of genetic linkage. A score of 15.65 means that the odds are $10^{15.65}$ to 1 that linkage exists. In humans, genetic linkage is considered established if the LOD Score exceeds 3.00.

Heterozygote Detection

Numerous studies demonstrated a mild deficiency of 21-OH in the obligate heterozygote parents of children with 21-OH deficiency.[14,15] However, similar studies in the sibs of patients with CAH were difficult to interpret because of the inability to ascertain which sibs were carriers of the gene and which were unaffected. HLA genotyping makes it possible to predict which sibs are carriers and which sibs are genetically unaffected (as shown in FIGURE 9, Family 16). The validity of the prediction of heterozygosity by HLA genotyping was supported by hormonal studies.[16,17] The response of 17-OHP to ACTH stimulation was higher in family members predicted by HLA genotyping to be heterozygotes than in family members predicted to be unaffected.[16] No other hormonal measurement was as useful in discriminating heterozygotes from normals.[16,18] Administration of dexamethasone before ACTH stimulation has been reported to improve the distinction of CAH heterozygotes from unaffected subjects in the adult population.[17,19]

Nonclassical 21-OH Deficiency

Nonclassical symptomatic (also called late-onset, attenuated, or acquired) 21-OH deficiency is a syndrome characterized by virilization, menstrual disturbances, and endocrinological features consistent with 21-OH deficiency that occur in later childhood or adolescence.[20,21] Unlike females with classical 21-OH deficiency, females with late-onset 21-OH deficiency demonstrate no evidence of *in utero* virilization and are born with normal vaginal and urethral orifices and no labial fusion. Some may have fusion of the labia minora only, however. Both types of patients respond similarly to glucocorticoid treatment. The late presentation of a biochemical defect has raised the question as to whether this is the same inherited disorder as CAH with delayed presentation or an acquired disorder distinct from CAH. Although initial studies suggested that the late-onset disorder was not HLA-linked,[22,23] more recent reports have provided evidence that this disorder is, in fact, also genetically linked to HLA.[24,25] Also, it has been proposed that classical and late-onset 21-OH deficiencies are allelic variants.[21,25,26] Pedigrees of families with symptomatic nonclassical 21-OH deficiency are shown in FIGURE 10.

Our HLA studies of families with classical 21-OH deficiency have also led us to uncover a new, cryptic form of 21-OH deficiency. In the course of HLA genotyping and hormonal testing, we encountered family members whose biochemical profiles were characteristic of a mild 21-OH deficiency, although the clinical hallmarks

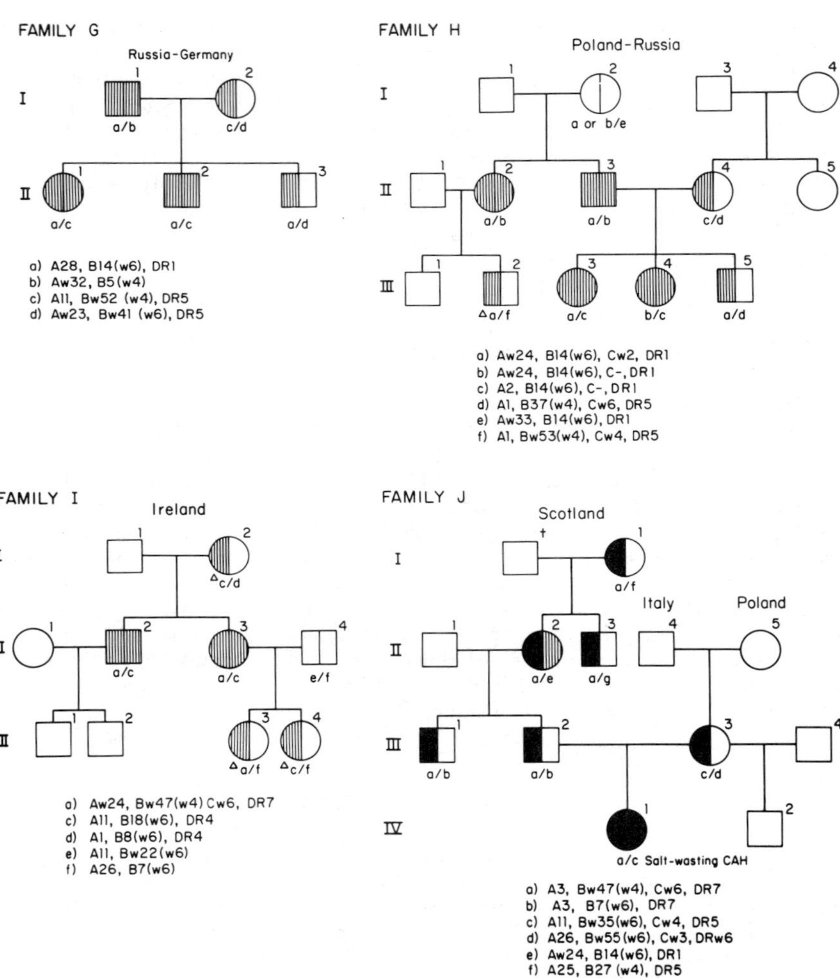

FIGURE 10. Pedigrees of families G–J with nonclassical symptomatic 21-hydroxylase deficiency:

- Family members with mild (nonclassical) 21-hydroxylase deficiency, which is less severe than classical 21-hydroxylase deficiency.
- Heterozygous carriers of a mild (nonclassical) deficiency gene. The assignment is based on HLA genotyping and 17-OHP response to ACTH stimulation.
- Δ Assignment of heterozygosity by HLA genotyping alone.
- Infant with classical 21-hydroxylase deficiency.
- Heterozygous carriers of the severe deficiency gene for CAH.
- Patient with nonclassical symptomatic 21-hydroxylase deficiency having both a severe 21-hydroxylase deficiency (classical) gene and a mild deficiency (nonclassical) gene. (From Kohn *et al.*,[22] by permission of The Endocrine Society.)

commonly accompanying the disorder (e.g., virilization, abnormal puberty and growth, infertility) were absent. The hormonal abnormalities which these nonclassical, asymptomatic patients demonstrated were similar to those observed in the nonclassical, symptomatic form of 21-OH deficiency.[27]

After further investigation, we concluded that these asymptomatic family members were genetic compounds having 21-OH deficiency as a result of inheriting two recessive genetic defects: (1) a severe 21-OH deficiency gene shared with the index case with classical CAH (21-OH^{SEVERE}), and (2) a mild, nonclassical 21-OH deficiency gene (21-OH^{MILD}). LOD score analysis established close genetic linkage between HLA and the gene for the mild 21-OH deficiency.[27]

Zachmann and Prader[28,29] also reported family members of patients with classical 21-OH deficiency whose hormonal responses are consistent with cryptic 21-OH deficiency. They proposed that these family members were unusually manifesting heterozygotes for the classical gene ($21\text{-OH}^{SEVERE}/21\text{-OH}^{NORMAL}$) rather than genetic compounds as we have proposed. Our demonstration that family members predicted by HLA genotyping to be heterozygous for the variant gene ($21\text{-OH}^{MILD}/21\text{-OH}^{NORMAL}$) have a 17-OHP response to ACTH identical to that of family members predicted to be heterozygous for the classical gene ($21\text{-OH}^{SEVERE}/21\text{-OH}^{NORMAL}$) argues against the suggestion of Zachmann and Prader.[30]

Hormonal Standards for Genotyping 21-OH Deficiency

We have developed a series of nomograms relating the baseline and ACTH-stimulated levels of 17-OHP, Δ^4, DHEA, the ratio DHEA/Δ^4, and testosterone. These nomograms provide hormonal standards to use in the assignment of the 21-OH deficiency genotype; i.e., patients whose hormonal values fall on the regression line within a defined group are assigned to that group (FIGURES 11 and 12). Clinical symptoms and signs distinguish between the symptomatic and asymptomatic forms of nonclassical 21-OH deficiency, which are biochemically indistinguishable. Heterozygotes for classical and nonclassical 21-OH deficiency demonstrate a similar hormonal response and are, thus, phenotypically indistinguishable.

The distribution of hormonal responses along a regression line suggests that there is a spectrum of enzymatic deficiency in these groups: patients with classical CAH have the most severe deficiency, patients with nonclassical adrenal hyperplasia have a less severe deficiency, while heterozygotes have an even milder deficiency unmasked only upon ACTH stimulation. Hormonal values in family members predicted by HLA genotyping to be unaffected for 21-OH deficiency fall at the lowest point of the regression line and serve as the best control population for normal 21-OH activity. It should be noted, however, that there is considerable overlap between the general population and the heterozygote population. This overlap of values is of great interest. It may be due to the high frequency of the 21-OH deficiency gene in the general population (see section on INCIDENCE OF 21-OH DEFICIENCY, below).

Allelic Variants

We propose that there are allelic variants at the 21-OH locus which produce different degrees of 21-OH deficiency resulting in the various forms of classical and nonclassical 21-OH deficiencies. We further suggest that the nonclassical forms represent either a genetic compound of the classical (or more severe) 21-OH deficiency

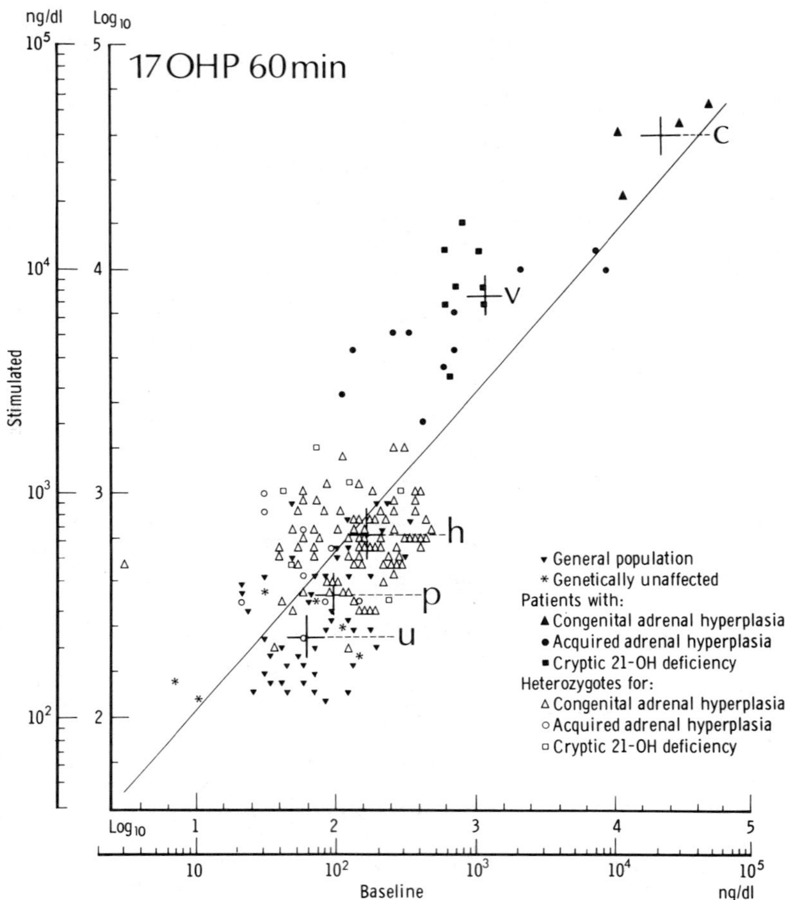

FIGURE 11. Nomogram relating baseline and 60-minute ACTH-stimulated serum 17-OHP concentrations. The mean values for each group are indicated as follows: c, CAH; v, patients with nonclassical symptomatic or asymptomatic (acquired, late-onset, or cryptic) 21-hydroxylase deficiency; h, heterozygotes for classical CAH, nonclassical symptomatic CAH (acquired or late-onset adrenal hyperplasia), and nonclassical asymptomatic CAH (cryptic 21-hydroxylase deficiency); u, family members predicted by HLA genotyping to be unaffected; and p, general population (not HLA genotyped).

allele and the mild, nonclassical 21-OH deficiency allele, or homozygosity for the mild, nonclassical allele.[21]

Genetic Linkage Disequilibrium

In genetic studies, it is important to distinguish between genetic linkage (which we have described for the HLA loci and the 21-OH deficiency locus) and genetic linkage

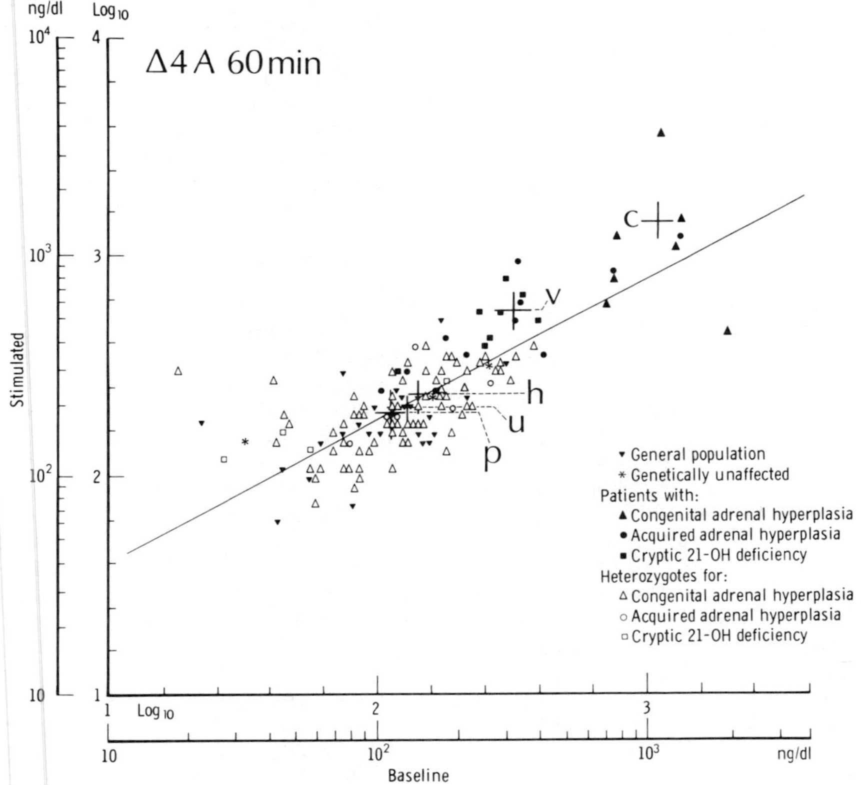

FIGURE 12. Nomogram relating baseline and 60-minute ACTH-stimulated serum Δ^4-androstenedione concentrations. The mean values for each group are indicated as described in FIGURE 11.

disequilibrium, which is the nonrandom association of particular alleles of different genetic loci. In patients with classical and nonclassical 21-OH deficiencies, certain HLA antigens appear with either a significantly increased or decreased frequency relative to their frequency in the general population. Thus, not only are 21-OH deficiency and HLA genetically linked, but there is genetic linkage disequilibrium between the 21-OH deficiency and specific HLA alleles.

The most significant association for classical 21-OH deficiency is with HLA-Bw47, where the combined relative risk is 15.4.[31] Slight increases have also been reported for Bw51, Bw53, Bw60, and DR7. A review of the Bw47-positive haplotypes in patients with 21-OH deficiency reveals that this antigen frequently occurs on one particular haplotype: A3, Cw6, Bw47, DR7. Although the gene frequency for Bw47 in different Caucasian populations is always very low (<0.005), international studies have found that Bw47 appears with remarkable frequency among 21-OH CAH patients; in one particular region of England, the Bw47 frequency among CAH patients is nearly 50 percent. Several studies have also demonstrated that A1, B8, and DR3 are consistently decreased among 21-OH deficient patients.[12,17,31]

It has recently been reported that some of these HLA antigen associations are characteristic of the simple virilizing form of the disease; others are characteristic of the salt-wasting form, and some are common to both forms. Dupont *et al.* studied the complete HLA genotypes in 150 international Caucasian families of patients with CAH.[32] The HLA antigens of the patients were compared with the HLA antigens on the unaffected parental HLA haplotypes. The B5 and B14 antigens were selectively increased among patients with the simple virilizing form; B40 was increased in the salt-wasting form. As previously reported, Bw47 was increased in both forms, and B8 was decreased in both forms. These data strongly suggest that simple virilizing and salt-wasting 21-OH deficiency CAH are due to different mutational defects in the 21-OH locus.

Genetic linkage disequilibrium has also been reported for the nonclassical forms of 21-OH deficiency.[15,24] Our studies document the significantly increased frequency of HLA-B14, DR1, and complement factor BfS in both late-onset and cryptic 21-OH deficiency.[21,25] These alleles tend to appear together on the haplotype HLA-B14, DR1, Bfs in patients with late-onset or cryptic 21-OH deficiency, suggesting that the haplotype segment—rather than the individual alleles—is highly associated with the nonclassical form of 21-OH deficiency.

Our analysis of genetic linkage disequilibrium provides provocative evidence that some patients with late-onset 21-OH deficiency, similar to patients with cryptic 21-OH deficiency, may have genetic compounds possessing both a classical and a nonclassical 21-OH deficiency gene. In a few late-onset patients, we detected HLA-Bw47, an allele occurring with strikingly high frequency in association with the gene for classical 21-OH deficiency. Conclusive proof for the presence of a classical 21-OH deficiency gene rests on the eventual appearance of a patient with classical 21-OH deficiency in the family pedigree. Indeed, in studying the family of a patient recently diagnosed as having congenital salt-wasting 21-OH deficiency (Family J, FIGURE 10), we found that the paternal grandmother has a hormonal response and medical history consistent with late-onset 21-OH deficiency. Particularly noteworthy, is the fact that the HLA haplotype which the grandmother shares with the patient with classical CAH contains Bw47, while the grandmother's other HLA haplotype contains B14, the HLA antigen most highly associated with nonclassical 21-OH deficiency.

It appears that late-onset adrenal hyperplasia may also result from a homozygous genotype for the mild, nonclassical 21-OH deficiency gene. Within the family of one late-onset patient (Family H, FIGURE 10), there were three family members who shared only one haplotype with the affected patient, yet nevertheless demonstrated hormonal levels in the range of nonclassical 21-OH deficiency patients. Analysis of this family pedigree reveals that the affected family members are all homozygous for haplotype B14, DR1, BfS.

The finding that both the late-onset and cryptic 21-OH deficiency genes are associated with the same haplotype segment (B14, DR1, BfS) suggests that these two variant genes are highly related. It is possible that individuals with nonclassical 21-OH deficiency alleles may have different clinical manifestations of their 21-OH defect, expressing either the late-onset or the cryptic disorder. In addition, the clinical difference between the late-onset and cryptic disorders no longer appears as distinct as initially appreciated. We have found marked variability in the symptoms of the late-onset patients we have recently examined, ranging from precocious pubic hair growth without subsequent symptoms of virilization to more severe degrees of hirsutism, acne, menstrual disturbances, and stunted growth. In one patient, the symptoms of late-onset 21-OH deficiency, including pubic hair and advanced growth, were evident at six months of age. The symptoms subsequently disappeared, but the biochemical abnormality persisted.

As a result of these studies, the concept of allelic variants at the 21-OH locus has evolved. In TABLE 1, we propose a scheme to clarify the concept of allelic variants of steroid 21-OH deficiency and to provide a glossary of terms by which to describe these disorders with consistency. Previously, the term "cryptic" 21-hydroxylase deficiency has been used to describe both the phenotype (asymptomatic) and genotype (21-OH deficiency^{SEVERE}/21-OH deficiency^{MILD}). The term "late-onset" 21-OH deficiency was used to refer to symptomatic patients. Thus, the term "late-onset" 21-OH deficiency described the phenotype and did not define the genotype. The new glossary of terms classifies patients based on phenotype (symptomatic or asymptomatic). The new terms recognize that these two nonclassical forms of 21-OH deficiency may have either of two genotypes (21-hydroxylase deficiency^{SEVERE}/21-OH deficiency^{MILD} or 21-OH deficiency^{MILD}/21-OH deficiency^{MILD}). Thus, nonclassical asymptomatic and symptomatic 21-OH deficiency may be discovered in the course of studies of families in which the index case is either a patient with classical (severe) or nonclassical (mild) 21-OH deficiency. Indeed, we predict that patients with nonclassical, asymptomatic 21-OH deficiency will be found in the course of screening the general population by hormonal studies.

Screening for CAH and Future Population Studies

In 1977, neonatal screening for CAH became possible by the development of a microfilter paper method for measuring 17-OHP.[33] This method utilizes a heel stick blood specimen which is spotted onto filter paper and then analyzed for 17-OHP by radioimmunoassay. This will be discussed in the paper by Pang *et al.* in this volume.[34]

Incidence of 21-OH Deficiency

In Alaska, the incidence of salt-wasting 21-OH deficiency in Yupik-speaking Eskimos is unusually high at one out of 684.[34] The incidence of classical 21-hydroxylase deficiency in a homogeneous Caucasian population has been estimated by screening to be 1/5000.[35,36] The corresponding carrier frequency is calculated to be 1/35.[37] However, this estimate of heterozygote frequency is only for the carriers of the classical gene defect. A recent comprehensive analysis of hormonal and HLA data gleaned from our studies has revealed a surprisingly high frequency of the gene for nonclassical 21-OH deficiency, making it the most common autosomal recessive genetic disorder in man.[38] The methods employed in the analysis are as follows: In patients with an autosomal recessive disorder, the parents are obligate heterozygotes. The haplotypes not transmitted to the affected offspring may be considered *a priori* a random sample of normal and affected haplotypes in the general population. Cortrosyn testing will unequivocally identify parents who are themselves patients with 21-OH

TABLE 1. Glossary of Terms for Disorders of Steroid 21-Hydroxylase (21-OH) Deficiency

Form of 21-OH Deficiency	Clinical Phenotype	Hormonal Phenotype	21-OH Deficiency Genotype[a]
Classical	Virilized prenatally and symptomatic	Markedly elevated serum 17-OHP and Δ^4-A concentration	$\frac{\text{21-OH deficiency}^{\text{SEVERE}}}{\text{21-OH deficiency}^{\text{SEVERE}}}$
Nonclassical	Symptomatic: virilized postnatally and symptomatic Asymptomatic: not virilized prenatally or postnatally, and asymptomatic	Modestly elevated serum 17-OHP and Δ^4-A concentration	$\frac{\text{21-OH deficiency}^{\text{SEVERE}}}{\text{21-OH deficiency}^{\text{MILD}}}$ or $\frac{\text{21-OH deficiency}^{\text{MILD}}}{\text{21-OH deficiency}^{\text{MILD}}}$

[a]The allelic variant which transmits severe 21-OH deficiency is in genetic linkage disequilibrium with HLA-Bw47. The allelic variant which transmits mild 21-OH deficiency is in genetic linkage disequilibrium with HLA-B14.

deficiency. Thus, in these parents, we can detect a 21-OH deficiency gene on the "random haplotype." By counting the number of parents in ethnically homogenous groups who are patients relative to the total number of random haplotypes, we are able to arrive at a disease gene frequency (q) for a given ethnic population. The Hardy-Weinberg Law will then allow us to calculate the respective normal gene (p), heterozygote ($2pq$), and affected homozygote (q^2) frequencies. In cases where one might not be certain whether the random haplotype contained a classical or nonclassical disease gene (as in FIGURE 13, D and E), we employed data from HLA-B associations (TABLES 2A and 2B). For instance, if an Ashkenazi Jewish parent was tested hormonally as a patient with nonclassical 21-OH deficiency and the random haplotype contained HLA-B40, we identified that 21-OH deficiency gene as nonclassical since B40 appeared in *no* Jewish patients with classical disease and was not decreased in frequency in nonclassical patients.

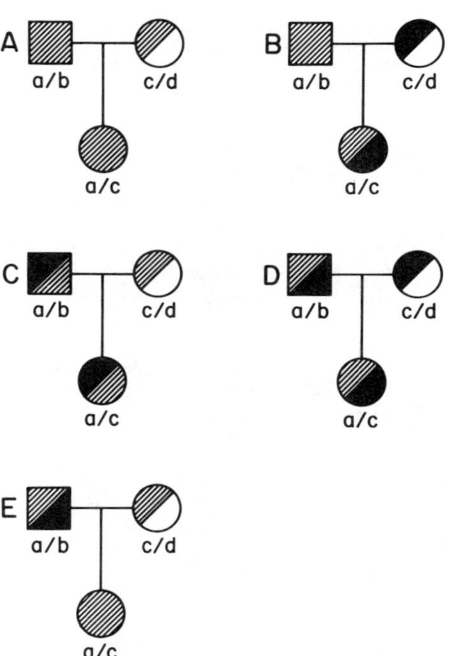

FIGURE 13. Possible 21-OH deficiency genotypes in families in which both an offspring and a parent have been diagnosed as patients with nonclassical 21-OH deficiency upon hormonal testing.

▨ refers to nonclassical 21-OH deficiency haplotype (21-OH def^{MILD}).

◩ refers to classical 21-OH deficiency haplotype (21-OH def^{SEVERE}). (From Speiser *et al.*,[38] by permission of the *American Journal of Human Genetics*.)

Relative risk for nonclassical 21-OH deficiency in the presence of the HLA-B14 marker was 31.9 overall. There is a minimally elevated risk of nonclassical disease with the Bw47 marker, which is reflective of those families in whom no patient with the classical 21-OH deficiency had (as yet) been identified. The risk of classical 21-OH deficiency in association with Bw47 was greatest (22.7) in Caucasian patients, among whom 48% had at least one Anglo-Saxon ancestor. Finally, we were able to establish linkage disequilibrium between HLA-B14 and nonclassical 21-OH deficiency in Ashkenazi Jewish, Hispanic, and Italian patients, but not in Yugoslavs or other Caucasians, suggesting that the latter groups may have a separate mutation.

TABLE 2A. HLA Gene Frequency Analysis in Nonclassical Patients by Ethnic Group

Population	Number of Haplotypes	Gene Frequency					
		B14	Bw47	B35	B8	B5	B40
Controls[a]							
1. Ashkenazi Jewish	254	0.120[b]	0.004	0.150	0.043	0.085	0.020
2. Hispanic	444	0.033	0.003	0.051	0.044	0.144	0.039
3. Yugoslav	152	0.007	0.040	0.166	0.090	0.090	0.011
4. Italian	1044	0.037	0.004	0.169	0.062	0.143	0.023
Sum of Ethnic Control Groups 1–4	1894	0.045	0.006	0.138	0.058	0.131	0.025
Nonclassical Patients[c]							
1. Ashkenazi Jewish	56	0.696[f]	0	0.071	0.018	0.089	0.018
2. Hispanic	9	0.444[f]	0	0	0	0.111	0.111
3. Yugoslav	8	0	0.125[d,e]	0.125	0	0.125	0
4. Italian	12	0.667[f]	0	0	0.083	0	0.167[j]
5. "Other Caucasians"	17	0.353	0.176[d]	0	0	0.059	0
Sum of Ethnic Patient Groups 1–4	85	0.600[f]	0.012	0.059[k]	0.024	0.082	0.047

[a]From Baur and Danilovs. *In* Histocompatibility Testing, p. 959 (see reference 51).

[b]Dupont, B. Unpublished data. New York City Ashkenazi B14 frequency is the same as the worldwide Ashkenazi B14 frequency.

[c]Ethnic background was homogeneous in each category, and was compared to ethnically matched controls. The group, "Other Caucasians," includes non-Jewish persons of German, French, Polish, Russian, Hungarian, Greek, Anglo-Saxon, and Nordic origin (41% had at least one Anglo-Saxon ancestor). Significance is not reported for "Other Caucasians" for lack of an appropriate control group. Haplotypes known to be associated with a classical 21-hydroxylase deficiency gene were omitted from this analysis, which explains the odd number of haplotypes.

[d]In these families, there were no known patients with the classical disorder; thus, the nonclassical patients might be genetic compounds carrying both a classical and nonclassical gene, but this could not be proven. The HLA Bw47;DR7 segment was never associated with a proven nonclassical gene.

[e]Occasionally, a Bw47 variant is seen in association with DR2.

[f–k]*Underscore* denotes statistical significance: f) $p \ll 0.001$, g) $p < 0.001$, h) $p < 0.005$, i) $p < 0.01$, j) $p < 0.025$, k) $p < 0.05$.

The nonclassical 21-OH deficiency gene and disease frequencies for the ethnic groups studied appear in TABLE 3. Amongst Ashkenazi Jewish subjects, one out of 27 is affected with the disease, one out of 3 may be a heterozygote. The gene and disease frequencies were also high in Hispanics, Yugoslavs, and Italians. Considering the population as a whole, one out of 111 individuals are affected with the disease, and one out of 6 may be a heterozygote. Given this amazingly high frequency of disease, we sought to confirm our findings by a separate method. We employed the sib-pair analysis of Thomson and Bodmer[39] for a recessive disease gene linked to HLA. The results are virtually the same (TABLE 3, B) as found by using the parental random haplotype method. FIGURE 14 depicts the relative frequencies of nonclassical 21-OH deficiency, classical 21-OH deficiency, and other common autosomal recessive diseases.

PRENATAL DIAGNOSIS OF CONGENITAL ADRENAL HYPERPLASIA

21-OH Deficiency

Since the report by Jeffcoate *et al.*[40] of the prenatal diagnosis of 21-OH deficiency by elevated concentrations of 17-ketosteroids and pregnanetriol in the amniotic fluid of the affected fetus, several investigators have attempted the prenatal diagnosis by measurement of various hormones. Most recently, elevated levels of 17-OHP and Δ^4 in the amniotic fluid of fetuses affected with CAH due to 21-OH deficiency have been used for prenatal diagnosis.[41] HLA genotyping of amniotic cells has provided an additional method for prenatal diagnosis of 21-OH deficiency in a pregnancy at risk, and has made possible the prediction of a heterozygous fetus.[31] When HLA genotyping of amniotic cells reveals that the fetus is HLA-identical to the affected sib, the fetus is predicted to be affected. The HLA prediction of CAH genotype should always be

TABLE 2B. HLA Gene Frequency Analysis in Classical Patients by Ethnic Group

Population	Number of Haplotypes			Gene Frequency					
	Total	SW	SV	B14	Bw47	B35	B8	B5	B40
Controls (See TABLE 2A)									
American Black	368			0.039	0.039	0	0.011	0.116	0.027
Sum of All Controls (plus N Amer Cauc)	2670			0.040	0.006	0.122	0.059	0.119	0.033
Classical Patients[a]									
1. Ashkenazi Jewish	12	6	6	0.167	0	0.417[g]	0	0.333[f]	0
2. Hispanic	23	14	9	0	0	0.087	0.043	0.261	0.043
3. Yugoslav	32	16	16	0	0.063	0.219	0.031	0.125	0
4. Italian	102	39	63	0.049	0.010	0.186	0.010[f]	0.157	0.059[g]
5. "Other Caucasians"	100	66	34	0.010	0.100	0.130	0.050	0.060	0.130
6. Black	11	9	2	0	0.091	0.091[g]	0	0.273	0
All Classical (CAH)[h]	324			0.025	0.059[b]	0.164[g]	0.031[g]	0.130	0.080[b]
Salt-wasters	176			0.006[f]	0.085[b]	0.170	0.028	0.108	0.114[b]
Simple Virilizers	148			0.047	0.027[f]	0.155	0.034	0.155	0.041

[a]Ethnic background was homogeneous in each category, and was compared with ethnically matched controls. The group, "Other Caucasians," includes non-Jewish persons of German, French, Polish, Russian, Hungarian, Greek, Anglo-Saxon, and Nordic origin (48% had at least one Anglo-Saxon ancestor). Significance is not reported in "Other Caucasians" for lack of an appropriate control group. If the ethnic background was nonhomogeneous (e.g., Italian-Polish), that haplotype was excluded from the analysis, which explains the odd number of haplotypes.

[b–g]*Underscore* denotes statistical significance: b) $p \ll 0.001$, c) $p < 0.001$, d) $p < 0.005$, e) $p < 0.01$, f) $p < 0.025$, g) $p < 0.05$.

[h]This classification includes additional patients who are of mixed ethnic origin. The "Sum of All Controls (plus N Amer Cauc)" served as a control.

TABLE 3. Nonclassical 21-Hydroxylase (NC 21-OHD) Gene and Disease Frequencies

A. *Hormonal Criteria*[a]

Ethnic Group[b]	Parental Haplotypes: Total No.	Parental Haplotypes: No. Random	No. Random Parental Haplotypes Nonclassical 21-OHD by ACTH	Gene Freq (q)	Disease Freq (q^2) (95% confidence limits)	Het Freq ($2q(1-q)$)
Ashkenazi Jewish	94	47	9	0.191 (0.092 – 0.333)	0.037 (0.008 – 0.111)	0.309 (0.166 – 0.445)
Hispanic	44	22	3	0.136 (0.029 – 0.349)	0.019 (0.001 – 0.122)	0.235 (0.056 – 0.454)
Yugoslavs	80	40	5	0.125 (0.042 – 0.268)	0.016 (0.002 – 0.072)	0.219 (0.080 – 0.392)
Italian	208	104	6	0.058 (0.022 – 0.121)	0.003 (0.0005 – 0.015)	0.109 (0.042 – 0.213)
"Other Caucasians"[b]	112	56	2	0.036[c] (0.004 – 0.123)	0.001 (0.00002 – 0.015)	0.069 (0.009 – 0.216)
Black	14	7	0	—[c]		
American Indian	4	2	0	—[c]		
Sum of Above Ethnic Patient Groups	556	278	25	0.090 (0.059 – 0.130)	0.009 (0.003 – 0.017)	0.164 (0.111 – 0.226)

B. *Sib Pair Analysis*

	Total Haplotypes	Sib Pairs Sharing One HLA Haplotype	q (ref 39)	q^2	$2q(1-q)$
Mixed Ethnic Group[d]	18	3 (0.648 – 7.452)	0.100 (0.009 – 0.324)	0.010 (0.00008 – 0.105)	0.180 (0.018 – 0.438)

[a]Only parents who had undergone ACTH testing were included.

[b]Ethnic background was homogeneous in each category except "Other Caucasians," which includes non-Jewish persons of German, French, Polish, Russian, Hungarian, Greek, Anglo-Saxon, and Nordic origin.

[c]Because of low disease frequency in these ethnic groups, more families must be studied.

[d]Including Ashkenazi Jews, Anglo-Saxons, Italians, Hispanics, Germans, and Native American Indians.

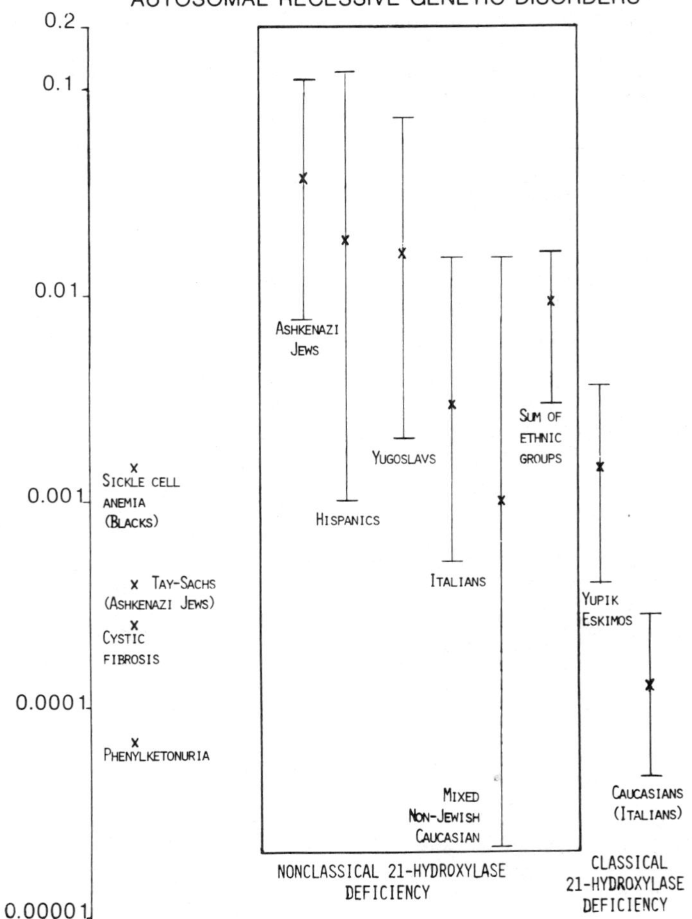

FIGURE 14. The disease frequencies of nonclassical 21-OH deficiency and classical 21-OH deficiency relative to other common autosomal recessive disorders. The latter frequencies are derived from Behrman *et al.*[50] Bars represent the 95% confidence limits. The lower confidence limit for "mixed non-Jewish Caucasians" approximates zero. (From Speiser *et al.*,[38] by permission of the *American Journal of Human Genetics*.)

corroborated by hormonal measurement of 17-OHP and Δ^4 in amniotic fluid. Caution must be exercised in interpreting results if multiple births are expected or if there is antigen sharing in the parents.

Our laboratory has had experience with 29 cases at risk for 21-OH deficiency referred for prenatal diagnosis by hormonal measurement and HLA genotyping. These data are discussed by Pang *et al.* in this volume.[42]

TREATMENT OF CONGENITAL ADRENAL HYPERPLASIA

The female pseudohermaphroditism caused by 21-OH deficiency can be remedied by surgical correction of the ambiguous genitalia, but the decision to do so must take into account an assessment of the patient's potential for future sexual function and fertility. The aim of surgical repair should be to remove the redundant erectile tissue, preserve the sexually sensitive glans clitoris, and provide a normal vaginal orifice that will function adequately for menstruation and intromission. When there is early therapeutic intervention in these patients (and because of the normal internal genitalia), normal puberty, fertility, and childbearing are possible.

The aim of endocrine therapy in CAH is to provide replacement of the deficient hormones. Glucocorticoid administration both replaces the deficient cortisol and suppresses ACTH overproduction. Plasma renin activity (PRA) has long been recognized to be elevated in the simple virilizing form, as well as in the salt-wasting form.[8,43] However, aldosterone levels have not been reported as being deficient in the simple virilizing form of 21-OH deficiency.[8] Rosler *et al.*[44] demonstrated that the addition of salt-retaining hormone to glucocorticoid therapy in simple virilizing patients with elevated PRA improves the hormonal control of the disease. Rosler showed that in patients with CAH due to 21-OH deficiency, the PRA was closely correlated to the ACTH level. The addition of salt-retaining steroids to the therapeutic regimen often made possible a decrease in the glucocorticoid dose. Normalization of PRA also resulted in improved statural growth.[8] In the past, urinary 17-ketosteroids and pregnanetriol excretion have been the biochemical monitors of hormonal control. Serum 17-OHP and Δ^4 levels provide a sensitive index of biochemical control. The serum testosterone is useful in females and prepubertal males, but not in newborn and pubertal males.

The combined laboratory determinations of PRA, 17-OHP, and serum androgens, and the clinical assessment of growth and pubertal status, must all be considered in adjusting the dose of glucocorticoid and salt-retaining steroid.

Sex Assignment in Congenital Adrenal Hyperplasia

Ambiguous sex in the newborn infant is a medical emergency. The decision as to sex assignment at birth has obvious long-term implications. A rational approach permits a careful assessment of the problem and choice of sex assignment.

Rarely, female infants have been so virilized as to have a penile urethra. In these infants, surgical correction of the external genitalia to conform to the female sex presents a much greater challenge to the skill and experience of the surgeon.[45,46] Should there be serious doubt as to the eventual outcome of corrective surgery as far as urinary continence is concerned, based upon anatomic considerations and the ready availability of surgical expertise, then due consideration should be given to a male sex assignment in such a newborn with a prominent phallus and a completely formed male-type urethra. Of course, fertility would be sacrified in the female given the male sex assignment, since a complete hysterectomy and oophorectomy would have to be done to complete such a sex reversal. The hysterectomy and oopherectomy are necessary in genetic females raised as males in order to avoid menstrual flow at puberty, which often occurs as cyclical hematuria.

Treatment with hydrocortisone is necessary in both sex assignments for growth and prevention of early epiphyseal fusion, while treatment with testosterone is necessary after puberty to induce male secondary sex characteristics in the castrated genetic female raised as a male. In the case of female sex assignment, the surgery for phallic correction must be carried out in early infancy; in the case of male sex assignment, hysterectomy and oophorectomy, which are relatively simple procedures, can be delayed until prepuberty. The function of the urethra is not compromised in the male assignment, while it may be in the female. The certain loss of fertility in the case of male sex assignment and the ease of surgical correction must be balanced against possible urological complications in the potentially difficult surgical correction when the baby is assigned to the female sex. A rational and judicious choice of sex assignment is a critical aspect of treatment, since the sex assignment has lifelong implications.

REFERENCES

1. FINKELSTEIN, M. & J. M. SHAEFER. 1979. Physiol. Rev. **59:** 353–406.
2. NEW, M. I. & M. P. SEAMAN. 1970. J. Clin. Endocrinol. Metab. **30:** 361–371.
3. JOST, A. Embryonic sexual differentiation. *In* Hermaphroditism, Genital Anomalies and Related Endocrine Disorders. (2nd edit., pp. 16–64). H. W. JONES & W. W. SCOTT, Eds.: 91–109. William and Wilkins. Baltimore.
4. JOSSO, N., J. Y. PICARD & D. TRAN. 1977. Recent Prog. Horm. Res. **33:** 117–168.
5. SYDNOR, K. L., V. C. KELLY, R. B. RAILE, R. S. ELY & G. SAYERS. 1953. Proc. Soc. Exp. Biol. Med. **82:** 695–697.
6. WILKINS, L. 1962. Arch. Dis. Child. **37:** 231–241.
7. KUHNLE, U., D. CHOW, R. RAPAPORT, S. PANG, L. S. LEVINE & M. I. NEW. 1981. J. Clin. Endocrinol. Metab. **52:** 534–544.
8. NEW, M. I., B. DUPONT, S. PANG, M. S. POLLACK & L. S. LEVINE. 1981. Recent Prog. Horm. Res. **37:** 105–181.
9. CHILDS, B., M. M. GRUMBACH, & J. J. VAN WYK. J. Clin. Invest. **35:** 213–222.
10. COHEN, J. M. 1969. Pediatrics 44: 621–622.
11. DUPONT, B., S. E. OBERFIELD, E. M. SMITHWICK, T. D. LEE & L. S. LEVINE. 1977. Lancet **2:** 1309–1311.
12. LEVINE, L. S., M. ZACHMANN, M. I. NEW, A. PRADER, M. S. POLLACK, G. J. O'NEILL, S. Y. YANG, S. E. OBERFIELD & B. DUPONT. 1978. N. Engl. J. Med. **299:** 911–915.
13. DUPONT, B., M. S. POLLACK, L. S. LEVINE, G. J. O'NEILL, B. R. HAWKINS & M. I. NEW. 1980. Congenital adrenal hyperplasia. *In* Histocompatibility Testing 1980. P. I. Terasaki, Ed.: 693–706. UCLA Tissue Typing Laboratory. Los Angeles.
14. BONGIOVANNI, A. M. 1953. Bull. Johns Hopkins Hosp. **92:** 244–251.
15. KNORR, D., F. BIDLINGMAIER, O. BUTENANDT, K. V. SCHNAKENBURG & W. WAGNER. 1977. Test for heterozygosity of congenital adrenal hyperplasia. *In* Congenital Adrenal Hyperplasia. P. A. Lee, L. P. Plotnick, A. A. Kowarski & C. J. Migeon, Eds.: 495–500. University Park Press. Baltimore.
16. LORENZEN, F., S. PANG, M. I. NEW, M. S. POLLACK, S. E. OBERFIELD, B. DUPONT, D. CHOW, B. SCHNEIDER & L. S. LEVINE. 1980. J. Clin. Endocrinol. Metab. **50:** 572–579.
17. GROSSE-WILDE, H. J., J. WEIL, E. ALBERT, S. SCHOLZ, F. BIDLINGMAIER, W. G. SIPPEL & D. KNORR. 1979. Immunogenetics **8:** 41–49.
18. LORENZEN, F., S. PANG, M. I. NEW, B. DUPONT, M. S. POLLACK, D. CHOW & L. S. LEVINE. 1979. Pediatr. Res. **13:** 1356–1370.
19. LEJEUNE-LENAIN, C., F. CANTRAINE, M. DUFRASNES, F. PREVOT, R. WOLTER & J. R. M. FRANCKSON. 1980. Clin. Endocrinol. **12:** 525–535.
20. DECOURT, J., M. F. JAYLE & E. BAULIEU. 1957. Ann. Endocrinol. **18:** 416–422.
21. KOHN, B., L. S. LEVINE, M. S. POLLACK, S. PANG, F. LORENZEN, D. LEVY, A. LERNER, G. F. RONDANINI, B. DUPONT & M. I. NEW. 1982. J. Clin. Endocrinol. Metab. **55:** 817–827.
22. NEW, M. I., F. LORENZEN, S. PANG, P. GUNCZLER, B. DUPONT & L. S. LEVINE. 1979. J. Clin. Endocrinol. Metab. **48:** 356–359.
23. MORILLO, E. & L. I. GARDNER. 1979. Lancet **2:** 202–203.
24. LARON, Z., M. S. POLLACK, R. ZAMIR, A. ROITMAN, Z. DICKERMAN, L. S. LEVINE, F.

LORENZEN, G. J. O'NEILL, S. PANG, M. I. NEW & B. DUPONT. 1980. Hum. Immunol. **1:** 55–66.

25. POLLACK, M. S., L. S. LEVINE, G. J. O'NEILL, S. PANG, F. LORENZEN, B. KOHN, G. F. RONDANINI, G. CHIUMELLO, M. I. NEW & B. DUPONT. 1981. Am. J. Hum. Genet. **33:** 540–550.
26. MIGEON, C. J., Z. ROSENWAKS, P. A. LEE, M. D. URBAN & W. B. BIAS. 1980. J. Clin. Endocrinol. Metab. **51:** 647–649.
27. LEVINE, L. S., B. DUPONT, F. LORENZEN, S. PANG, M. S. POLLACK, S. E. OBERFIELD, B. KOHN, A. LERNER, E. CACCIARI, F. MANTERO, A. CASSIO, C. SCARONI, G. CHIUMELLO, G. F. RONDANINI, L. GARGANTINI, G. GIOVANNELLI, R. VIRDIS, E. BARTOLOTTA, C. MIGLIORI, C. PINTOR, L. TATO, F. BARBONI & M. I. NEW. 1980. J. Clin. Endocrinol. Metab. **51:** 1316–1324.
28. ZACHMANN, M. & A. PRADER. 1978. Acta Endocrinol. **87:** 557–565.
29. ZACHMANN, M. & A. PRADER. 1979. Acta Endocrinol. **95:** 542–546.
30. LEVINE, L. S., B. DUPONT, F. LORENZEN, S. PANG, M. S. POLLACK, S. E. OBERFIELD, B. KOHN, A. LERNER, E. CACCIARI, F. MANTERO, A. CASSIO, C. SCARONI, G. CHIUMELLO, G. F. RONDANINI, L. GARGANTINI, G. GIOVANNELLI, R. VIRDIS, E. BARTOLOTTA, C. MIGLIORI, C. PINTOR, L. TATO, F. BARBONI & M. I. NEW. 1981. J. Clin. Endocrinol. Metab. **53:** 1193–1198.
31. POLLACK, M. S., L. S. LEVINE, M. ZACHMANN, A. PRADER, M. I. NEW, S. E. OBERFIELD & B. DUPONT. 1979. Transplant. Proc. **XI:** 1315–1316.
32. DUPONT, B., R. VIRDIS, A. J. LERNER, M. S. POLLACK, C. NELSON & M. I. NEW. 1984. Program and Abstracts of the 7th International Congress of Endocrinology, Quebec. Excerpta Med. Int. Congr. Ser. **652:** 543–566.
33. PANG, S., J. HOTCHKISS, A. L. DRASH, L. S. LEVINE & M. I. NEW. 1977. J. Clin. Endocrinol. Metab. **45:** 1003–1008.
34. PANG, S., D. A. SPENCE & M. I. NEW. 1985. Ann. N. Y. Acad. Sci. **458:** 90–102.
35. CACCIARI, E., A. BALSAMO, A. CASSIO, S. PIAZZI, F. BERNARDI, S. SALARDI, A. CICOGNANI, P. PIRAZZOLI, F. ZAPPULLA, M. CAPELLI, B. PAOLINI & L. MAZZANTI. 1983. Neonatal screening for congenital adrenal hyperplasia using a microfilter paper method for 17-α-hydroxyprogesterone RIA. *In* Recent Progress in Pediatric Endocrinology. G. Chiumello & M. Sperling, Eds.: 221–224. Raven Press. New York.
36. NATOLI, G., L. MOSCHINI, P. ACCONCIA, G. ALBINO, P. COSTA & G. PANSA. 1983. Neonatal screening by microassay of 17-α-hydroxyprogesterone in congenital adrenal hyperplasia. *In* Recent Progress in Pediatric Endocrinology. G. Chiumello & M. Sperling, Eds.: 285–290. Raven Press. New York.
37. FRISTROM, J. W. & P. T. SPIETH. 1980. Principles of Genetics. p. 533. Chiron Press. New York.
38. SPEISER, P. W., B. DUPONT, P. RUBINSTEIN, A. PIAZZA, A. KASTELAN & M. I. NEW. 1985. Am. J. Hum. Genet. **37:** 650–667.
39. THOMPSON, G. & W. BODMER. 1977. The genetic analysis of HLA and disease associations. *In* HLA and Disease. J. Dausset & A. Svejgaard, Eds.: 84–93. Williams and Wilkins Co. Baltimore.
40. JEFFCOATE, T. N. A., J. R. H. FLIEGNER, S. H. RUSSELL, J. C. DAVIS & A. P. WADE. 1965. Lancet **2:** 553–555.
41. PANG, S., L. S. LEVINE, L. L. CEDERQVIST, M. FUENTES, V. M. RICCARDI, J. H. HOLCOMBE, H. M. NITOWSKY, G. SACHS, C. E. ANDERSON, M. A. DUCHON, R. OWENS, I. MERKATZ & M. I. NEW. 1980. J. Clin. Endocrinol. Metab. **51:** 223–229.
42. PANG, S., M. S. POLLACK, M. LOO, O. GREEN, R. NUSSBAUM, G. CLAYTON, B. DUPONT & M. I. NEW. 1985. Ann. N. Y. Acad. Sci. **458:** 111–129.
43. BARTTER, F. C., R. I. HENKIN & G. T. BRYAN. 1968. J. Clin. Invest. **47:** 1742–1752.
44. ROSLER, A., L. S. LEVINE, B. SCHNEIDER, M. NOVOGRODER & M. I. NEW. 1977. J. Clin. Endocrinol. Metab. **45:** 500–512.
45. LEVINE, L. S. & M. I. NEW. 1984. Steroid 21-hydroxylase deficiency. *In* Adrenal Diseases in Childhood, vol. 13. M. I. New & L. S. Levine, Eds.: 1–46. *In* Pediatric and Adolescent Endocrinology. Z. Laron, Ed. Karger AG. Basel, Switzerland.
46. MININBERG, D., L. S. LEVINE & M. I. NEW. 1982. Current concepts in congenital adrenal

hyperplasia. *In* Genetic Disorders, Syndromology, and Prenatal Diagnosis, vol. 5. T. V. N. Persaud, Ed.: 181–196. Alan R. Liss, Inc. New York.

47. NEW, M. I. & L. S. LEVINE. 1973. Congenital adrenal hyperplasia. *In* Advances in Human Genetics. vol. 4. H. Harris & K. Hirschhorn, Eds.: 251–326. Plenum Press. London.

48. NEW, M. I. & R. E. PETERSON. 1966. Pediatr. Clin. North Am. **13:** 43–58.

49. NEW, M. I., B. DUPONT, K. GRUMBACH, & L. S. LEVINE. 1982. Congenital adrenal hyperplasia and related conditions. *In* The Metabolic Basis of Inherited Disease. 5th edit. J. B. Stanbury, J. B. Wyngaarden, D. S. Fredrickson, J. F. Goldstein & M. S. Brown, Eds.: 973–1000. McGraw-Hill. New York.

50. HOLMES, L. B. 1983. Prenatal disturbances: genetic factors. *In* Nelson Textbook of Pediatrics. 12th edit. R. E. Behrman & V. C. Vaughan, III, Eds.: 284. W. B. Saunders Company. Philadelphia.

51 BAUR, M. P. & J. A. DANILOVS. 1980. Population analysis of HLA-A,B,C,DR and other genetic markers. *In* Histocompatibility Testing 1980. P. I. Terasaki, Ed. UCLA Tissue Typing Laboratory. Los Angeles.

Extended MHC Haplotypes in Salt-Losing 21-Hydroxylase Deficiency[a]

CHESTER A. ALPER,[b,c,f] ELLEN FLEISCHNICK,[b,c,f]
ZUHEIR AWDEH,[b,d] DONALD RAUM,[b,e,g]
JOHN F. CRIGLER, JR.,[c,f]
PARK S. GERALD,[c,f]
AND EDMOND J. YUNIS[b,d,h]

[b]Center for Blood Research;
Departments of [c]Pediatrics, [d]Pathology, and [e]Medicine
Harvard Medical School;
[f]Children's Hospital;
[g]Beth Israel Hospital;
and
[h]Dana-Farber Cancer Institute
Boston, Massachusetts 02115

Our interest in 21-hydroxylase deficiency (21-OHD) is not only for its own sake, but as a model for major histocompatibility complex (MHC)-linked diseases of less well-defined etiology and inheritance, particularly type I diabetes mellitus (IDDM). In our view, IDDM shares with 21-OHD a recessive inheritance of disease or susceptibility genes[1–3] and MHC linkage. This linkage is demonstrable in family studies,[4,5] and is observed as the increased incidence of some MHC alleles and the decreased incidence of others in patient populations.[6–9] 21-OHD, on the other hand, has a number of features that make its inheritance much easier to understand than that of IDDM. Penetrance is complete so that all homozygous susceptibles have disease, and most heterozygotes can be detected by appropriate chemical testing.[10] In IDDM, penetrance is between 0.2 and 0.5,[11] and there is no sure test for either healthy susceptibles or carriers. Genetic heterogeneity is clear-cut in 21-OHD[12–14] with its salt-losing, non-salt-losing, late-onset, and cryptic forms. In IDDM, although claims have been made that there are distinct forms marked by different MHC alleles,[15] this is not clear. Finally, 21-OHD is rare and IDDM is common, and since both are recessive disorders, this is even more true of their susceptibility or disease genes.

Our approach to these diseases has been through the study of MHC haplotypes determined from the typing of patients and their families of HLA-A, B, C, DR, C2, BF, C4A, C4B, and GLO.[16,17] The four complement genes (C2, BF, C4A, C4B) are considered a single genetic unit termed a "complotype."[18] Combinations on single chromosomes of specific HLA-B, complotype, and HLA-DR alleles that show significant linkage disequilibrium are called "extended MHC haplotypes." There are about a dozen extended haplotypes in Caucasians with a frequency of about 1% or more (TABLE 1), and they comprise nearly 30% of Caucasian chromosomes.[19] They show limited variation in HLA-A alleles and it is likely that some of them embrace more of the short arm of chromosome 6 than is revealed by current genetic markers.

[a]These studies were supported by NIH grant nos. AM–26844, HL–29583, CA–19589, CA–20531, CA–06516, and HD–17461.

Most of the extended haplotypes appear to have been "frozen" at some point during human evolution, and this is currently manifested by the relative fixity of alleles, even those at widely separated genetic loci on the short arm of chromosome 6.

In theory, if a disease susceptibility gene is embedded in or is closely linked to an extended haplotype, the alleles in that haplotype will be increased among patients.[20] Conversely, if a common extended haplotype does not carry the susceptibility gene, the alleles of that haplotype will be decreased in patients and appear to be "protective."

In this paper, we report further studies of MHC haplotypes in salt-losing 21-OHD and extend our previous analyses.[17] Based on these studies, we attempt to define the characteristics of extended and other MHC haplotypes in MHC-linked diseases, whether fully penetrant or not.

METHODS

Patients and Samples

The study group includes the previously reported families,[17] which now consists of individuals in 31 Caucasian families for whom typing of multiple MHC loci was performed. There were 36 patients with the salt-losing form of 21-OHD in 31 families, including 15 males and 21 females. Of the ten new families, six were followed by the endocrine division of the Children's Hospital and four at the Massachusetts General Hospital.

TABLE 1. Extended Haplotypes among Caucasian Chromosomes

Haplotype	Frequency
[B8, DR3, SC01]	0.093
[B7, DR2, SC31]	0.060
[Bw44, DR7, FC31]	0.037
[Bw44, DR4, SC30]	0.034
[Bw57, DR7, SC61]	0.030
[Bw35, DR1, FC3, 20]	0.012
[Bw38, DR4, SC21]	0.012
[Bw61, DRw6, SC02]	0.011
[B14, DR1, SC22]	0.011
[Bw62, DR4, SC33]	0.010
[B18, DR3, F1C30]	0.010
[B18, DR2, S042]	0.010

Blood for typing of BF, C2, and C4 was collected into EDTA. Plasma obtained by centrifugation was frozen immediately and stored at −80°C. Portions were thawed immediately before analysis. Blood for HLA typing was drawn into a heparinized syringe containing approximately 1 ml of sodium heparin (1000 U/ml) and diluted with equal volumes of RPMI 1640. Lymphocytes were separated using Ficoll-Hypaque, frozen, and stored in liquid nitrogen until analysis.

BF, C2, and C4 Typing

Typing of BF was by agarose gel electrophoresis at pH 8.6 and immunofixation with goat antihuman B (Atlantic Antibodies; Scarborough, Maine), as described previously.[21] For C2 typing,[22] plasma samples were subjected to isoelectric focusing in polyacrylamide gel, and C2 patterns were developed in an agarose gel overlay containing antibody-sensitized sheep red cells and fresh normal human serum diluted 1:90. C4 types were obtained by electrophoresis in agarose gels of desialated plasma and immunofixation with anti-C4 (Atlantic Antibodies), as described previously.[23] For detection of half-null haplotypes, desialated plasma samples were subjected to crossed immunoelectrophoresis.[24]

HLA Typing

HLA-A, B, and C antigens were assigned by the standard NIH lymphocyte microcytotoxicity assay. HLA-DR specificities were assigned on testing B lymphocytes separated from frozen lymphocytes on nylon columns.[25]

GLO I Typing

Hemolysates were subjected to electrophoresis on cellulose acetate for GLO I typing. Patterns were developed with suitable reagents.[2]

Analysis of Data

MHC haplotypes occurring in patients were considered "disease" haplotypes, and those occurring in healthy family members, but not patients, were considered family normal.[16] Both sets of chromosomes or haplotypes were compared with a much larger population of normal haplotypes ($n = 448$) from Caucasians residing, for the most part, in the Boston area. All haplotypes were analyzed for distribution of HLA-DR specificities and of extended MHC haplotypes.[19] Individual complotypes were designated, in arbitrary order, by their BF, C2, C4A, and C4B markers. Significance of differences in distribution between populations of chromosomes were tested in contingency tables and in χ^2 analysis with correction for continuity where appropriate.

RESULTS

There were 64 chromosomes from patients with the salt-losing form of 21-OHD, and 65 normal chromosomes in these same families. FIGURE 1A shows the distribution of patient haplotypes (left-hand bars) in comparison with normal haplotypes in these families (center bars) and the general normal Caucasian haplotypes obtained from other studies (right-hand bars), all arranged according to HLA-DR types. The contribution of specific extended haplotypes (letters in clear portions of the bars) and nonextended haplotypes (crosshatched areas) to total haplotypes of each DR type is also shown. Significant differences were seen in HLA-DR3-bearing haplotypes (decreased in patient chromosomes, particularly when compared with family normal haplotypes, $p < 0.025$) and in HLA-DR7 (increased in patient chromosomes when compared with general control haplotypes, $p < 0.04$). There were no significant differences in HLA-DR frequencies between family normal and general control haplotypes except that HLA-DR3 was increased among family normal chromosomes ($p < 0.02$).

From FIGURE 1A, it is evident that all of the increase in HLA-DR7 haplotypes among patient chromosomes was the result of the increase in the extended haplotype [HLA-Bw47, DR7, FC0,31]. This haplotype occurs in normal chromosomes at a frequency of no more than 0.001 (one instance in 1322) and has a relative risk of 370 in patient chromosomes. Of the two common extended haplotypes with HLA-DR7

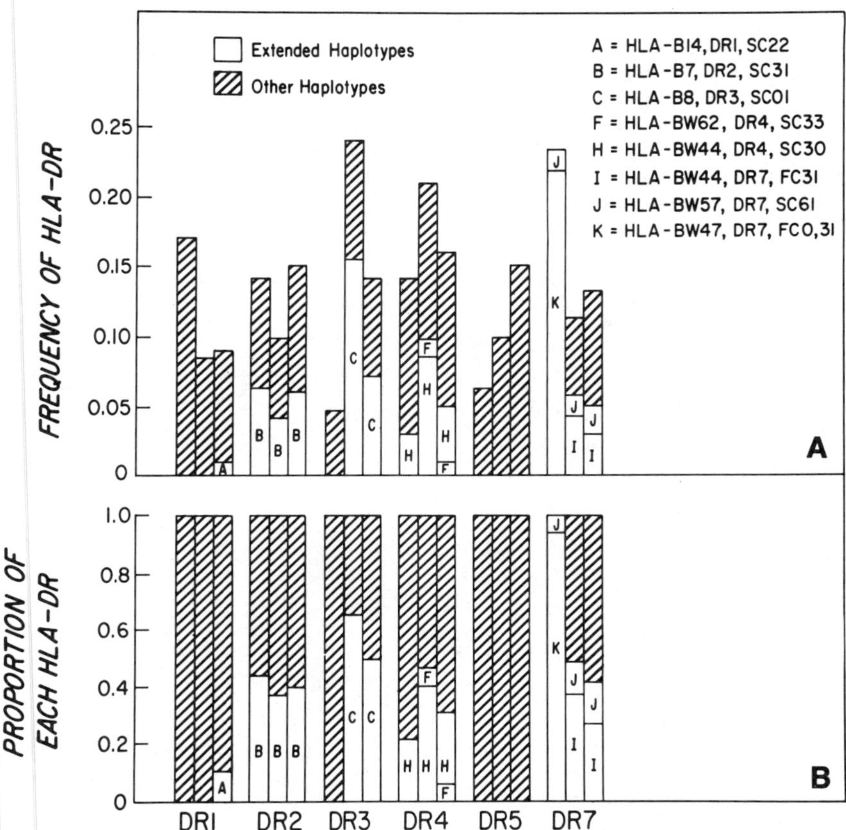

FIGURE 1. In FIGURE 1A, the frequencies of haplotypes in patients with salt-losing 21-OHD are shown by their DR types (left-hand bars) for comparison with normal chromosomes for the same families (middle bars) and general normal Caucasian chromosomes (right-hand bars). Extended haplotypes are shown within each bar as clear areas, with letters defined in the upper right portion of the figure. Randomly associating, nonextended haplotypes are shown in the upper portions of each bar as crosshatched areas. In FIGURE 1B, the same data are shown normalized to the frequency of each HLA-DR allele to permit direct assessment of contribution of each extended haplotype to its HLA-DR allele frequency.

among normal chromosomes, there was one instance of [HLA-Bw57, DR7, SC61] and none of [HLA-Bw44, DR7, FC31]. The most common of all extended haplotypes [HLA-B8, DR3, SC01] was not found at all among patient chromosomes, despite its being increased among family normal chromosomes ($p = 0.001$).

Of other extended haplotypes, [HLA-B7, DR2, SC31], [HLA-Bw57, DR7, SC61], and [HLA-Bw44, DR4, SC30] were present in normal absolute percentages. [HLA-Bw44, DR7, FC31] was not found among patient chromosomes, but only two instances were expected. All other extended haplotypes were found among normal chromosomes at too low a frequency to be expected with certainty among patient chromosomes.

FIGURE 1B permits the assessment of the proportion of specific extended haplotypes in both patient and normal chromosomes of specific HLA-DR types. It is clear that there was a striking increase in the proportion of [HLA-Bw47, DR7, FC0,31] among patient haplotypes with HLA-DR7. Similarly, HLA-DR3-bearing haplotypes are severely depleted of [HLA-B8, DR3, SC01].

Over 20% of chromosomes 6p from patients with salt-losing 21-OHD carried the rare MHC haplotype [HLA-Bw47, DR7, FC0,31]. HLA-A specificities and GLO types on these haplotypes are given in TABLE 2. Over 60% of these haplotypes carried HLA-A3 (frequency in normal chromosome population = 13%), two carried HLA-A28 (15.4 versus 4.5%), and there were single instances of HLA-A11 (7.7 versus 5%), Aw24 (7.7 versus 9%), and A2 (7.7 versus 27%). Over 80% of these haplotypes bore GLO 1, whereas 39% of our general normal Caucasian MHC haplotypes carried GLO 1.

TABLE 2. HLA-A and GLO Types in [HLA-Bw47, DR7, FC0,31] Haplotypes from Patients with Salt-Losing 21-OHD

HLA-A	Proportion	Decimal Fraction	Normal Frequency	GLO	Proportion	Decimal Fraction	Normal Frequency
A3	8/13	0.615	0.144	1	13/15	0.87	0.39
A28	2/13	0.154	0.041	2	2/15	0.13	0.61
A11	1/13	0.077	0.055				
Aw24	1/13	0.077	0.076				
A2	1/13	0.077	0.270				

DISCUSSION

The most striking differences in MHC haplotypes between normal chromosomes and those from patients with salt-losing 21-OHD are the increase in [HLA-Bw47, DR7, FC0, 31] and the decrease in [HLA-B8, DR3, SC01]. This is in keeping with our concept that if a disease susceptibility gene is on an extended haplotype with its relatively fixed alleles, it will be on most instances of that extended haplotype, and, if it is not, that haplotype and its alleles will be reduced in frequency on patient chromosomes and appear to be "protective" against the disease.[20] The recessive nature of the inheritance of 21-OHD is consistent with the presence of [HLA-Bw47, DR7, FC0,31] homozygotes[17] and the complete absence of [HLA-B8, DR3, SC01] from patient chromosomes. As befits a rare disorder, the extended haplotype which is increased markedly in patients is also rare. That most, if not all such marker extended haplotypes carry susceptibility genes is borne out by detection of the carrier state for 21-OHD in healthy heterozygotes with no family history of disease for a haplotype found in Australia in 21-OHD patients.[27]

The fact that three common extended haplotypes are found among patient chromosomes ([HLA-B7, DR2, SC31], [HLA-Bw44, DR4, SC30], and [HLA-Bw57, DR7, SC61]) in normal or near-normal numbers raises the possibility that some of these combinations of alleles sometimes do and sometimes do not mark extended haplotypes.

The haplotype [HLA-Bw47, DR7, FC0,31] has many features of an extended haplotype even though our information on it is mostly derived from study of patients with salt-losing 21-OHD. The limited variation in HLA-A (over 60% HLA-A3) and the striking and highly significant association with GLO 1[17] strongly suggest "fixity" of at least 5 centiMorgans of chromosome 6. Unlike [B8, DR3, SC01, GLO2], which shows striking segregation distortion from the male of 90% in disease (IDDM and gluten enteropathy) families, [HLA-Bw47, DR7, FC0,31] shows no distortion of transmission from fathers in the present families (data not shown).

Comparison of patient haplotypes with family normal haplotypes[16] has the disadvantage of providing relatively few control haplotypes compared with general controls. On the other hand, population stratification effects are minimized. The increased incidence of [HLA-B8, DR3, SC01] in family normal chromosomes is in sharp contrast to its absence in patient chromosomes, and this may well reflect the Northern European distribution of both this haplotype[28] and the mutation(s) responsible for salt-losing 21-OHD.

In both 21-OHD and IDDM, the increase of certain extended haplotypes results in increases in their specific alleles in the patient populations.[16] Also, just as [HLA-Bw47, DR7, FC0,31] appears to carry a single mutant gene which probably arose in the British Isles, the extended haplotypes marking IDDM appear to have arisen in specific populations for which the extended haplotypes are markers: [HLA-B8, DR3, SC01] in Northern Europe, [HLA-B18, DR3, F1C30] in the Pyrenees and Mediterranean, [HLA-B15, DR4, SC33] possibly also in Northern Europe, and [HLA-Bw38, DR4, SC21] in Ashkenazi Jews. The difference is that [HLA-Bw47, DR7, FC0,31] is rare, and the extended haplotypes marking IDDM are all relatively common, particularly in their respective primary populations. Both diseases have a decrease in those extended haplotypes that presumably lack susceptibility genes [HLA-B8, DR3, SC01] in 21-OHD and [HLA-B7, DR2, SC31] in IDDM. The normal frequency of the latter extended haplotype in salt-losing 21-OHD is, however, not explainable in these terms. Finally, consistent with recessive inheritance of disease susceptibility, the decrease in "protective" haplotypes or alleles is extreme. Pure dominant inheritance predicts 50% reduction in a 100% "protective" marker or haplotype.

It is our hope that further study of MHC markers in 21-OHD will provide additional clues about IDDM and other mysterious diseases of unclear etiology, pathogenesis, and genetic determination. It has already provided tantalizing analogies and insights.

ACKNOWLEDGMENTS

We thank Miss Deborah Marcus, Miss Catherine Ramaika, Miss Carroll Goldsmith, Mrs. Sharon Martin Alosco, and Mrs. Rosanne Stein for expert technical assistance; Dr. Dorothy Villee, Mrs. Stephanie deVos, and Ms. Kris Olsen for help in obtaining patient samples; and Drs. John Crawford, Alia Antoon, and Robert Richie, Jr. for permission to study their patients.

REFERENCES

1. Bongiovanni, A. M. 1978. Congenital adrenal hyperplasia and related conditions. *In* Metabolic Basis of Inherited Disease. J. Stanbury, J. B. Wyngaarden & D. S. Frederickson, Eds.: 868–893. McGraw–Hill. New York.

2. RUBINSTEIN, P., N. SUCIU-FOCA & J. F. NICHOLSON. 1977. Genetics of juvenile diabetes mellitus: A recessive gene closely linked to HLA-D and with 50 percent penetrance. N. Engl. J. Med. **297:** 1036–1040.
3. RAUM, D., Z. AWDEH & C. A. ALPER. 1981. BF types and the mode of inheritance of insulin-dependent diabetes mellitus (IDDM). Immunogenetics **12:** 59–74.
4. CUDWORTH, A. G. & J. C. WOODROW. 1976. Genetic susceptibility in diabetes mellitus: Analysis of the HLA association. Br. Med. J. **7:** 846–848.
5. DUPONT, B., S. E. OBERFELD, E. M. SMITHWICK, T. D. LEE & L. S. LEVINE. 1977. Close genetic linkage between HLA and congenital adrenal hyperplasia (21-hydroxylase deficiency). Lancet **2:** 1309–1311.
6. SINGAL, D. P. & M. A. BLAJCHMAN. 1973. Histocompatibility antigens, lymphocytotoxic antibodies, and tissue antibodies in patients with diabetes mellitus. Diabetes **22:** 429–432.
7. ZAPPACOSTA, S., M. DEFELICE, M. MINOZZI, G. LOMBARDI, R. VALENTINO & G. VANACORE. 1978. HLA and congenital adrenal hyperplasia. Lancet **2:** 524.
8. POLLACK, M. S., L. LEVINE, M. ZACHMANN, A. PRADER, M. NEW, S. OBERFELD & B. DUPONT. 1979. Possible genetic linkage disequilibrium between HLA and the 21-hydroxylase deficiency gene (congenital adrenal hyperplasia). Transplant. Proc. **11:** 1315–1316.
9. KLOUDA, P. T., R. HARRIS & D. A. PRICE. 1980. Linkage and association between HLA and 21-hydroxylase deficiency. J. Med. Genet. **17:** 337–341.
10. GUTAI, J. P., A. A. KOWARSKI & C. J. MIGEON. 1977. Detection of the heterozygous state in siblings of patients with congenital adrenal hyperplasia due to 21-hydroxylase deficiency. J. Pediatr. **94:** 770–772.
11. CAHILL, G. F., JR. 1979. Current concepts of diabetic complications with emphasis on hereditary factors: A brief review. *In* Genetic Analysis of Common Diseases: Applications to Predictive Factors in Coronary Heart Disease, 1979. C. F. Sing & M. Skolink, Eds.: 113–129. Alan R. Liss. New York.
12. MCKUSICK, V. A. 1975. Mendelian Inheritance in Man. 4th edit. Johns Hopkins University Press. Baltimore.
13. NEWMARK, S., R. DLUHY, G. WILLIAMS, P. POCHI & L. ROSE. 1977. Partial 11- and 21-hydroxylase deficiencies in hirsute women. Am. J. Obstet. Gynecol. **127:** 594–598.
14. LEVINE, L. S., B. DUPONT, F. LORENZEN & S. PANG. 1980. Cryptic 21-hydroxylase deficiency in families of patients with classical congenital adrenal hyperplasia. J. Clin. Endocrinol. Metabl. **51:** 1316–1324.
15. ROTTER, J. I. & D. L. RIMOIN. 1978. Heterogeneity of diabetes mellitus—update, 1978. Diabetes **27:** 599–605.
16. RAUM, D., Z. AWDEH, E. J. YUNIS, C. A. ALPER & K. H. GABBAY. 1984. Extended major histocompatibility complex haplotypes in type I diabetes mellitus. J. Clin. Invest. **74:** 449–454.
17. FLEISCHNICK, E., Z. L. AWDEH, D. RAUM, J. GRANADOS, S. M. ALOSCO, J. F. CRIGLER, JR., P. S. GERALD, C. M. GILES, E. J. YUNIS & C. A. ALPER. 1983. Extended MHC haplotypes in 21-hydroxylase-deficiency congenital adrenal hyperplasia: Shared genotypes in unrelated patients. Lancet **i:** 152–156.
18. ALPER, C. A., D. RAUM, S. KARP, Z. L. AWDEH & E. J. YUNIS. 1983. Serum complement "supergenes" of the major histocompatibility complex in man (complotypes). Vox Sang. **45:** 62–67.
19. AWDEH, Z. L., D. RAUM, E. J. YUNIS & C. A. ALPER. 1983. Extended HLA/complement allele haplotypes: Evidence for T/t-like complex in man. Proc. Natl. Acad. Sci. USA **80:** 259–263.
20. ALPER, C. A., Z. L. AWDEH, D. D. RAUM & E. J. YUNIS. 1982. Extended major histocompatibility complex haplotypes in man: Role of alleles analogous to murine t mutants. Clin. Immunol. Immunopathol. **24:** 276–285.
21. ALPER, C. A., T. BOENISCH & L. WATSON. 1972. Genetic polymorphism in human glycine-rich beta-glycoprotein. J. Exp. Med. **135:** 68–80.
22. ALPER, C. A. 1976. Inherited structural polymorphism in human C2: Evidence for genetic linkage between *C2* and *Bf*. J. Exp. Med. **144:** 1111–1115.
23. AWDEH, Z. L. & C. A. ALPER. 1980. Inherited structural polymorphism of the fourth component of human complement. Proc. Natl. Acad. Sci. USA **77:** 3576–3580.

24. AWDEH, Z. L., D. RAUM & C. A. ALPER. 1979. Genetic polymorphism of human complement C4 and detection of heterozygotes. Nature **282:** 205–207.
25. DANILOVS, J. A., G. AYOUB & P. I. TERASAKI. 1980. B lymphocyte isolation by thrombin-nylon wool. *In* Histocompatibility Testing, 1980. P. I. Terasaki, Ed.: 287–288. UCLA Tissue Typing Laboratory. Los Angeles.
26. KÖMPF, J., S. BISSBORT, S. GUSSMANN & H. RITTER. 1975. Polymorphism of red cell glyoxalase I (E.C. 4.4.1.5). A new genetic marker in man. Humangenetik **27:** 141–143.
27. MCCLUSKEY, J., P. H. KAY, M. STUCKEY, F. T. CHRISTIANSEN, R. L. DAWKINS & G. WILSON. 1983. MHC "supratype" predicting heterozygous 21-hydroxylase deficiency. Lancet **i:** 764–765 (letter).
28. BODMER, W. & G. THOMPSON. 1977. Population genetics and evolution of the HLA system. *In* HLA and Disease. J. Dausset & A. Svejgaard, Eds.: 280–295. Munksgaard. Copenhagen.

The HLA Associations in Congenital Adrenal Hyperplasia due to 21-Hydroxylase Deficiency in a Yugoslav Population

A. KAŠTELAN, LJ. BRKLJAČIĆ-ŠURKALOVIĆ, AND M. DUMIĆ

Tissue Typing Centre Department of Urology
and
Department of Pediatrics
Medical Faculty University of Zagreb
41000 Zagreb, Yugoslavia

Since the findings by Dupont *et al.* (1977)[1] on the close genetic linkage between HLA and congenital adrenal hyperplasia (CAH) due to 21-hydroxylase deficiency (21-OH def.), numerous studies have examined HLA-linked genetic markers in many different populations.

In Zagreb, Yugoslavia, we have studied 47 families of patients with congenital adrenal hyperplasia due to 21-hydroxylase deficiency. Thirty-eight of these families were highly informative for the genetic studies. There were 21 unrelated patients with salt-wasting (SW) form, 13 unrelated patients with simple-virilizing (SV) form, and 10 patients with the nonclassical form of the disease. The diagnosis of CAH was based on clinical, hormonal, and HLA studies.

The hormonal profile of family members consisted of baseline (Pang *et al.*, 1977[2]) and ACTH-stimulated levels of 17-hydroxyprogesterone (17-OHP) in 60 or 360 minute tests (Lorenzen *et al.*, 1979;[3] Laron *et al.*, 1980[4]).

HLA typing for the HLA-A, B, C, and DR antigens was performed by standard techniques using VIIIth and IXth Histocompatibility Workshop reagents and additional highly selected local antisera. The HLA antigens in the patient groups were compared with HLA antigens found in the nonaffected parental HLA haplotypes and with HLA antigens in the normal population.

RESULTS AND CONCLUSIONS

The gene frequencies for HLA antigens in SW and SV patients are compared with the HLA antigens found on the nonaffected parental haplotypes in TABLE 1. As can be seen, the most striking increase in gene frequency was found for antigen HLA-DR7, which was detected in 29 percent of haplotypes in SW patients and only in 4 percent of haplotypes in SV patients ($\chi^2 = 4.85; p = 0.02$). The increase of this antigen was even higher in SW patients when compared with the healthy parental haplotypes ($\chi^2 = 9.0$; $p = 0.002$). No significant difference was found in antigen HLA-Bw47 between the two patient groups, but the difference was significant between SW patients and control ($\chi^2 = 3.64; p = 0.05$). Two antigens, HLA-Bw16 and HLA-DR3, seem to be "protective" for the classical 21-OH deficiency since their gene frequencies showed a

TABLE 1. The Gene Frequencies for HLA Antigens with Deviation among the CAH Patients[a]

HLA	Total CAH	SW	SV	Control	Population
Bw47	0.088 (6/68)	0.119 (5/42)	0.038 (1/26)	0.015 (1/68)	0.004
DR7	0.191 (13/68)	0.286 (12/42)	0.038 (1/26)	0.059 (4/68)	0.069
Bw16 (38, 39)	0.0 (0/68)	0.0 (0/42)	0.0 (0/26)	0.118 (8/68)	0.067
B8	0.044 (3/68)	0.048 (2/42)	0.038 (1/26)	0.132 (9/68)	0.077
DR1	0.074 (5/68)	0.048 (2/42)	0.115 (3/26)	0.118 (8/68)	0.115
DR3	0.029 (2/68)	0.048 (2/42)	0.0 (0/26)	0.147 (10/68)	0.092

[a]Significant deviations are underlined.

significant decrease in the total CAH patient group when compared with the control ($\chi^2 = 6.51; p = 0.01$ for the former antigen; $\chi^2 = 4.47; p = 0.03$ for the latter antigen). Deviation for HLA-B8 antigen did not reach statistical difference.

In five out of ten SW patients in whom at least one of the 21-OH deficiency genes was associated with antigen HLA-DR7, the 21-OH deficiency gene was transmitted with the complete or partial part of the HLA-A3, Cw6, Bw47, BfF, DR7 complotype (TABLE 2). The other five SW patients inherited the 21-OH deficiency gene in linkage disequilibrium with HLA-DR7 antigen only. The HLA-Bw47 antigen in these patients

TABLE 2. The High Risk Genes Bw47 and DR7 in SW CAH Patients

KAS-41	Aw24	Cw6	Bw47	BfF	DR7
	Aw24	Cw6	B7	BfS	DR2
KAS-39	A3	Cw6	Bw47	BfF	DR7
	A2	Cw5	B12	BfS	DR5
KAS-43	A2	CwX	Bw47	BfF	DR7
	A9	CwX	B5	BfS	DR2
KAS-44	A3	Cw6	Bw47	—	DR7
	A29	Cw2	B7	—	DRX
KAS-64	A3	—	Bw47	—	DR7
	A1	—	B18	—	DR4
KAS-51	A2	Cw6	Bw21	BfS	DR7
	A2	CwX	B5	BfS	DRX
KAS-65	A11	—	Bw35	BfF	DR7
	A1	—	Bw35	BfS	DRw6
KAS-57	Aw32	—	B5	—	DR7
	A1	—	Bw22	—	DR5
KAS-5	A2	—	B40		DR7
	A2	—	B27	—	DR7
KAS-66	A2	—	B13	—	DR7
	A2	—	Bw21	—	DR7

TABLE 3. The HLA and 21-OH Genotypes of Ten Individuals with the Mild 21-OH Deficiency Gene

Family I.D.	Disease Form	HLA-Genotype	21-OH Genotype
KAS-47	CRY	A3,Cw6,Bw39,DR6/Aw30,Cw6,B13,DR5	21-OH^{CRY}/21-OH^{CAH}
KAS-35	CRY	A28,Cw4,Bw35,DR8/A2,Cw5,Bw44,DR2	21-OH^{CRY}/21-OH^{CAH}
KAS-40	CRY	A25,B18,DR5/A3,B18,DR1	21-OH^{CRY}/21-OH^{CAH}
[a]KAS-41	CRY	A28,B14,DRw6/A11,B5,DRw6	21-OH^{CRY}/21-OH^{CAH}
KAS-50	CRY	Aw33,Cw8,B14,DR1/A11,Cw4,Bw53,DRw6	21-OH^{CRY}/21-OH^{CAH}
KAS-50	NONCLASSICAL?	A25,Cw5,B12,DR5/A11,Cw4,Bw53,DRw6	21-OH^{NONCLASS}/21-OH^{CAH}
KAS-45	CRY	A2,Cw6,B18,DR5/A2,Cw5,Bw50,DR7	unknown
KAS-45	AAH	A2,Cw6,B18,DR5/A2,CwX,Bw44,DR2	unknown
KAS-32	AAH	A2,Bw44,DRw6/AX,B18,DR5	unknown
[b]KAS-51	AAH	Aw33,Cw8,B14,DR1/A2,CwX,Bw35,DRX	unknown

[a]Hungarian origin.
[b]German origin.

TABLE 4. The Clinical and Hormonal Profile of Family Members KAS-50[a]

	Danijela 21-OH^{CAH} (Child)		Marino 21-OH$^{NON\text{-}CLASS?}$ (Child)				Dušan 21-OH^{CRY} (Father)	
Age	5.5 Yr		4 Mo		5.5 Mo		32 Yr	
Sex	F		M				M	
Ht(cm)	103.5 (10%)		63 (50%)		67 (50%)		177	
Clitoromegaly	Yes		NA				NA	
Labial fusion	Yes		NA				NA	
17-OHP/ng/dl/0' and 60'/	3850	NT	2139	2790	3590	20140	500	3010
DHEA /ng/dl/0'and 60'/	25	NT	540	1021	411	622	NT	
Δ^4/ng/dl/0' and 60'/	425	NT	439	553	159	530	200	565
PRA/ng/ml/	42.80		135.8		106.0		NT	
Na/mmol/l	134.0		138–140					
K/mmol/l	4.5		5.8–6.3					

[a]NA, Not applicable; NT, Not tested.

was missed. In two patients (KAS-5 and KAS-66), both parental HLA-DR7 associated 21-OH deficiency genes segregated together.

TABLE 3 shows the HLA and 21-OH deficiency genotypes in ten patients with the nonclassical form of CAH. The patients with nos. 51, 32, and 45 were "late-onset," while others were "cryptic." Family no. 45 had two nonclassical patients; a daughter with late-onset and a father with the "cryptic" form of 21-OH deficiency. Family no. 50 also had two nonclassical patients; a father who was definitely "cryptic" and a son who probably displayed the "nonclassical" form of the disease.

As can be seen in TABLE 3, the HLA-B14, DR1 haplotype was found in only two patients (KAS-50 and KAS-51). One was of non-Yugoslav origin. All but one of the "cryptic" patients (KAS-45) had a severe classical 21-OH deficiency gene in addition to a mild "cryptic" 21-OH deficiency allele.

TABLE 4 shows the clinical and hormonal profile of the three family members KAS-50 (cf. also TABLE 3). The index case was a 5.5 year-old girl (Danijela) with classical 21-OH deficiency. Her father (Dušan) displayed a "cryptic" hormonal profile, while her younger brother (Marino), who showed no clinical symptoms of

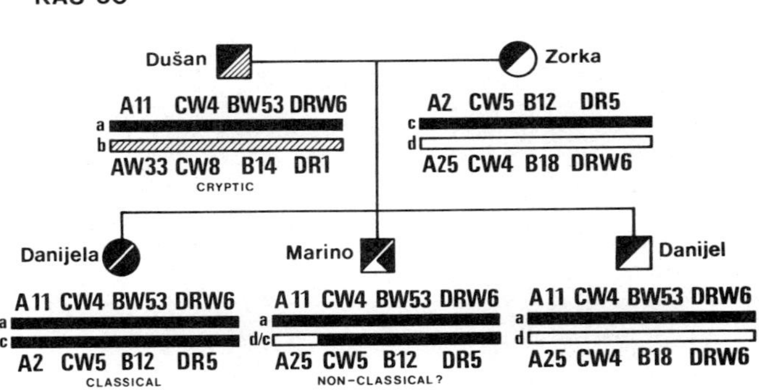

FIGURE 1. The HLA typing results of family members KAS-50.

CAH, displayed two different patterns of 17-OHP levels after ACTH stimulation. Both were characteristic for the nonclassical form of 21-OH deficiency.

The HLA typing of this family revealed (FIGURE 1) that this child was identical with his index case sister for the HLA-B, C, DR loci and presumably also for the classical 21-OH deficiency allele. However, these two siblings differed at the HLA-A chromosome segment since Marino inherited a maternal HLA-A/C recombinant haplotype. This may give rise to a "mild" clinical and hormonal expression of the disease in this patient.

From these data, we can conclude that: (1) in the Yugoslav population, the SW, but not the SV form of CAH is strongly associated with antigen HLA-DR7 (no association to any particular HLA antigen is found in the SV form of CAH); (2) the HLA-B8, DR3 haplotype and the HLA-Bw16 antigen are protective for CAH; (3) the nonclassical 21-OH deficiency in the Yugoslav population seems not to be associated with the HLA-B14, DR1 haplotype; and (4) the clinical and hormonal expression of 21-OH deficiency may be associated with the "A" locus of the HLA system.

REFERENCES

1. Dupont, B., S. E. Oberfield, E. M. Smithwick, T. D. Lee & L. S. Levine. 1977. Close genetic linkage between HLA and congenital adrenal hyperplasia (21-hydroxylase deficiency). Lancet **2:** 1309–1312.
2. Pang, S., J. Hotchkiss, A. L. Drash, L. S. Levine & M. I. New. 1977. Microfilter paper method for 17-hydroxyprogesterone radioimmunoassay: Its application for rapid screening for congenital adrenal hyperplasia. J. Clin. Endocrinol. Metab. **45:** 1003–1008.
3. Lorenzen, F., S. Pang, M. I. New, B. Dupont, M. S. Pollack, D. M. Chow & L. S. Levine. 1979. Hormonal phenotype and HLA-genotype in families of patients with congenital adrenal hyperplasia (21-hydroxylase deficiency). Pediatr. Res. **13:** 1356–1360.
4. Laron, Z., M. S. Pollack, R. Zamir, A. Roitman, Z. Dickerman, L. S. Levine, G. Lorenzen, G. J. O'Neill, S. Pang, M. I. New & B. Dupont. 1980. Late onset 21-hydroxylase deficiency and HLA in the Ashkenazi population: A new allele at the 21-hydroxylase locus. Hum. Immunol. **1:** 55–66.

HLA Associations in 21-Hydroxylase Deficiency (Congenital and Late-Onset Adrenal Hyperplasia) in France

P. COUILLIN,[a] R. RAPPAPORT,[b] F. KUTTENN,[c] J. HORS,[d]
J. FEINGOLD,[e] J. BOUÉ,[a] AND A. BOUÉ[a]

[a]*Unité de Recherches de Biologie prénatale*
INSERM U.73
75016 Paris, France

[b]*Service d'Endocrinologie Pédiatrique*
Hôpital des Enfants Malades

[c]*Service d'Endocrinologie de la Reproducion*
Hôpital Necker

[d]*Unité de Recherches d'Immunogénétique de*
la Transplantation Humaine
INSERM U.93, Hôpital Saint-Louis

[e]*Unité de Recherches de Génétique Epidémiologique*
INSERM U.155
Paris, France

INTRODUCTION

It was demonstrated in 1977[1] that congenital adrenal hyperplasia due to 21-hydroxylase (21-OH) deficiency behaved as an autosomal genetic trait in close linkage with HLA. Nonrandom gametic associations for the 21-OH deficiency trait and HLA loci have been studied by several investigators.[2,3] More recently, there has been evidence that the trait responsible for the late-onset form of 21-OH deficiency is also linked to HLA loci.[4,5]

The aim of this study was to investigate the gene frequency patterns among French patients affected with the congenital form or the late-onset form of 21-OH deficiency.

MATERIALS AND METHODS

Patients with 21-OH deficiency were carefully selected on the basis of clinical examination and biochemical investigation. The study included 109 French families, from various parts of France, in which the index case had a congenital adrenal hyperplasia, and 44 families in which the index case had a late-onset form of the 21-OH deficiency. HLA typing was performed by microlymphocytotoxicity.[6] DR specificities were recognized on B lymphocyte enriched suspensions obtained by rosetting. Bf variants were determined by immunofixation electrophoresis.[7]

Gene frequencies were calculated after direct counting and were compared with those obtained from the parents' haplotypes not bearing the 21-OH deficient gene and with those from a control reference panel (Dausset, unpublished data). Statistical significance was evaluated after two by two comparisons and a χ^2 test.

TABLE 1. Frequencies of HLA-A and HLA-B Alleles in Abnormal and Normal Haplotypes of 109 French Families with 21-OH Deficiency, Congenital Form

Markers	Abnormal Haplotype	Normal Haplotype	Controls 209 Families	p values 1-2	p values 1-3
HLA-A					
1	9.4	16	12	0.05	
2	25.8	26.6	23.6		
3	19.7	11.2	13.2	0.02	0.02
9	10.8	15.4	14.5		
10	9.8	2.7	6.9	0.01	
11	7.0	6.9	6.7		
w 19	3.3	4.2	6.8		0.1
28	5.6	5.8	3.7		
29	4.7	5.8	5.4		
w 32	1.9	4.3	3.5		
w 33	0.4	1.1	1		
—	1.4	0	2.7		
HLA-B					
5	8.4	7.9	7.9		
7	13.1	11.6	9.6		
8	2.8	8.5	8.3	0.02	0.01
12	12.2	17.5	16.1		
13	0.9	0	2.9		
14	3.8	4.2	5.9		
15	8.4	4.8	5.4		0.1
w 16	5.2	5.8	3.6		
17	2.3	7.4	3.8	0.02	
18	3.8	6.3	6.6		
w 21	3.8	2.7	3.7		
w 22	1.9	2.7	2.7		
27	5.2	2.7	4.3		
w 35	11.3	9.5	9.8		
37	0.9	1.6	0.6		
40	8.9	5.8	5.2		0.05
w 41	0.5	0.5	0.5		
w 47	3.8	0	0.4	0.02	0.001
—	2.8	0.5	2.7		
	$n = 213$	$n = 189$	$n = 874$		

RESULTS

Families with Congenital Adrenal Hyperplasia

Only two significant differences were observed between the frequency of HLA alleles in the patients and in the controls (TABLE 1): a negative association with B8 and a positive association with Bw47, but they were of low incidence. All the Bw47 were associated on the same haplotype with A3, C6, DR7. This association may explain the slight increase of the frequency of HLA-A3 in the patients.

TABLE 2. Frequencies of DR Alleles

	1 Congenital Form	2 Late-Onset Form	3 Controls	p values 2-3
HLA-DRw				
1	7.8	17.9	4.3	0.01
2	23.4	3.6	14.9	0.5
3	3.1	0	9.6	0.5
4	10.9	3.6	6.7	
5	6.3	19.6	15.9	
6	12.5	39.3	6.7	0.001
7	14.1	8.9	11.1	
—	21.9	7.1	30.8	0.001
	$n = 64$	$n = 56$	$n = 208$	

No significant differences were observed in the frequencies of DR and Bf alleles (TABLES 2 and 3).

Families with Late-Onset 21-OH Deficiency

A strong positive association was observed with HLA–B14, which was present in 30 index cases (68.2 per cent) of the 44 families studied (TABLE 4). In 16 families, the haplotype was Aw33, B14. In the 14 families in which late-onset 21-OH deficiency was not associated to B14, we noticed that either B12 or B35 was present in the patients (TABLE 4).

Significant differences were also observed for DR alleles: positive associations were found with Drw1 ($p < 0.01$) and with DRw6 ($p < 0.001$) (TABLE 2).

DISCUSSION

This study has confirmed the close genetic linkage between 21-OH deficiency trait and HLA-B locus in the congenital forms and in the late-onset forms of the disease. Sums of LOD-scores were maximum for $\Theta = 0.00$, and reached 32.113 for the congenital form (82 informative families) and 11.495 for the late-onset form (20 informative families).

There are differences, though, in the frequencies of the different HLA alleles associated with the 21-OH deficient gene in the two clinical forms. In the congenital form, in more than 90 per cent of the French index cases, the 21-OH deficiency trait was randomly associated to HLA loci. A low incidence of associations was observed, and these associations confirmed the findings of other studies[2,3]: a negative association

TABLE 3. Frequencies of Bf Alleles in Abnormal and Normal Haplotypes of French Families with 21-OH Deficiency, Congenital Form

Markers	Abnormal Haplotype	Normal Haplotype	Controls	p values 1-2	p values 1-3
BfS	63.6	79.3	81.1	NS	0.05
BfF	36.4	20.7	18.2	NS	0.05
	$n = 33$	$n = 29$	$n = 187$		

with B8 and a positive association with haplotype A3, Cw6, Bw47, DR7 (7 out of 109 index patients). The seven French index patients with A3, Cw6, Bw47, DR7 belong to families from the northern half of France. In addition, the association with Bw47 has been observed in central England[8] and also in patients having British or Irish ancestry.[3]

We have studied some other families with the congenital form of 21-OH deficiency which have been referred from foreign countries. In twenty families from Western and Central Europe, seven (35%) had the haplotype A3, Bw47 associated to the trait (families from Denmark, Belgium, and Czechoslovakia), but in sixteen families from North Africa, no Bw47 was found. From these results, it appears that the association between the haplotype A3, Cw6, Bw47 and 21-OH deficiency is mainly distributed in Northern Europe and may have a common origin.

On the other hand, a nonrandom gametic association for the late-onset 21-OH deficient trait was clearly demonstrated with B14 and with the HLA haplotype segment Aw33, C-,B14. In the French population, B14 is in genetic linkage disequilibrium with Aw33 (the observed incidence of Aw33, B14 is 0.6% against 0.06%

TABLE 4. Forty-four Families with Late-Onset 21-OH Deficiency Genotype of the Index Case

HLA-B Markers[a]	No. of Cases	Total No.	Total %
B14/B14	4	30	68.2%
B14/B12	2		
B14/B35	2		
B14/By	22		
B12/By	7	14	31.8%
B35/By	6		
B12/B35	1		
By/By	0		

[a]HLA-By is a HLA-B different from B12,B14, and B35.

calculated). However, the degree of genetic linkage disequilibrium is higher in the patients with late-onset 21-OH deficiency (19.3%).

With the numbers of observations of late-onset 21-OH deficiency that are available, it is not possible to know if there are associations with B12 and B35. It will be important to collect more data to elucidate this problem.

ACKNOWLEDGMENTS

The authors wish to acknowledge the invaluable contribution of all the physicians without whom this study would not have been possible, and especially Drs. P. Canlorbe, F. Girard, J. C. Job, and J. P. Luton.

REFERENCES

1. Dupont, B., E. M. Smithwick, S. E. Oberfield, T. D. Lee & L. S. Levine. 1977. Close genetic linkage between HLA and congenital adrenal hyperplasia (21-hydroxylase deficiency). Lancet ii: 1309–1312.

2. DUPONT, B., M. POLLACK, L. LEVINE, G. O'NEILL, B. HAWKINS & M. NEW. 1980. Congenital adrenal hyperplasia and HLA: Joint report from the Eighth International Histocompatibility Workshop. *In* Histocompatibility Testing 1980. P. I. Terasaki, Ed.
3. FLEISCHNICK, E., D. RAUM, S. M. ALOSCO, P. S. GERALD, E. J. YUNIS, J. F. AWDEH, J. F. CRIGLER, C. M. GILES & C. A. ALPER. 1983. Extended MHC haplotypes in 21-hydroxylase deficiency congenital adrenal hyperplasia: shared genotypes in unrelated patients. Lancet **i:** 152–156.
4. COUILLIN, P., R. RAPPAPORT, F. KUTTENN, P. CANLORBE, J. HORS, A. MARCELLI-BARGE, J. FEINGOLD, M. C. GRISARD, J. BOUÉ. 1982. HLA and 21-hydroxylase deficiency in the French population. Tissue Antigens **19:** 100–107.
5. POLLACK, M. S., L. S. LEVINE, G. O'NEILL, S. PANG, F. LORENZEN, B. KOHN, G. P. RONDANINI, G. CHIUMELLO, M. I. NEW & B. DUPONT. 1981. HLA linkage and B14, DR1, BfS haplotype association with the genes for late-onset and cryptic 21-hydroxylase deficiency. Am. J. Genet. **33:** 540–550.
6 MITTAL, K. K., M. R. MICKEY, D. P. SINGAL & P. I. TERASAKI. 1968. Refinement of microdroplet lymphocyte cytotoxicity test. Transplantation **6:** 913–927.
7. APER, C. A. & A. M. JOHNSON. 1969. Immunofixation electrophoresis: A technique for the study of protein polymorphism. Vox Sang. **17:** 445–452.
8. PUCHOLT, V., J. S. FRITZSIMMONS, K. GELSTHORPE, M. A. REYNOLDS & R. D. G. MILNER. 1980. Location of the gene for 21-hydroxylase deficiency. J. Med. Genet. **17:** 447–452.

The Association Between Congenital Adrenal Hyperplasia and HLA in Southern Italy

SERAFINO ZAPPACOSTA,[a] MICHELE MAIO,[a]
MARIO DE FELICE,[a] AND ROSSELLA VALENTINO[b]

[a]*Cattedra di Immunologia*
Dipartimento di Biologia e Patologia Cellulare e Molecolare
and
[b]*Istituto di Endocrinologia*
2a Facoltà di Medicina e Chirurgia
Università di Napoli
80131 Napoli, Italy

In this paper, we report on our studies carried out on families from Southern Italy having one or more children suffering from congenital adrenal hyperplasia (CAH) due to 21-hydroxylase deficiency. The work was initially aimed at the confirmation of the linkage of the enzyme deficiency to the HLA complex in our population,[1] after the discovery of Dupont *et al.* in 1977.[2,3] The number of typed families was subsequently extended to focus on the possible association of the disease with HLA alleles and to correlate different clinical forms of the disease to HLA patterns. Aside from the usefulness of HLA typing for clinical purposes, the study seemed of interest because of the genetic peculiarities of Southern Italian populations, which had been shown to diverge from other European Caucasoid ethnic groups.[4]

Our study included 13 families with a total of 72 members, 19 of which were afflicted with the so-called "classical" form of CAH. In all patients, the disease was ascertained by clinical and biochemical means. All family members were typed for nearly all known HLA specificities (HLA-A, -B, and -C loci), as recognized at the 7th and 8th International Histocompatibility Testing Workshops. In some families, DR, Bf, and C2 phenotypes were also determined by standard methods.

In each of the families studied, the affected siblings are HLA-identical. In 3 out of 13 families, a male affected child, clinically apparently normal, was detected after HLA-typing and assay of serum 17-hydroxyprogesterone (17-OH-P) (FIGURE 1a). These individuals should probably not be assimilated to "cryptic" patients as indicated in the literature.[5] They actually are HLA-identical to symptomatic patients in the same family, and are, apparently, homozygotes for the pathologic trait.

As can be seen from TABLE 1, in which all HLA haplotypes, including complement markers, are reported for the affected subjects, there seems to exist no predominance of any HLA allele except a slight increase of Aw24 (7/13) and of Bw35 (6/13). When the gene frequencies of these alleles are compared to those of the same alleles in the control population, there are no significant deviations. In fact, it can be noticed from TABLE 2 that the uncorrected p value for Aw24 is close to 0.025.

As far as the HLA-Bw47 antigen is concerned, a strong association has been reported with the classical form of CAH.[6,7] This finding does not seem to receive support from our data (TABLE 2), in which only one patient shows the antigen in a heterozygous status. Haplotypes normally recognized in our local Caucasoids, as for instance the gene pairs Aw24, Bw35; Aw33, B14; Bw35, Cw4; Bw51, Cw5; B12, Cw5,

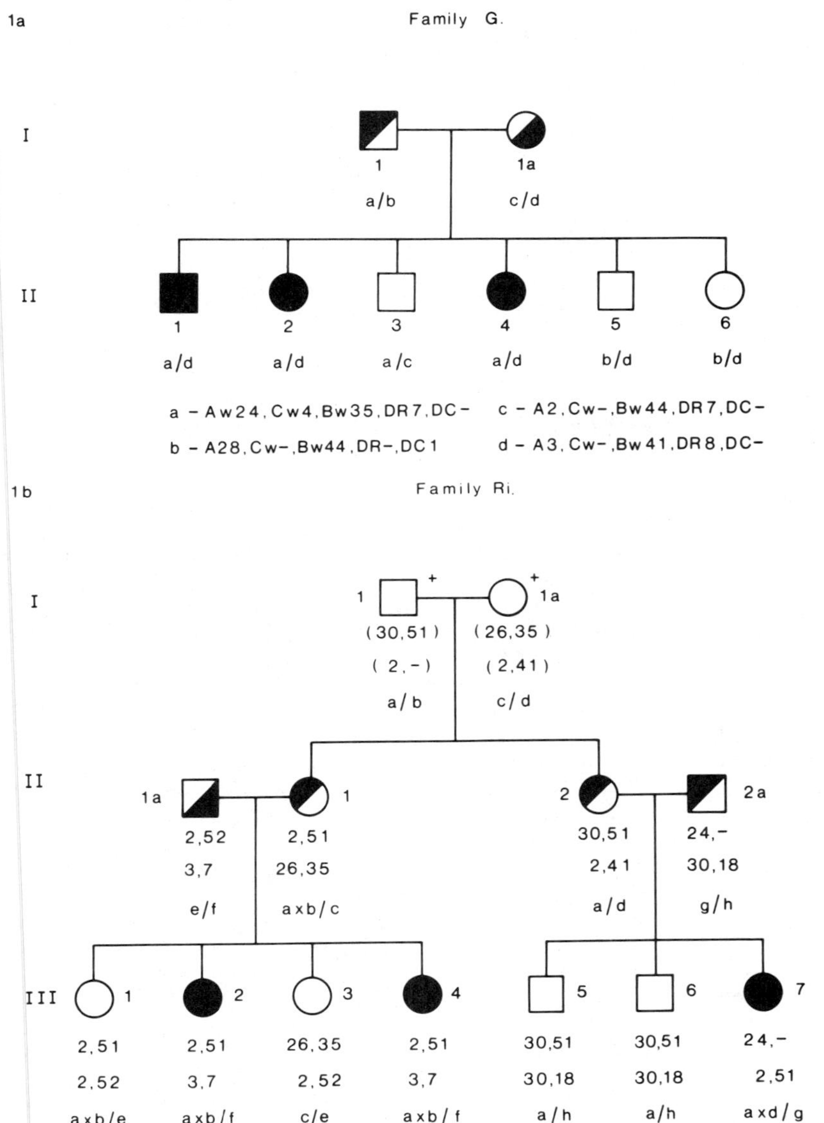

FIGURE 1. Pedigrees of two families with CAH. In family G, (1a), the asterisk indicates a clinically apparent normal child with high 17-OH-P serum levels. In family Ri, (1b), subject III_7 and possibly subject II_1 are HLA-A:B recombinants. Haplotypes of generation I are deduced from offspring. For discussion, see text.

TABLE 1. HLA Haplotypes of 13 Unrelated Patients with CAH

Patient Identification	Haplotypes											
Sco.	Aw24	Bw35	Cw-		BfS	C2C	/A2	B18	Cw-		BfS	C2C
Gri.	Aw24	B18	Cw4	DR7			/A3	Bw41	Cw-	DR8		
Ro. I	A2	B18	Cw2				/A2	B12	Cw5			
Ro. II	Aw24	Bw35	Cw4	DR-	BfS	C2C	/Aw24	B18	Cw-	DR5	BfS	C2C
Mo.	Aw31	B15	Cw-	DR5	BfS	C2C	/Aw31	B15	Cw-	DR-	BfS	C2C
Su.	A3	B7	Cw5		BfS	C2C	/A28	Bw35	Cw4		BfS	C2C
Sa.	A28	Bw44	Cw-		BfS	C2C	/A28	B14	Cw-		BfS	C2C
Pa.	A3	Bw35	Cw4	DR5	BfS	C2C	/Aw24	B15	Cw3	DR-	BfS	C2C
Ri.	A2	Bw51	Cw5				/A3	B7	Cw-			
Pu.	A-	B18	Cw-				/Aw24	B18	Cw-			
Ma.	Aw33	B14	Cw-		BfS		/A2	B-	Cw2		BfS	
To.	A1	B14	Cw4				/Aw24	Bw35	Cw4			
Ga.	Aw24	Bw47	Cw6				/Aw23	Bw38	Cw5			

are represented in our patient group, but none of them seem to prevail over the other ones. On the other hand, the small number of patients does not allow a statistical analysis of haplotypes.

Our data, thus, favor the absence, at least in our population, of a definite "classical" association between 21-OH deficiency and HLA alleles, and, given the localization of the gene, suggest the possibility of associations with any allele of the system, particularly at the B and DR loci.

Phenotypes of the complement factor Bf, the gene of which has been precisely localized between the HLA-B and -D/DR loci,[8] show also a divergent behavior from those reported in the literature for Caucasoid populations.[7] All individuals of our patient group are homozygotes for the BfS allele, even in families carrying the BfF allele. As is shown in TABLE 3, in which the haplotypes of both parents of each sick child are indicated, the haplotypes carrying the 21-OH deficiency gene are associated in all instances with the BfS allele. Should we attempt a rough statistical analysis of these data, given the small number of cases, it emerges that the frequency of SS homozygotes is almost doubled as compared to the expected one.

FIGURE 1b shows the pedigree of a large family with 21-hydroxylase deficiency in which one or possibly two recombinants were detected after HLA-typing. The first generation haplotypes were deduced from offspring. In the hypothesis that subject I_1 is a homozygote for Bw51, i.e., possesses the A2, Bw51 haplotype, it is unnecessary to consider subject II_1 as a recombinant. In this instance, however, I_1 must have been affected from the disease, which cannot be proved. The other recombinant is member

TABLE 2. Significance (Uncorrected p) of Gene Frequencies for HLA-Aw24 and HLA-Bw47 in CAH Patients

	Patients ($N = 13$)	Controls ($N = 327$)	Significance[a]
Aw24	0.321	0.123	$p < 0.025$
Bw47	0.040	0	$p = 0.038$[b]

[a]Uncorrected p values.
[b]Fisher exact test.

TABLE 3. Parental HLA Haplotypes in CAH Families[a]

Family Ro. II	F	A11	B14	DR1	BfS	Aw24	Bw35	DR-	BfS
	M	Aw24	B18	DR5	BfS	Aw24	Bw39	DR7	BfF
Family Sco.	F	Aw24	Bw35		BfS	Aw30	Bw38		BfF
	M	A2	B18		BfS	A28	Bw35		BfF
Family Mo.	F	Aw31	B15	DR5	BfS	Aw24	B13	DR7	BfS
	M	Aw24	B18	DR5	BfS	Aw31	B15	DR-	BfS
Family Su.	F	A3	B7		BfS	Aw24	B5		BfS
	M	A2	Bw35		BfS	A26	B12		BfF
Family Sa.	F	A2	B27		BfF	A28	Bw44		BfS
	M	A28	B14		BfS	Aw32	Bw38		BfS
Family Pa.	F	A2	B12	DR3	BfF	A3	Bw35	DR5	BfS
	M	Aw24	B18	DR5	BfS	Aw24	B15	DR-	BfS
Family Ma.	F	A1	B15		BfF	Aw33	B14		BfS
	M	A2	Bw51		BfS	A2	B-		BfS

[a]Haplotypes carrying the 21-OH deficiency gene are boxed.

III_7. In either interpretation, the recombinant(s) is (are) A:B and the pathologic trait travels with the B locus, as has been repeatedly reported by others.[9]

To discuss the local implications of our observations, we should perhaps refer to our previous studies on the genetic structure of Southern Italian populations.[4,10] By comparing HLA gene and haplotype frequencies of a population from the Campania region with other European and Mediterranean populations, a typical Mediterranean structure of the Campanian population emerges, confirming other genetic and historical findings. The comparison of the Campanian data with similar data from a Northern Italian population (Bergamo) indicates profound differences in the frequencies of some HLA antigens, as reported in TABLE 4. Thus, when genetic distances are measured between the Campanian population and other European and Mediterranean ethnic groups for the HLA-A and -B loci on the basis of data available in the literature (TABLE 5), striking features emerge. From all comparisons made, it became clear that

TABLE 4. HLA Antigen Frequencies in a Northern (Bergamask) and a Southern (Campanian) Italian Population

	Antigen Frequency[a]		
Antigen	Bergamo (321)[b]	Campania (254)[b]	Significance[c]
A1	.246 (79)	.150 (38)	**
A9	.218 (70)	.331 (84)	**
A11	.140 (45)	.039 (10)	***
B5	.308 (99)	.205 (52)	**
B7	.103 (33)	.051 (13)	*
B14	.037 (12)	.087 (22)	*
B40	.022 (7)	.075 (19)	***
Cw1	.103 (33)	.039 (10)	**
Cw6	.031 (10)	.228 (58)	***

[a]Number of positive subjects are given in parentheses.
[b]Numbers tested.
[c]*, **, and *** indicate 5%, 1%, and 0.1% significance level of chi-square, respectively.

the people from Campania are genetically quite distant from the populations of Northern Europe (English, German, and Danish), while the distance from the French and the Spanish people are less pronounced, the same as that from Bergamo. While keeping some Middle Eastern features, the Campanians place themselves as far from Northern Italian as from the Spanish and the French populations. Among all groups compared, including Turkish and Lebanese, the largest distance is measured from the English, the Danish, and the German populations. The reason for such a finding is probably related to the historical background of Southern Italy, and is reflective of the migratory phenomena the population faced in previous centuries. One factor is certainly caused by the extensive network of relationships between countries bordering on the Mediterranean Sea, allowing genetic admixture. The invasions by the Normans, the Angevins, and the Spanish may not have contributed so much to the gene pool, though, since these were small dominant groups. Lastly, in more recent years, the one-way migration towards the North has never received a valid counterpart, and this has contributed in keeping the gene pool more homogeneous.

Reviewing our 21-hydroxylase deficiency data in the light of these considerations, we can provide points of discussion to attempt to explain the differences observed with

TABLE 5. Genetic Distances[a] for HLA-A and -B Locus Genes Between the Campanian and Other European and Mediterranean Populations[b]

Population	Locus A	Locus B	Average
English	.0184	.0263	.0223
German	.0148	.0196	.0172
Danish	.0094	.0225	.0159
French	.0059	.0150	.0109
Spanish	.0088	.0103	.0095
Bergamask	.0106	.0080	.0093
Lebanese	.0109	.0141	.0125
Turkish	.0150	.0106	.0128
Arab	.0255	.0247	.0251

[a]Genetic distances were calculated by the formula of Cavalli-Sforza.[11]
[b]Sources of compared gene frequencies are indicated in reference 4.

the allele associations. The fact that the Campanian population is divergent from other Northern European (and Northern American) Caucasoid populations (in which most of the studies of the disease were carried out) can, perhaps, explain the lack of significant associations between B locus alleles and 21-hydroxylase deficiency. On the other hand, it should be borne in mind that the 21-hydroxylase deficiency is, together with idiopathic hemochromatosis, the only disease of which the genetic linkage with HLA has been proved.

Therefore, the way of considering this enzyme deficiency as any other "statistically" associated disorder may not be appropriate. This remark, as well as the observation of a close linkage with the B locus, would imply an association with a given allele much stronger than that found with Bw47 in all populations except perhaps the English. It seems, thus, that the initial hypothesis of Levine *et al.*[3] could be reconsidered as a random association between alleles of the HLA-B locus and the 21-hydroxylase locus. The origin of the association is based on the assumption of multiple unrelated mutations and of a rather high frequency of these mutations.

In conclusion, it is still difficult to put forward a convincing interpretation of the B locus and 21-hydroxylase deficiency association. Only molecular genetics studies of the HLA complex, now rapidly progressing, will give a conclusive answer to this question.

REFERENCES

1. Zappacosta, S., M. de Felice, M. Minozzi, G. Lombardi, R. Valentino & G. Vanacore. 1978. HLA and congenital adrenal hyperplasia. Lancet **2:** 524.
2. Dupont, B., S. C. Oberfield, E. M. Smithwich, T. D. Lee & L. S. Levine. 1977. Close genetic linkage between HLA and congenital adrenal hyperplasia (21-hydroxylase deficiency). Lancet **20:** 1309.
3. Levine, L. S., M. Zachmann, M. I. New, A. Prader, M. S. Pollack, G. J. O'Neill, S. Y. Yang, S. E. Oberfield & B. Dupont. 1978. Genetic mapping of the 21-hydroxylase deficiency gene within the HLA linkage group. N. Engl. J. Med. **299:** 911.
4. Zappacosta, S., M. de Felice, M. Fiore & G. B. Ferrara. 1980. The HLA system in the Campania region: A genetic study. Tissue Antigens **16:** 286.
5. New, M. I., B. Dupont, S. Pang, M. Pollack & L. S. Levine. 1983. An update of congenital adrenal hyperplasia. Recent Progr. Horm. Res. **37:** 105.
6. Pucholt, V., J. S. Fitzsimmons, K. Gelsthorpe, M. A. Reynolds & R. D. G. Milner. 1980. Location of the gene for 21-hydroxylase deficiency. J. Med. Genet. **17:** 447.
7. Dupont, B., M. S. Pollack, L. S. Levine, G. J. O'Neill, B. R. Hawkins & M. I. New. 1980. Congenital adrenal hyperplasia. Joint report. *In* Histocompatiblity Testing 1980. P. I. Terasaki, Ed: 693. UCLA Tissue-Typing Laboratory. Los Angeles.
8. Schreuder, I., T. Meo, A. Termijtelen & P. Meera Khan. 1980. The Bf locus is between HLA-B and D/DR. *In* Histocompatibility Testing 1980. P. I. Terasaki, Ed.: 931. UCLA Tissue-Typing Laboratory. Los Angeles.
9. Hawkins, D. R., J. A. Danilovs & G. J. O'Neill. 1980. Analysis of recombinant families. Joint report. *In* Histocompatibility Testing 1980. P. I. Terasaki, Ed.: 148. UCLA Tissue-Typing Laboratory. Los Angeles.
10. Zappacosta, S., M. de Felice, M. Fiore & G. B. Ferrara. 1980. The pattern of HLA antigens in two Italian regions. La Ricerca Clin. Lab. **10:** 405.
11. Cavalli-Sforza, L. L. 1969. Human diversity. Proc. Int. Congr. Genet. **3:** 405.

HLA Associations in Late-Onset 21-Hydroxylase Deficiency in Israel

Z. LARON, A. ROITMAN, A. PERTZELAN,
H. KAUFMAN,[a] AND R. ZAMIR[b]

Institute of Pediatric and Adolescent Endocrinology
[a]Steroid Laboratory
and
[b]Tissue Typing Unit
Beilinson Medical Center
49 100 Petah Tikva, Israel
and
Sackler School of Medicine, Tel Aviv University
Tel Aviv, Israel

In the mid-1960's and early 70's, we observed in our clinic a considerable number of pubertal females who had normal genitalia at birth, but, with the onset of puberty, developed marked acne and hirsutism accompanied by irregular menstruation, with advanced bone age in the younger patients. In many, there was also obesity, and, in some, clitoromegaly.The only definite pathological laboratory finding was an elevated urinary excretion of pregnanetriol (3–4 mg/24 hr) often accompanied by elevated 17-ketosteroids. In contrast to the large number of females with this disorder, there were only two such males showing precocious adrenarche and having high urinary pregnanetriol levels. These subjects were all diagnosed as having "postpubertal" congenital adrenal hyperplasia (CAH),[1] i.e., 21-hydroxylase deficiency (21-OH deficiency), and some were treated with glucocorticoids for various lengths of time. At that time, it was not clear whether this metabolic derangement was of hereditary or acquired nature. Other investigators described in adults a nontypical form of 21-OH deficiency of late-onset[2–4] characteristically found in female patients with normal genitalia and a normal onset of puberty, followed by the appearance of mild to severe hirsutism, irregular menstruation, and infertility.

After Dupont *et al.*[5] observed that CAH due to 21-OH deficiency behaved as an autosomal genetic trait in close linkage with HLA, we decided to investigate the linkage between HLA and the late-onset form of 21-OH deficiency. A collaborative study (carried out by teams from Petah Tikva and New York) of five Jewish Ashkenazic families in Israel, in which the index case had been diagnosed as having late-onset 21-OH deficiency, revealed a relationship between the biochemical abnormality in this syndrome and the HLA haplotype of B14, DR1, BfS.[6] This association was also reported for a number of Caucasian adults who were probably of non-Jewish origin.[7–9] Subsequently, we were able to investigate additional families of probands diagnosed to have this syndrome either prepubertally or during puberty

Forthwith, we present a review of all the patients in Israel known by us to have late-onset 21-OH deficiency, including all our published[6,10] and unpublished cases, a patient reported from a nearby hospital,[11] and eight cases brought to our attention by colleagues in Haifa,[12] with particular emphasis on the HLA haplotypes and 21-OH deficiency in various ethnic groups.

SUBJECTS AND METHODS

Diagnostic Criteria for Late-Onset 21-OH Deficiency in Childhood

The typical clinical signs, chief complaints, and main reasons for referral are as follows:

Prepubertal Age

Girls: precocious adrenarche, advanced bone age, occasional clitoromegaly, no labio-scrotal fusion (FIGURE 1).
Boys: precocious adrenarche, pre–acne, enlarged penis (FIGURE 2), advanced bone age, tall stature.

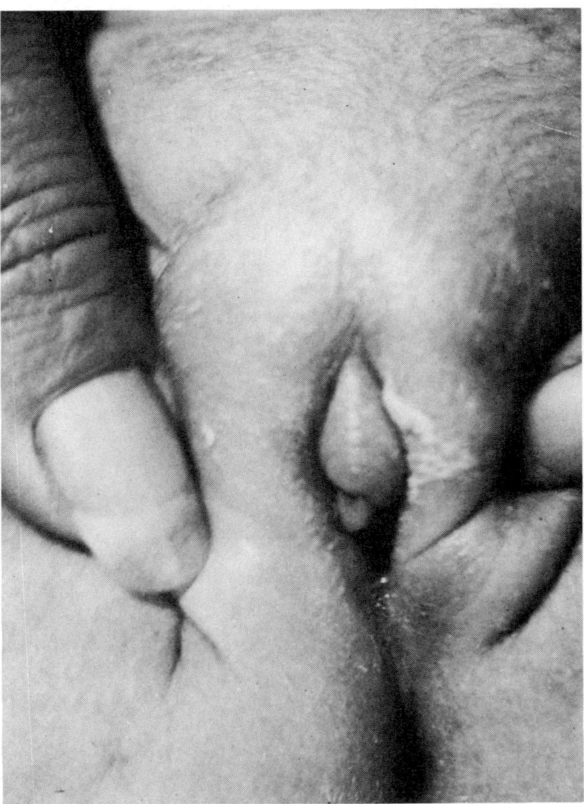

FIGURE 1. Clitoromegaly and pubic hair fuzz in a 3½ year-old girl with late-onset 21-OH deficiency.

Pubertal Age

Girls: hirsutism of varying degree frequently associated with marked acne (this is usually the main complaint), menstrual disorders (late menses, oligomenorrhea, secondary amenorrhea), occasionally small breasts, clitoromegaly, and advanced bone age (FIGURES 3 and 4).

Boys: advanced adrenarche, acne, muscular body-build, macrogenitosomia with normal gonads (FIGURE 5).

If untreated, subjects of both sexes have a final height which is usually shorter than expected due to early closure of the bone epiphyses (FIGURE 6).

Laboratory Findings

Elevated urinary pregnanetriol and normal or high basal levels of plasma 17-OHP, but an excessive response to a single i.v. bolus of 0.25 mg $ACTH^{1\text{-}24}$ (plasma sampling

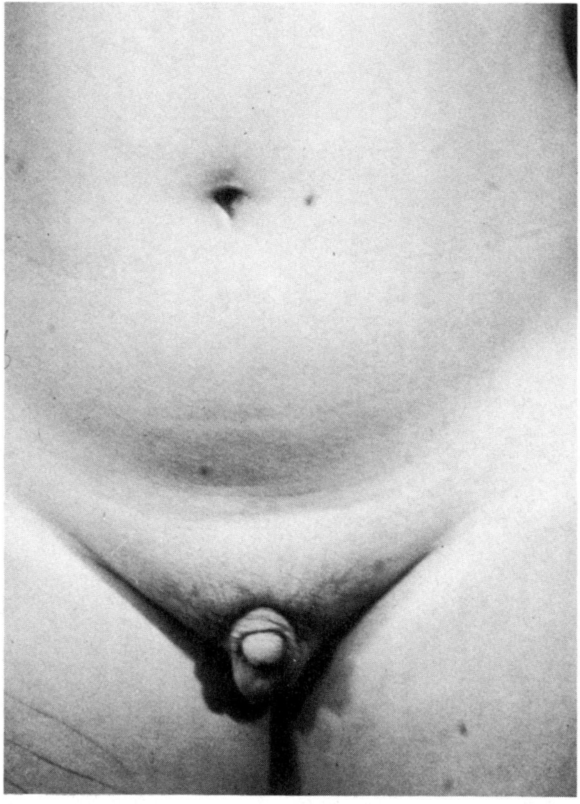

FIGURE 2. Precocious adrenarche in a seven year-old boy with late-onset type 21-OH deficiency.

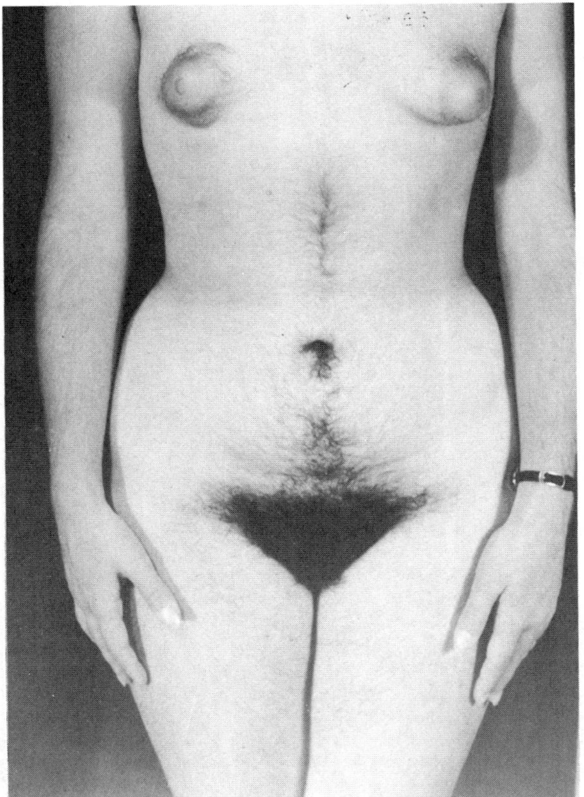

FIGURE 3. Hirsutism and small breasts in an 18 year-old girl with late-onset type 21-OH deficiency.

at 0, 30, 60 mins.),[6] which is more exaggerated than in congenital 21-OH deficiency; there is no obvious cortisol deficiency as measured by the plasma 11-OHCS response to ACTH.

Family History

This is usually positive; examination of the parents and siblings or comprehensive anamnesis often yields a history of hirsutism, irregular menstruation, and reproductive difficulties.

SUBJECTS

Among the 46 known cases of late-onset 21-OH deficiency in Israel, 37 index cases and their families are presently under observation in our clinic. Of these, 33 are

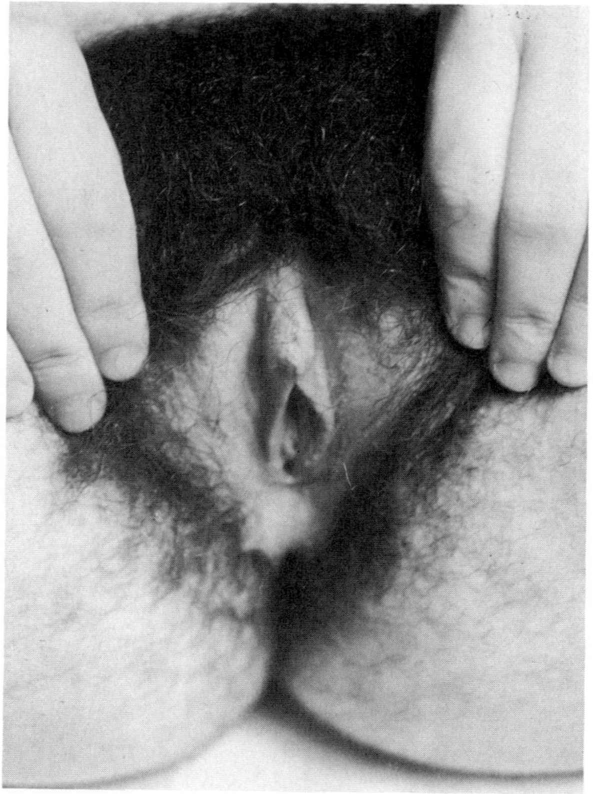

FIGURE 4. Heavy pubic hair and clitoromegaly in a 16 year-old girl with late-onset type 21-OH deficiency.

females and 4 males; 26 are postpubertal and 11 prepubertal. A comprehensive investigation has been carried out in eight of these families, seven of which are described in two publications.[6,10] (Further reports are in preparation, but the overall study of the family members has not yet been completed). In addition, there is the single prepubertal boy with precocious adrenarche and his family, who have been described by Kaushanski *et al.*,[11] and the eight families with prepubertal index cases (two of them boys) presently being investigated in Haifa.[12] Investigation of all these families revealed numerous abnormal findings, both clinical and in the laboratory results, in many of the family members.

METHODS

Endocrine Studies

Twenty-four hour urine collections were used for the determination of 17-ketosteroids,[13] pregnanetriol, and pregnanetriolone.[14] Plasma 17-OHP, androstenedione, testosterone, and aldosterone were measured by radioimmunoassay using commercial kits and 11-OHCS fluorometrically.[15]

HLA Typing

For HLA typing (A, B, and C specificities), the standard NIH technique[16] was employed. HLA-DR typing was performed on enriched B lymphocytes using a 2-aminoethylisothiouronium bromide (AET)-treated sheep red blood cell preparation (Sigma, USA). The procedure was the NIH two-stage technique with prolonged incubation.[17]

RESULTS

Forthwith, we present some examples of the family studies performed, including values of plasma 17-OHP and HLA genotypes.

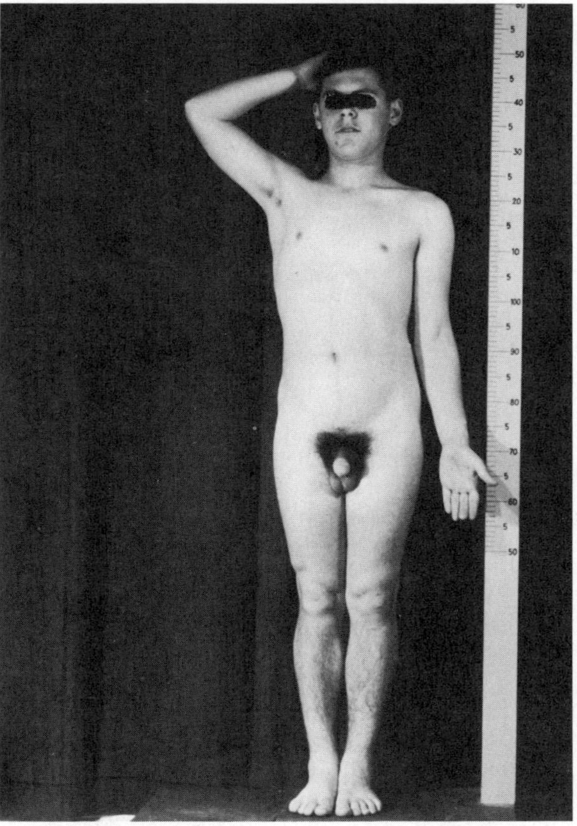

FIGURE 5. Advanced sexual development and musculature in a 12 year-old boy with late-onset type 21-OH deficiency.

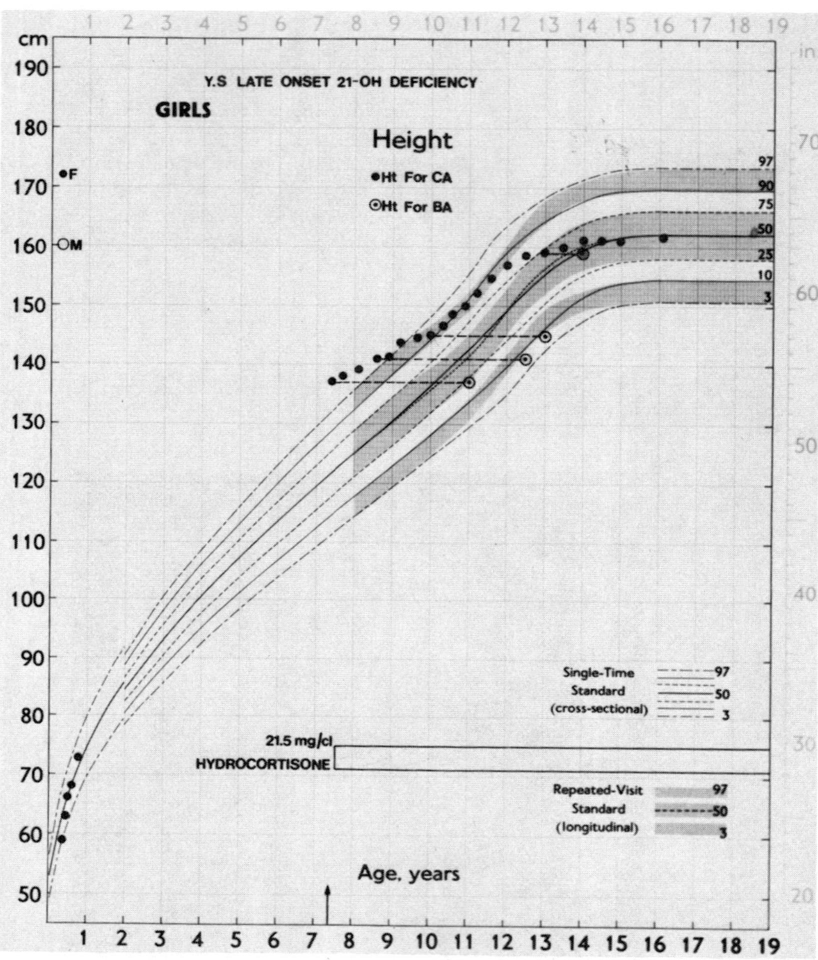

FIGURE 6. Linear growth chart of a girl with late-onset type 21-OH deficiency. Note advanced bone age and shortening of expected final height despite therapeutic trial. (Chart prepared by J.M. Tanner and R.H. Whitehouse, University of London Institute of Child Health, for the Hospital of Sick Children.)

FIGURE 7 shows Family M of Ashkenazi origin. Both parents have a normal basal level of plasma 17-OHP and a normal response to ACTH, and both are considered to be heterozygotes for the late-onset 21-OH deficiency gene (marked L). The index case, who was born prematurely in the sixth month of gestation and had a birth weight of 620 g, was slightly short and had an enlarged clitoris when referred to us at the age of 2½ years. Basal levels of her plasma 17-OHP were elevated with an exaggerated response to ACTH, and she is considered to be homozygous for this trait. The link to

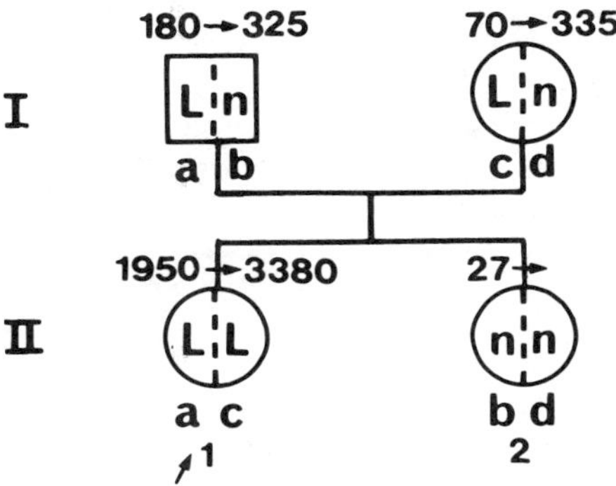

FIGURE 7. Pedigree and proposed genotype based on plasma 17-OHP and HLA studies in Family M of Jewish Ashkenazi origin. The index case is homozygous for the 21-OH deficiency gene. The 21-OH genes are marked: n = normal; L = late-onset 21-OH deficiency. A plasma 17-OHP response below 400 ng/dl is considered normal. The numbers above the subjects represent plasma levels of 17-OHP (ng/dl): basal → 60 min after ACTH. HLA haplotypes: a: A28 B14 DR1; b: A1 B5 DR1; c: Aw24 B14 Cw2 DR1; d: Aw32 B27 DRw6. (Reproduced with permission from *Isr. J. Med. Sci.*[10])

HLA B14 and DR1 haplotype is obvious, with the proband having two, one from each parent.

FIGURE 8 shows the pedigree of Family IN in which the proband MJ is a heterozygote for the 21-OH deficiency gene. The family is of Ashkenazi origin, and the girl was referred at the age of 2½ years with an apparently enlarged clitoris to undergo

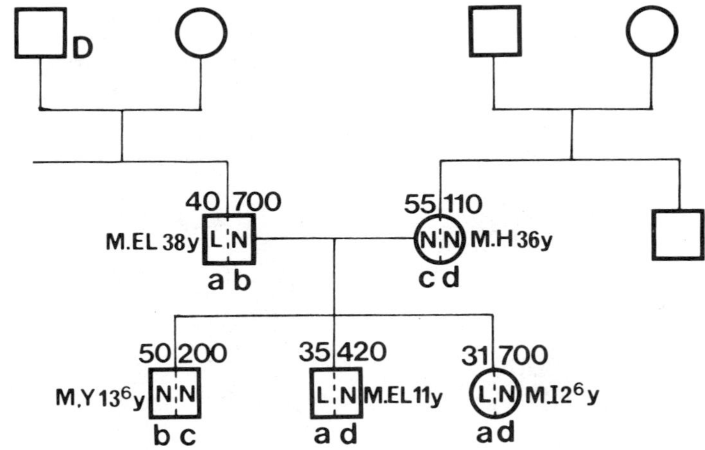

FIGURE 8. Family IN: Late-onset 21-OH deficiency. Pedigree and proposed genotype based on plasma 17-OHP determination and HLA studies in Family IN of Jewish Ashkenazi origin, in which the father and the index case are heterozygotes for the 21-OH deficiency gene. D denotes deceased. HLA haplotypes: a: A3 B14 DR1; B: A11 Bw51; c: A2 B39; d: Aw31 Bw35 Cw4.

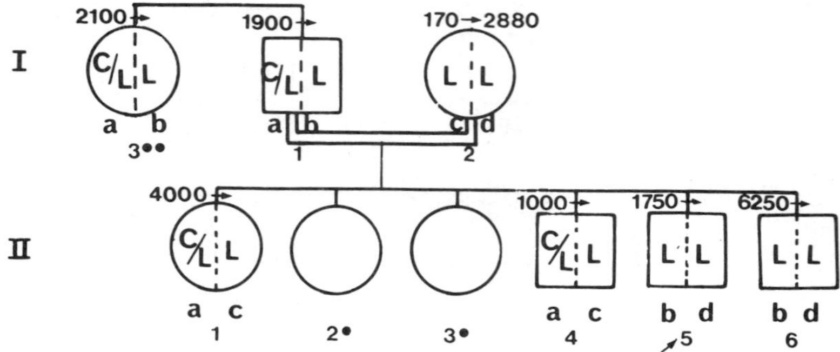

FIGURE 9. Proposed genotype based on 17-OHP determination and HLA studies of an inbred Arab Moslem family in which both parents and all children are affected. ● denotes deceased; ●● denotes DR antigens not typed in this patient. C/L denotes double heterozygote for the late-onset 21-OH deficiency gene (L) and the congenital type 21-OH def. gene (C). For explanation, see text. The numbers above the subjects represent plasma levels of 17-OHP (ng/dl): basal → 60 min after ACTH. HLA haplotypes: a: A3 B15 DR5; b: A1 B14 DRw9; c: A1 B14 DR7; d: A1 B14 DR2. (Reproduced with permission from *Isr. J. Med. Sci.*[10])

evaluation of her growth. She was found to have normal basal levels of 17-OHP and a slightly elevated response to ACTH, similar to the findings in her father. Both have one haplotype of HLA B14, DR1. One brother has a slightly abnormal rise of 17-OHP; it is intended to repeat this test. Having the same haplotypes as the proband, he is also considered heterozygous for the late-onset 21-OH deficiency gene.

The next family (FIGURE 9) is of inbred Arab Moslem origin, and both parents and all the children (probably including two deceased children) are affected. All of the children and the father had elevated basal levels of plasma 17-OHP; the mother had only an exaggerated response to ACTH. These subjects were considered to be either homozygotes for the late-onset 21-OH deficiency gene or double heterozygotes for two

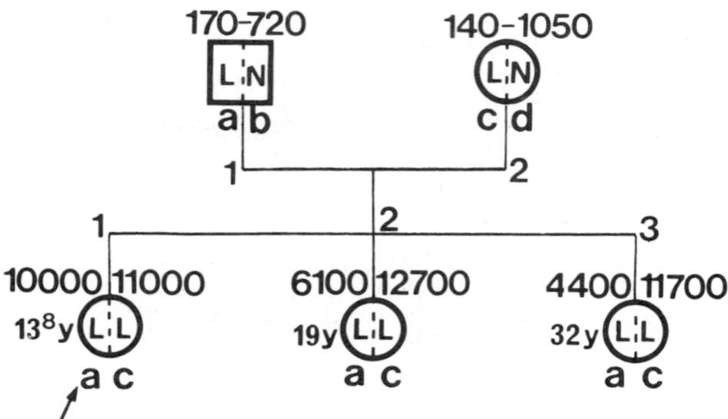

FIGURE 10. Pedigree and proposed genotype based on plasma 17-OHP levels and HLA studies in a Jewish family of Yemenite origin (Family Sh). It is assumed that all children are homozygous for the late-onset 21-OH deficiency gene (L). HLA haplotypes: a: Aw24 Bw44; b: A2 Bw35; c: Aw24 Bw44; d: A2 Bw35. Data kindly obtained from Drs. Schneer and Kleinhaus.[12]

distinct allelic genes, the congenital type gene and the late-onset gene (to be discussed later). Both the mother and the father had one HLA B14 haplotype, but no DR1. It is of note that HLA B14 was also found in Jewish children with late-onset 21-OH deficiency whose parents originated in Morocco.

In two families of Jewish Yemenite origin (FIGURES 10 and 11), both parents carry the late-onset 21-OH deficiency gene (L) and all the children in one family and at least one other in the second family are affected. The HLA types found, which are probably linked to the gene deficiency, are Bw44 and, in one parent, Bw38. Thus, it would appear that in the Yemenite Jews, who constituted a population isolated for almost 2000 years, there is a different linkage between the late-onset 21-OH deficiency gene and the HLA system. DR and complement groups have not yet been determined in these families.

Finally all of the children found to be homozygous for the late-onset 21-OH deficiency gene showed a clinical manifestation typical of this disorder.

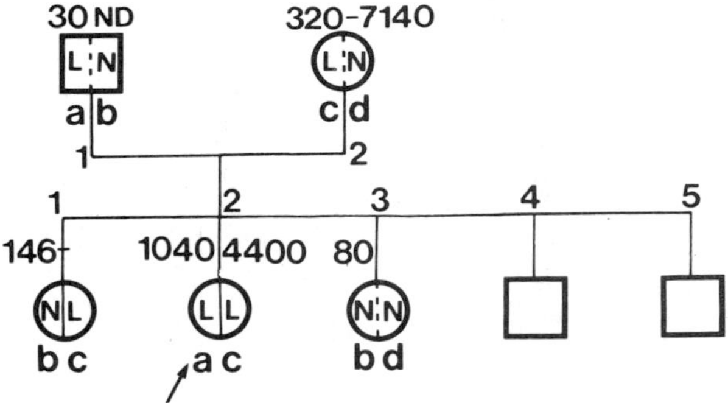

FIGURE 11. Pedigree and proposed genotype based on plasma 17-OHP levels and HLA studies in a Jewish family of Yemenite origin. (Family SY). Two children were not available for investigation and the father did not undergo an ACTH test. It is assumed that the index child is homozygous and one sibling heterozygous for the late-onset 21-OH deficiency gene. HLA haplotypes: a: A9 Bw38; b: A3 B5; c: A9 Bw44; d: A29 B17.

DISCUSSION

The mosaic of responses registered regarding the severity of clinical signs, the number of affected family members, and the levels and degree of response of plasma 17-OHP, has led us to the following interpretation of the etiology and expression of the late-onset form of 21-OH deficiency and its relationship with the congenital form.

We assume there are two distinct allelic genes for 21-OH deficiency, one for the congenital (classical) type (C) and the other for the late-onset type (L) (TABLE 1). The patients with clinical manifestations of late-onset 21-OH deficiency are either homozygous for the specific late-onset gene or double heterozygous for both genes. It is possible that the subjects described as "unusual heterozygotes"[18] or as having "cryptic" 21-OH deficiency[19] are actually double heterozygotes for the two genes. Some relatives of patients with congenital 21-OH deficiency having only mild symptoms or none at all were found to have definite biochemical evidence of 21-OH deficiency and are, thus, similar to individuals with late-onset 21-OH deficiency or some of the relatives of the latter. It is of interest that we were also able to detect

TABLE 1. Classification of 21-Hydroxylase Deficiency (Two Allelic Gene Hypothesis)

Diagnostic Group	21-OHase Locus Genes	Symptoms	Plasma 17-OHP ng/dl Basal	Plasma 17-OHP ng/dl Post ACTH[a]	Assoc. HLA
Normal	n n	–	< 200	< 400	
Congenital heterozygote	C n	–	< 200	< 1000	C: Bw47
Congenital homozygote	C C	+	> 5-10000		
Late-onset heterozygote	L n	–	< 200	< 1000	L:B14 DR1 (Ashk)
Late-onset homozygote	L L	+/–	< 200-3000	> 2000	Bw44 (Yemenite)
Late-onset double heterozygote "cryptic"	C L	+/–	500-10000	< 2000	

[a]60 min after $ACTH^{1\text{-}24}$ 0.25 mg i.v.

heterozygotes for late-onset 21-OH deficiency having a slight, but definite biochemical abnormality.

Another finding of interest in our population was the association of the late-onset 21-OH deficiency gene with the B14 and DR1 HLA groups in the Ashkenazi Jews, in the North African Jews, and, even, in the Israeli Arabs (who had only one B14) (similar to that reported in a probably non-Jewish group of Caucasian European origin), whereas in the Yemenite Jews, the association was with HLA Bw44 and possibly also Bw38. Further studies on the population genetics of 21-OH deficiency and its linkage to HLA are clearly indicated.

In view of the relatively large number of index cases diagnosed in two clinics and the identification of several homozygous or heterozygous individuals in each of their families (although, studies are not yet complete), it may be assumed that there is a high overall incidence of late-onset 21-OH deficiency in Israel. In our clinic alone, there are registered 37 late-onset 21-OH deficiency cases compared to 29 index cases with the congenital type. Also, the apparent sex ratio seems to be different, with 33 females to 4 males as compared to 20 females to 9 males. Particularly since this condition of pathological androgenization is treatable, it is felt that greater attention should be directed towards discovering additional cases. Thus, any child showing even slight signs of androgenization, including bone age advancement, should be submitted to have an ACTH test and his plasma 17-OHP determined. An abnormal finding should be followed by the investigation of as many family members as possible.

The difference in linkage to HLA groups found between Ashkenazi and Yemenite Jews should stimulate further studies in the population genetics of 21-OH deficiency.

REFERENCES

1. LARON, Z. 1968. Ten years of a metabolic-endocrinologic unit for children and adolescents. Harefuah **75:** 117–118.
2. NEWMARK, S., R. DLUHY, G. WILLIAMS, P. POCHI & L. ROSE. 1977. Partial 11- and 21-hydroxylase deficiencies in hirsute women. Am. J. Obstet. Gynecol. **127:** 594–598.
3. ZACHMANN, M. & A. PRADER. 1978. Unusual heterozygotes of congenital adrenal hyperplasia due to 21-hydroxylase deficiency. Acta Endocrinol. (Copenhagen) **87:** 557–565.
4. BIRNBAUM, M. D. & L. J. ROSE. 1979. The partial adrenocortical hydroxylase deficiency syndrome in infertile women. Fertil. Steril. **32:** 536–541.
5. DUPONT, N., S. E. OBERFIELD, E. M. SMITHWICK, T. D. LEE & L. S. LEVINE. 1977. Close genetic linkage between HLA and congenital adrenal hyperplasia (21-hydroxylase deficiency). Lancet **i:** 1309–1312.
6. LARON, Z., M. S. POLLACK, R. ZAMIR, A. ROITMAN, Z. DICKERMAN, L. S. LEVINE, F. LORENZEN, G. J. O'NEILL, S. PANG, M. I. NEW & B. DUPONT. 1980. Late-onset 21-hydroxylase deficiency and HLA in the Ashkenazi population: A new allele at the 21-hydroxylase locus. Hum. Immunol. **1:** 55–66.
7. BLANKSTEIN, J., C. FAIMAN, F. I. REYES, M. L. SCHROEDER & J. S. D. WINTER. 1980. Adult-onset familial adrenal 21-hydroxylase deficiency. Am. J. Med. **68:** 441–448.
8. CHROUSOS, G. P., D. L. LORIAUX, D. L. MANN & G. B. CUTLER, JR. 1982. Late-onset 21-hydroxylase deficiency mimicking idiopathic hirsutism or polycystic ovarian disease: An allelic variant of congenital virilizing adrenal hyperplasia with a milder enzymatic defect. Ann. Intern. Med. **96:** 143–148.
9. CHROUSOS, G. P., D. L. LORIAUX, D. MANN & G. B. CUTLER, JR. 1982. Late-onset 21-hydroxylase deficiency is an allelic variant of congenital adrenal hyperplasia characterized by attenuated clinical expression and different HLA haplotype associations. Horm. Res. **16:** 193–200.

10. ROITMAN, A., M. STIVEL, R. ZAMIR, H. KAUFMAN, A. PERTZELAN & Z. LARON. 1982. Late-onset type of 21-hydroxylase deficiency in childhood. Isr. J. Med. Sci. **18:** 763–768.
11. KAUSHANSKY, A., H. KAUFMAN, R. ZAMIR & E. ELIAN. 1981. Late-onset adrenal hyperplasia (21-hydroxylase deficiency): 17-OH progesterone response to ACTH stimulation and HLA typing. Horm. Res. **14:** 73–78.
12. SCHNEER, J. & N. KLEINHAUS. Personal communication.
13. PETERSON, R. S. & C. E. PIERCE. 1960. *In* Lipids and Steroid Hormones in Clinical Chemistry. Sunderman & Sunderman, Eds.: 158–164. J. B. Lippincott Co. New York.
14. QUAZY, Q. H. & J. G. HILL. 1973. A modified method for estimation of pregnanediol, pregnanetriol, and seven common 17-ketosteroids in urine by gas-liquid chromatography. Steroids **22:** 311–325.
15. DE MOOR, P., O. STEENO, M. RASKIN & A. HENDRICKS. 1960. Fluorometric determination of free plasma 11-hydroxy corticosteroids in man. Acta Endocrinol. (Copenhagen) **33:** 297.
16. BRAND, D. L., J. G. RAY, D. B. HARE, D. D. KAYHOE & G. D. MELELLARED. 1970. Preliminary trials towards standardization of leukocyte typing. In Histocompatibility Testing. Terasaki, Ed.: 357. Munksgaard. Copenhagen.
17. VAN ROOD, J. J. 1980. Procedure for DRw typing on enriched B-lymphocyte suspension. Eighth International Histocompatibility Workshop Letter. Los Angeles.
18. ZACHMANN, M. & A. PRADER. 1979. Unusual heterozygotes of congenital adrenal hyperplasia due to 21-hydroxylase deficiency confirmed by HLA tissue typing. Acta Endocrinol. (Copenhagen) **92:** 542–546.
19. LEVINE, L. S., B. DUPONT, F. LORENZEN, S. PANG, M. POLLACK, S. OBERFIELD, B. KOHN, A. LERNER, E. CACCIARI, F. MANTERO, A. CASSIO, C. SCARONI, G. CHIUMELLO, G. RONDANINI, L. GARGANTINI, G. GIOVANELLI, R. VIRDIS, E. BARTOLOTTE, C. MIGLION, C. PINTOR, L. TATO, F. BARBONI & M. L. NEW. 1980. Cryptic 21-hydroxylase deficiency in families of patients with classical congenital adrenal hyperplasia. J. Clin. Endocrinol. Metab. **51:** 1316–1324.

HLA-B14 and Nonclassical 21-Hydroxylase Deficiency in a Heterogeneous New York Population[a]

LENORE S. LEVINE

St. Luke's-Roosevelt Hospital Center
College of Physicians and Surgeons
of Columbia University
New York, New York 10025

Symptomatic and asymptomatic nonclassical forms of 21-hydroxylase deficiency have been recognized and well-characterized genetically and hormonally.[1–4] Like classical congenital adrenal hyperplasia,[5,6] nonclassical 21-hydroxylase deficiency is an HLA-linked autosomal recessive disorder.[1–3] It has been suggested that nonclassical 21-hydroxylase deficiency and classical congenital adrenal hyperplasia are due to allelic variations at the steroid 21-hydroxylase locus.[1–4,7–11] An increased association of HLA-B14 and nonclassical 21-hydroxylase deficiency has been demonstrated in reports of several populations.[1,2,7–15]

This report summarizes the association of HLA-B14 and nonclassical 21-hydroxylase deficiency in a heterogeneous New York population studied at The New York Hospital-Cornell Medical Center.

SUBJECTS

The diagnosis of nonclassical 21-hydroxylase deficiency was made on the basis of elevated baseline and ACTH stimulated 17-hydroxyprogesterone level (17-OHP).[1–4] Symptomatic patients also received a dexamethasone suppression test and demonstrated suppression of elevated serum 17-OHP and androgens. In the course of HLA and hormonal studies of families of CAH patients, the diagnosis of nonclassical 21-OH deficiency was made in 11 individuals from 7 families (FIGURE 1). These included three HLA identical sisters of patients' fathers. These family members were designated as having "cryptic" 21-hydroxylase deficiency because they did not present with symptoms of androgen excess, but were diagnosed as a result of family studies.[2,3] All but one of these family members considered themselves asymptomatic, and although some female family members were hirsute, none had genitalia abnormalities, amenorrhea, infertility, or virilization. One family member (indicated by * in FIGURE 1), though, was symptomatic, with short adult height (4′8″), hirsutism, deep voice, and temporal recession. She had a history of amenorrhea between 14 and 16 years of age, but had subsequent regular periods and no difficulty with fertility, having had three uneventful pregnancies.

[a]This work was supported in part by a grant (RR-47) from the General Clinical Research Centers Program of the Division of Research Resources, NIH and NIH awards HD–00072, HD–15084, and CA–22507, and Core Grant CA–08748.

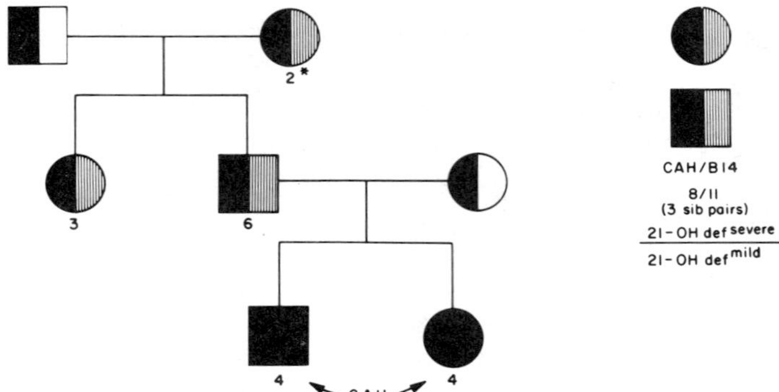

FIGURE 1. Nonclassical 21-hydroxylase deficiency in CAH families.

■ ● denotes index cases with CAH.

◧ ◐ denotes family members with nonclassical 21-hydroxylase deficiency having one severe deficiency gene and one mild deficiency gene.

◧ ◐ denotes heterozygote carriers of a severe deficiency gene for CAH.

*denotes one paternal grandmother of a CAH patient who was short and hirsute, and had temporal recession and amenorrhea from 14–16 years of age.

In addition, there were 28 patients who presented with signs of androgen excess ranging from precocious adrenarche with tall stature and advanced bone age to acne, hirsutism, and menstrual irregularities, and who were diagnosed to have symptomatic nonclassical 21-hydroxylase deficiency (TABLE 1). Two patients were free of symptoms and were untreated at the time of the family studies. Hormonal and HLA studies of the families of these 28 patients revealed 16 additional family members with nonclassical 21-hydroxylase deficiency; 8 were HLA identical sibs of the index cases (FIGURE 2). The majority of these family members were asymptomatic, though some had a history of precocious adrenarche, hirsutism, or short adult height.

MATERIALS AND METHODS

HLA genotyping was performed on peripheral blood lymphocytes for the well-characterized antigens of the HLA-A, -B, and -C loci by the standard NIH 2-stage

TABLE 1. Clinical Features in Patients with Symptomatic Nonclassical 21-OH Deficiency

Precocious Adrenarche (tall stature, advanced bone age)
Severe Acne
Temporal Balding
I° or II° Amenorrhea
Clitoromegaly
Short Adult Height

microcytotoxicity test using 142 selected antisera representing two or three antisera for each of the HLA-A, -B, and -C antigens.[16]

Typing for the HLA-DR antigen, expressed only on the B cell lymphocyte subpopulation, was performed by the modified complement-dependent cytotoxicity technique using nylon column-purified B cells.[17]

Gene frequency was calculated as follows:

$$\frac{\text{number of haplotypes with B14}}{\text{total number of haplotypes}}.$$

Antigen frequency was calculated as follows:

$$\frac{\text{number of individuals with B14}}{\text{total number of individuals}}.$$

For each of these calculations, HLA identical sibs were counted as one individual.

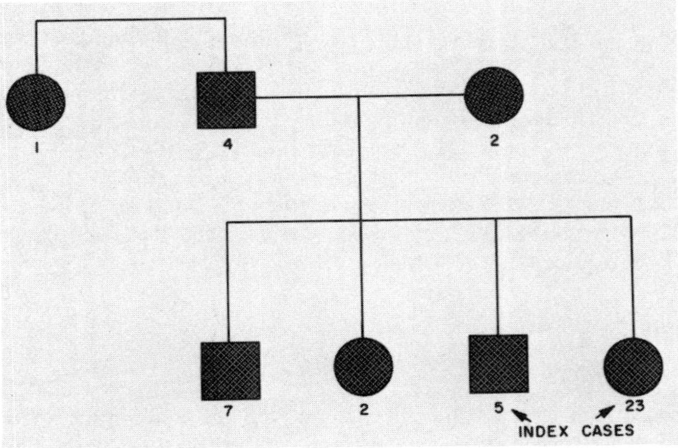

FIGURE 2. Index cases with symptomatic nonclassical 21-hydroxylase deficiency and their family members with symptomatic or asymptomatic nonclassical 21-hydroxylase deficiency.

RESULTS

Eight of the 11 CAH family members demonstrated to have nonclassical 21-hydroxylase deficiency had HLA-B14 in combination with the HLA-B antigen linked to the classical 21-hydroxylase deficiency gene found in the index case with CAH (FIGURE 1). Twenty-two of the 28 index cases with symptomatic nonclassical 21-hydroxylase deficiency had at least one HLA-B14 antigen; of these, eight were HLA-B14 homozygous and five had HLA-B14 in combination with an HLA-B antigen in genetic disequilibrium with classical CAH[18] [HLA-B47, B60(40), B51(5)] (FIGURE 3). Three additional patients were HLA-B35/B14. HLA-B35, although not in genetic disequilibrium with CAH in the New York population, has been reported to be increased in other populations.[19] Eight of the 16 affected family members of the symptomatic nonclassical 21-OH deficiency patients were HLA-B14 homozygous, and

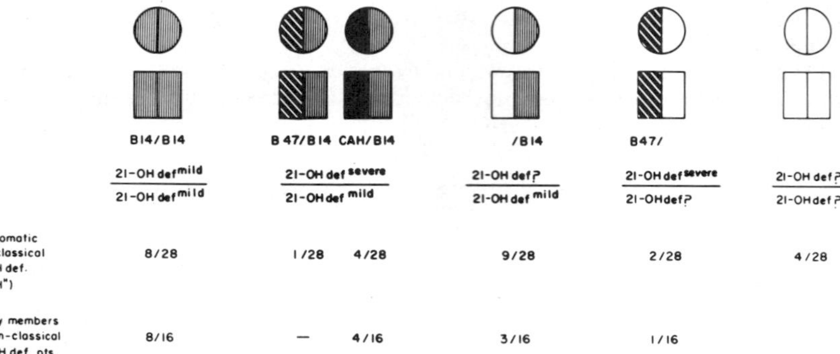

FIGURE 3. HLA-B and CAH genotypes in index cases with symptomatic nonclassical 21-hydroxylase deficiency and their family members with symptomatic or asymptomatic nonclassical 21-hydroxylase deficiency.

seven had one HLA-B14 antigen. Thus, 45 of the 55 individuals with nonclassical 21-hydroxylase deficiency had at least one HLA-B14 (FIGURE 4). Further, 40 of 43 of the HLA-B14 antigens were associated with DR1, confirming that nonclassical 21-OH deficiency is in genetic disequilibrium with a particular haplotype segment, HLA-B14 DR1, rather than with a single HLA determinant, HLA-B14.

Calculation of gene and antigen frequencies for HLA-B14 in nonclassical 21-hydroxylase deficiency gave a gene frequency of 0.537 and an antigen frequency of 80.5%, compared to 0.062 and 12.1%, respectively, in the New York Caucasian control population (TABLE 2).

DISCUSSION

The results of the gene frequency and antigen frequency determinations in the population with nonclassical 21-hydroxylase deficiency studied at The New York Hospital confirm the strong genetic disequilibrium between nonclassical 21-hydroxylase deficiency and HLA-B14 documented in previous reports.[1-3,12,13] We have previously postulated that classical congenital adrenal hyperplasia results from a

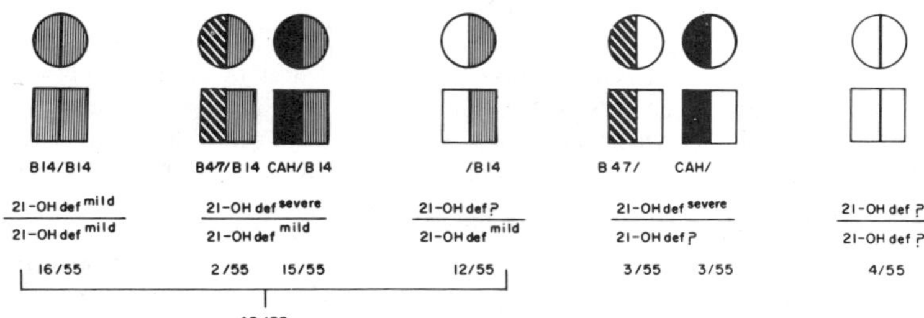

FIGURE 4. HLA-B and CAH genotypes in a total population of patients with nonclassical 21-hydroxylase deficiency.

combination of two severe deficiency genes (21-OH def^{severe}/21-OH def^{severe}), whereas the nonclassical disorder results from a combination of a severe and a mild deficiency gene (21-OH def^{severe}/21-OH def^{mild}) or a combination of two mild deficiency genes (21-OH def^{mild}/21-OH def^{mild}).[1] Patients with nonclassical 21-OH deficiency, in whom a HLA antigen in genetic linkage disequilibrium with classical CAH is present in combination with HLA-B14, would be postulated to have the CAH genotype of 21-OH def^{severe}/21-OH def^{mild}, whereas a patient who is HLA-B14 homozygous would have the 21-OH deficiency genotype of 21-OH def^{mild}/21-OH def^{mild}. Thus, in the 55 individuals in the New York population of symptomatic and asymptomatic nonclassical 21-hydroxylase deficiency, the 16 HLA-B14 homozygous patients would be postulated to have 21-OH def^{mild}/21-OH def^{mild}; 17 could be identified as having 21-OH def^{severe}/21-OH def^{mild} based upon an identifiable CAH haplotype in combination with HLA-B14.

Additional genetic linkage studies of classical and nonclassical 21-hydroxylase deficiency may permit the further definition of the CAH genotype in patients with nonclassical 21-hydroxylase deficiency.

TABLE 2. HLA-B14 in a New York Population of Patients with Nonclassical 21-OH Deficiency

	Gene Frequency[a]	Antigen Frequency[b]
Patients	0.537	80.5%
New York City Caucasian Controls	0.062	12.1%

[a]Number of haplotypes with B14/Total number of haplotypes.
[b]Number of patients with B14/Total number of patients × 100.

ACKNOWLEDGMENTS

The data presented herein is the result of the collaborative efforts of many, including Drs. Bo Dupont, Maria New, Marilyn Pollack and Songya Pang, and Chris Nelson.

REFERENCES

1. KOHN, B., L. S. LEVINE, M. S. POLLACK, S. PANG, F. LORENZEN, D. LEVY, A. LERNER, G. F. RONDANINI, B. DUPONT & M. I. NEW. 1982. Late-onset steroid 21-hydroxylase deficiency: A variant of classical congenital adrenal hyperplasia. J. Clin. Endocrinol. Metab. **55:** 817–827.
2. LEVINE, L. S., B. DUPONT, F. LORENZEN, S. PANG, M. S. POLLACK, S. E. OBERFIELD, B. KOHN, A. LERNER, E. CACCIARI, F. MANTERO, A. CASSIO, C. SCARONI, G. CHIUMELLO, G. F. RONDANINI, L. GARGANTINI, G. GIOVANNELLI, R. VIRDIS, E. BARTOLOTTA, C. MIGLIORI, C. PINTOR, L. TATO, F. BARBONI & M. I. NEW. 1980. Cryptic 21-hydroxylase deficiency in families of patients with classical congenital adrenal hyperplasia. J. Clin. Endocrinol. Metab. **51:** 1316–1324.
3. LEVINE, L. S., B. DUPONT, F. LORENZEN, S. PANG, M. S. POLLACK, S. E. OBERFIELD, B. KOHN, A. LERNER, E. CACCIARI, F. MANTERO, A. CASSIO, C. SCARONI, G. CHIUMELLO, G. F. RONDANINI, L. GARGANTINI, G. GIOVANNELLI, R. VIRDIS, E. BARTOLOTTA, C. MIGLIORI, C. PINTOR, L. TATO, F. BARBONI & M. I. NEW. 1981. Genetic and hormonal characterization of the cryptic 21-hydroxylase deficiency. J. Clin. Endocrinol. Metab. **53:** 1193–1198.

4. New, M. I., F. Lorenzen, A. J. Lerner, B. Kohn, S. E. Oberfield, M. S. Pollack, B. Dupont, E. Stoner, D. J. Levy, S. Pang & L. S. Levine. 1983. Genotyping steroid 21-hydroxylase deficiency: Hormonal reference data. J. Clin. Endocrinol. Metab. **57:** 320–326.
5. Dupont, B., E. M. Smithwick, S. E. Oberfield, T. D. Lee & L. S. Levine. 1977. Close genetic linkage between HLA and congenital adrenal hyperplasia (21-OH deficiency). Lancet **2:** 1309–1312.
6. Levine, L. S., M. Zachmann, M. I. New, A. Prader, M. S. Pollack, G. J. O'Neill, S. Y. Yang, S. E. Oberfield & B. Dupont. 1978. Genetic mapping of the 21-hydroxylase deficiency gene within the HLA linkage group. N. Engl. J. Med. **299:** 911–915.
7. Migeon, C. J., Z. Rosenwaks, P. A. Lee, M. D. Urban & W. B. Bias. 1980. The attenuated form of congenital adrenal hyperplasia as an allelic form of 21-hydroxylase deficiency. J. Clin. Endocrinol. Metab. **51:** 647–649.
8. Blankstein, J., C. Faiman, F. I. Reyes, M. L. Schroeder & J. S. D. Winter. 1980. Adult-onset familial adrenal 21-hydroxylase deficiency. Am. J. Med. **68:** 441–448.
9. Chrousos, G. P., D. L. Loriaux, D. L. Mann & G. B. Cutler. 1982. Late-onset 21-hydroxylase deficiency mimicking idiopathic hirsutism or polycystic ovarian disease: An allelic variant of congenital virilizing adrenal hyperplasia with a milder enzymatic defect. Ann. Intern. Med. **96:** 143–148.
10. Chrousos, G. P., D. L. Loriaux, D. Mann & G. B. Cutler. 1982. Late-onset 21-hydroxylase deficiency is an allelic variant of congenital adrenal hyperplasia characterized by attenuated clinical expression and different HLA haplotype associations. Horm. Res. **16:** 193–200.
11. Lee, P. A., Z. Rosenwaks, M. D. Urban, C. J. Migeon & W. D. Bias. 1982. Attenuated forms of congenital adrenal hyperplasia due to 21-hydroxylase deficiency. J. Clin. Endocrinol. Metab. **55:** 866–871.
12. Laron, Z., M. S. Pollack, R. Zamir, A. Roitman, Z. Dickerman, L. S. Levine, F. Lorenzen, G. J. O'Neill, S. Pang, M. I. New & B. Dupont. 1980. Late-onset 21-hydroxylase deficiency and HLA in the Ashkenazi population: New allele at the 21-hydroxylase locus. Hum. Immunol. **1:** 55–66.
13. Pollack, M. S., L. S. Levine, G. J. O'Neill, S. Pang, F. Lorenzen, B. Kohn, G. F. Rondanini, G. Chiumello, M. I. New & B. Dupont. 1981. HLA linkage and B14, DR1, BfS haplotype association with the genes for late-onset and cryptic 21-hydroxylase deficiency. Am. J. Hum. Genet. **33:** 540–550.
14. Zachmann, M. & A. Prader. 1979. Unusual heterozygotes of congenital adrenal hyperplasia due to 21-hydroxylase deficiency confirmed by HLA tissue typing. Acta Endocrinol. (Copenhagen) **99:** 166–173.
15. Lobo, R. A. & U. Goebelsmann. 1980. Adult manifestation of congenital adrenal hyperplasia due to incomplete 21-hydroxylase deficiency mimicking polycystic ovarian disease. Am. J. Obstet. Gynecol. **138**(6): 720–726.
16. Terasaki, P. I. & J. D. McClelland. 1964. Microdroplet assay for human cytotoxins. Nature **204:** 998.
17. Danilovs, J., P. I. Terasaki, M. S. Park & G. Ayoub. 1979. B lymphocyte isolation by thrombin-nylon wool. Eighth International Histocompatibility Workshop Newsletter, no. 6.
18. Dupont, B., M. S. Pollack, L. S. Levine, G. J. O'Neill, B. R. Hawkins & M. I. New. 1980. Congenital adrenal hyperplasia: Joint report from the Eighth International Histocompatibility Workshop. *In* Histocompatibility Testing. P. I. Terasaki, Ed.: 693–706. UCLA Tissue Typing Laboratory. Los Angeles.
19. Zappacosta, S., M. de Felive, M. Minozzi, G. Lombardi, R. Valentino & G. Vanacore. 1978. HLA and congenital adrenal hyperplasia. Lancet **2:** 524.

Different Gene Defects in the Salt-Wasting (SW), Simple Virilizing (SV), and Nonclassical (NC) Types of Congenital Adrenal Hyperplasia (CAH)

D. KNORR,[a] E. D. ALBERT,[b] F. BIDLINGMAIER,[a]
W. HÖLLER,[a] AND S. SCHOLZ[b]

[a]*Kinderklinik der Universität München*
Abteilung für Pädiatrische Endokrinologie
München, Federal Republic of Germany

[b]*Kinderpoliklinik der Universität München*
Labor für Gewebetypisierung
München, Federal Republic of Germany

INTRODUCTION

When the genetic basis of classical CAH was found to be placed near the HLA region on chromosome 6,[1,2] there was, unlike other diseases, an initial failure to find nonrandom associations with specific HLA alleles. Later, the HLA allele Bw47 was found to occur much more often among several Caucasian CAH individuals than in the respective general population.[3] Our group also found an overrepresentation of HLA B5 among German CAH patients.[2] From various reports, there was no doubt that HLA B14 is closely linked to the nonclassical CAH forms.[4-6]

The aim of this study was to investigate the HLA haplotype associations of the three clinical variants of 21-hydroxylase deficiency (salt-wasting (SW), simple virilizing (SV), and nonclassical (NC) form), and to find out if a genetic difference existed between its two classical forms.

SUBJECTS AND METHODS

CAH patients and relatives investigated are listed in TABLE 1.

Statistical analysis was done using the χ^2 method.

All p values were corrected according to Svejgaard and Ryder.[7]

As a biochemical test for heterozygosity, 17OH-progesterone levels were measured by a specific RIA before and after stimulation by a bolus injection of ACTH i.v.[8] Women were tested between days 2 and 8 of their menstrual cycle.

RESULTS

Looking at all known HLA-A and HLA-B alleles, the incidence of A3, A33, B5, B14, and Bw47 was found to be significantly higher in CAH patients than in the control group.[9] The incidence of Bw47, B5, and B14 among clinical types of CAH is

TABLE 1. Individuals Investigated by HLA Typing in This Study

General population $n = 4302$
CAH families investigated $n = 134$
CAH relatives $n = 333$
CAH patients $n = 163$
Salt-wasting CAH (SW) $n = 70$
Simple viril. CAH (SV) $n = 74$
Nonclassical CAH (NC) $n = 19$
late-onset $n = 13$
cryptic $n = 6$

listed in TABLE 2. There was a highly significant overrepresentation of Bw47 in SW cases, of B5 in SV patients, and of B14 in NC patients.

Since according to the data on CAH frequency[10] any group from the general population is supposed to contain 2% heterozygous gene carriers, we compared our data also with the 240 disease-unaffected haplotypes of heterozygous family members. Those haplotypes can be regarded as the ideal control group since there are *per definitionem* no CAH genes on them.

A comparison of the gene frequency patterns of patients and of this control group is shown in TABLE 3. Bw47 was found to be a strong indicator for the SW gene.

Among the HLA-A alleles, there was a striking elevation of A3 in the SW cases (TABLE 4).

Looking at the complete haplotypes, for carriers of the haplotype A3, Bw47, Cw6, DR7, there is an extremely high risk to be a carrier of the salt-wasting CAH gene. An even higher association could be found for the HLA haplotype Aw33, B14, DR1 with NC-CAH. In addition, for the first time, we could show that there is a significant association of the haplotypes A2, B5 and A3, B5 with SV-CAH.

The hormonal data of this study correspond to the genetic findings. According to our biochemical test, there is most probably heterozygosity of 21-hydroxylase deficiency when the increase in plasma 17OH-progesterone after ACTH stimulation surpasses 260 ng/dl.[8,9,11–20] The test results obtained from the patients' relatives demonstrated that HLA genotypically ascertained heterozygotes for SW-CAH as a group had higher increases in 17OH-progesterone than the relatives of patients with SV- and NC-CAH (FIGURE 1).

DISCUSSION

The close genetic linkage between SW-CAH and Bw47 and between SV-CAH and HLA B5 found in our study gives evidence, for the first time, of a genetic difference

TABLE 2. Frequency (%) of HLA-B Alleles in CAH Families

	Families n	Bw47	B5	B14
Salt-wasters	59	34.0	17.0	6.8
Simple viril. CAH	60	10.0	46.7	1.6
Late-onset CAH	11	9.0	9.0	100.0
General population	200	0.5	17.0	2.5
p versus General population		10^{-13}	0.001	10^{-30}

TABLE 3. Frequency (%) of HLA-B Alleles in CAH Haplotypes

	Number of Haplotypes	Bw47	B5	B14
Salt-wasting CAH	118	19.5	9.3	3.4
Simple viril. CAH	120	5.0	27.5	0.8
Nonclassical CAH	30	6.7	3.3	46.7
Unaffected haplotypes of relatives	240	0.0	6.3	0.0
p versus Unaffected haplotypes		10^{-10}	10^{-6}	10^{-25}

between the two classical CAH forms. Similarly, the genetic independence of NC-CAH was already established from its nearly total connection with B14,[4] which was confirmed by us.

The assumption of genetic differences among the three CAH types is underlined by the finding that the patients' HLA antigens Bw47 and B14 are mostly part of a more extended haplotype, namely A3, Bw47, DR7 and Aw33, B14, DR1, respectively, which normally occurs very rarely.

We conclude from our genetic data and our biochemical test results that not only is NC-CAH an independent genetic form of 21-hydroxylase deficiency, but also the two classical types of CAH can be divided into two separate genetic forms.

SUMMARY

HLA (human leucocyte antigens) alleles and plasma 17-hydroxyprogesterone levels after ACTH stimulation were studied in 134 German families of patients with salt-wasting (SW), simple virilizing (SV), and nonclassical (NC) late-onset forms of CAH. HLA typing revealed a genetic difference between the two classical disease forms. SW-CAH was strongly associated with Bw47 and SV-CAH was closely linked to B5.

The nearly complete connection of NC-CAH with B14 was confirmed. Bw47 and B14 were mostly components of the normally rare haplotypes A3, Bw47, DR7 and Aw33, B14, DR1, respectively. They did not occur in the families' disease-unaffected haplotypes.

The HLA linkage data were consistent with those obtained from the ACTH

TABLE 4. Incidence (%) of HLA Haplotypes Specifically Associated with CAH

	A3,Cw6,Bw47,DR7	A2,B5	A3,B5	Aw33,B14,DR1
Salt-wasting CAH $n = 118$	15.3	3.4	2.5	0.8
Simple viril. CAH $n = 120$	2.7	10.9	11.8	0.9
Nonclassical CAH $n = 30$	3.3	3.3	0.0	30.0
Unaffected haplotypes (controls) $n = 240$	0.0	1.7	1.3	0.0
p versus Controls	10^{-7}	0.005	0.0005	10^{-15}

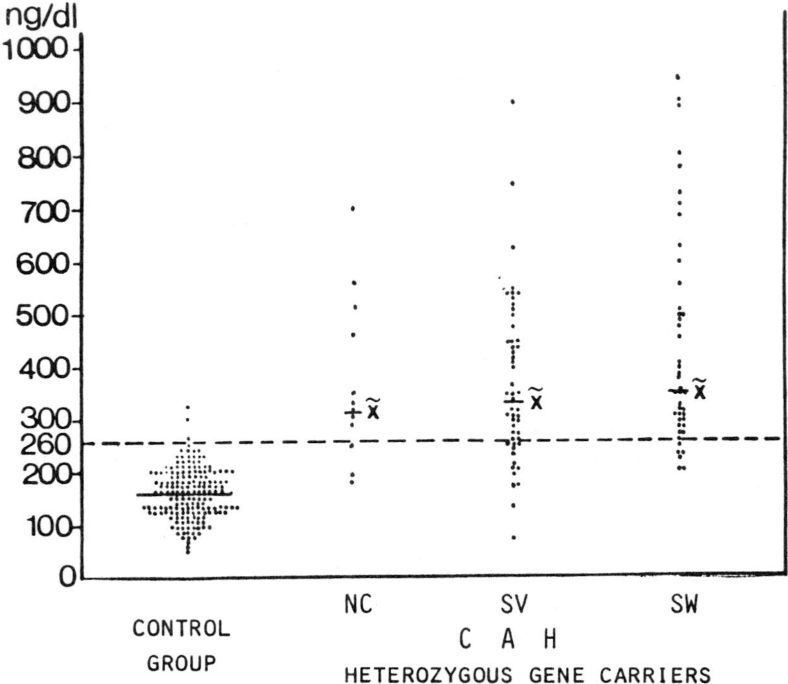

FIGURE 1. Increase of plasma 17OH-progesterone within 60 minutes after i.v. stimulation with 0.25 mg 1-24 ACTH (Synacthen®) in heterozygous gene carriers of 21-hydroxylase deficiency (parents and heterozygous affected siblings of patients with nonclassical CAH (NC), with simple virilizing CAH (SV), and with salt-wasting CAH (SW)).

stimulation test which showed a higher 17-hydroxyprogesterone increase in the group of genetically defined heterozygous relatives of SW patients than in the groups of heterozygous members of SV and NC families.

REFERENCES

1. Dupont, B., E. E. Oberfield, E. M. Smithwick, T. D. Lee & L. S. Levine. 1977. Close genetic linkage between HLA and congenital adrenal hyperplasia (21-hydroxylase deficiency). Lancet **2:** 1309.
2. Grosse-Wilde, H., J. Weil, E. Albert, S. Scholz, F. Bidlingmaier, W. G. Sippell & D. Knorr. 1979. Genetic linkage studies between congenital adrenal hyperplasia and the HLA blood group system. Immunogenetics **8:** 41.
3. Pollack, M. S., M. I. New, G. J. O'Neill, L. S. Levine, C. Callaway, S. Pang, E. Cacciari, F. Mantero, A. Cassio, C. Scaroni, G. Chiumello, G. F. Rondanini, L. Gargantini, G. Giovannelli, R. Virdis, E. Bartolotta, C. Migliori, C. Pintor, L. Tato, F. Barboni & B. Dupont. 1981. HLA-genotypes and HLA-linked genetic markers in Italian patients with classical 21-hydroxylase deficiency. Hum. Genet. **58:** 331.
4. Laron, Z., M. S. Pollack, R. Zamir, A. Roitman, Z. Dickerman, L. S. Levine, F.

LORENZEN, G. J. O'NEILL, S. PANG, M. I. NEW & B. DUPONT. 1980. Late-onset 21-hydroxylase deficiency and HLA in the Ashkenazi population: A new allele at the 21-Hydroxylase locus. Hum. Immunol. **1:** 55.

5. POLLACK, M. S., L. S. LEVINE, G. J. O'NEILL, S. PANG, F. LORENZEN, B. KOHN, G. F. RONDANINI, G. CHIUMELLO, M. I. NEW & B. DUPONT. 1981. HLA linkage and B14, DR1, BfS haplotype association with the genes for late-onset and cryptic 21-hydroxylase deficiency. Am. J. Hum. Genet. **33:** 540.
6. CHROUSOS, G. P., D. L. LORIAUX, D. MANN & G. B. CUTLER, JR. 1982. Late-onset 21-hydroxylase deficiency is an allelic variant of congenital adrenal hyperplasia characterized by attenuated clinical expression and different HLA haplotype associations. Horm. Res. **16:** 193.
7. SVEJGAARD, A. & L. L. RYDER. 1977. Associations between HLA and disease. Notes on methodology and a report from the HLA and Disease Registry. *In* HLA and Disease. J. Dausset & A. Svejgaard, Eds.: 46. Munksgaard. Copenhagen.
8. WEIL, J., F. BIDLINGMAIER, W. G. SIPPELL, O. BUTENANDT & D. KNORR. 1979. Comparison of two tests for heterozygosity in congenital adrenal hyperplasia (CAH). Acta Endocrinol. **91:** 109.
9. HÖLLER, W., S. SCHOLZ, D. KNORR, F. BIDLINGMAIER, E. KELLER & E. D. ALBERT. 1985. Genetic differences between the salt-wasting, simple virilizing, and nonclassical types of congenital adrenal hyperplasia. J. Clin. Endocrinol. Metab. **60:** 757.
10. MAUTHE, I., H. LASPE & D. KNORR. 1977. Zur Häufigkeit des Kongenitalen Adrenogenitalen Syndroms (AGS): München 1963–1972. Klin. Pädiatr. **189:** 172.
11. KNORR, D., F. BIDLINGMAIER, O. BUTENANDT, K. VON SCHNAKENBURG & W. WAGNER. 1975. A test for heterozygosity in congenital adrenal hyperplasia (CAH). Pediatr. Res. **9:** 681.
12. GAREIS, F. J. & P. A. LEE 1977. Heterozygous carriers of 21-hydroxylase deficiency adrenal hyperplasia. *In* Congenital Adrenal Hyperplasia. P. A. Lee, L. P. Plotnick, A. A. Kowarski & C. J. Migeon, Eds.: 509. University Park Press. Baltimore.
13. KNORR, D., F. BIDLINGMAIER, O. BUTENANDT, K. VON SCHNAKENBURG & W. WAGNER. 1977. Test for heterozygosity of congenital adrenal hyperplasia. *In* Congenital Adrenal Hyperplasia. P. A. Lee, L. P. Plotnick, A. A. Kowarski & C. J. Migeon, Eds.: 495. University Park Place. Baltimore.
14. GUTAI, J. P., A. A. KOWARSKI & C. J. MIGEON. 1977. The detection of the heterozygous carrier for congenital virilizing adrenal hyperplasia. J. Pediatr. **90:** 924.
15. KRENSKY, A. M., A. M. BONGIOVANNI, J. MARINO, J. PARKS & A. TENORE. 1977. Identification of heterozygote carriers of congenital adrenal hyperplasia by radioimmunoassay of serum 17-OH progesterone. J. Pediatr. **90:** 930.
16. LORENZEN, F., S. PANG, M. I. NEW, B. DUPONT, M. POLLACK, D. M. CHOW & L. S. LEVINE. 1979. Hormonal phenotype and HLA-genotype in families of patients with congenital adrenal hyperplasia (21-hydroxylase deficiency). Pediatr. Res. **13:** 1356.
17. LORENZEN, F., S. PANG, M. I. NEW, M. POLLACK, S. OBERFIELD, B. DUPONT, D. CHOW, B. SCHNEIDER & L. LEVINE. 1980. Studies of the C–21 and C–19 steroids and HLA genotyping in siblings and parents of patients with congenital adrenal hyperplasia due to 21-hydroxylase deficiency. J. Clin. Endocrinol. Metab. **50:** 572.
18. NEW, M. I., F. LORENZEN, A. J. LERNER, B. KOHN, S. E. OBERFIELD, M. S. POLLACK, B. DUPONT, E. STONER, D. J. LEVY, S. PANG & L. S. LEVINE. 1983. Genotyping steroid 21-hydroxylase deficiency: Hormonal reference data. J. Clin. Endocrinol. Metab. **57:** 320.
19. PETERSEN, K. E., A. SVEJGAARD, M. D. NIELSEN & J. DISSING. 1982. Heterozygotes and cryptic patients in families of patients with congenital adrenal hyperplasia (21-hydroxylase deficiency). HLA and glyoxalase I typing and hormonal studies. Horm. Res. **16:** 151.
20. MAUSETH, R. S., J. A. HANSEN, E. K. SMITH, E. R. GIBLETT & V. C. KELLEY. 1980. Detection of heterozygotes for congenital adrenal hyperplasia: 21-hydroxylase deficiency—a comparison of HLA typing and 17-OH progesterone response to ACTH infusion. J. Pediatr. **97:** 749.

The Coexistence of IgA Deficiency and 21-Hydroxylase Deficiency Marked by Specific MHC Supratypes

T. J. COBAIN, M. S. STUCKEY, J. McCLUSKEY[a],
A. N. WILTON, A. GEDEON, M. J. GARLEPP,
F. T. CHRISTIANSEN, AND R. L. DAWKINS

Department of Clinical Immunology
Royal Perth Hospital and Q.E.II Medical Centre
Perth, Western Australia

[a]Laboratory of Immunology
National Institute of Allergy and Infectious Diseases
National Institutes of Health
Bethesda, Maryland 20205

INTRODUCTION

Recent evidence has demonstrated that IgA deficiency is associated with two particular MHC supratypes.[1] One of these, A1, CW7, B8, C2C, BfS, C4AQ0, C4B1, DR3, DQ2 is readily recognizable from HLA-A, B, and DR typing alone.[2] The second important supratype was predicted on the basis of increased frequencies of B14 and A28 in patients with serum IgA deficiency. It is already known that one HLA-B14 containing supratype is associated with the phenotypically cryptic or attenuated late-onset form of 21-hydroxylase deficiency. We were, therefore, interested to determine whether this same supratype carried genes involved in IgA deficiency or whether different B14-containing supratypes were involved in these apparently unrelated disorders. Additionally, further genetic characterization of the susceptibility determinants of IgA deficiency might allow predictions on the mode of inheritance of this disorder.

METHODS

Histocompatibility and Complement Typing

HLA typing was performed on peripheral blood lymphocytes using a panel of standardized international workshop antisera in a microcytotoxicity assay. Complement typing was performed by immunofixation following agarose gel electrophoresis of EDTA plasma samples.[4]

Synacthen Testing

Following intravenous injection of synthetic ACTH (synacthen), blood samples were drawn at zero time and 30 minutes. Plasma 17-α-hydroxyprogestrone levels were measured by radioimmunoassay.[3]

IgA Levels

Serum IgA concentration was measured by rate nephelometry on a Beckman Automated Immunochemistry System. The 95% reference range is 0.65–3.4 gL^{-1}. IgA deficiency is defined as serum concentration <0.3 g/l.

RESULTS

Genotypes and Supratypes in IgA Deficiency

Thirteen families with IgA deficiency were studied (TABLE 1). Two supratypes, B14, C4A2, C4B1/2, BfS (denoted as 14-S-2-1/2) and A1, B8, C4AQ0, C4B1, BfS, DR3 (denoted as 8/3-S-0-1), occurred frequently. Each was present in the affected members of five families, and one patient possessed both supratypes. HLA-A28 was also frequent in this patient group, but this allele was not always part of the same supratype.

TABLE 1. Unrelated Haplotypes in IgA Deficient Subjects from 13 Families

	First Haplotype						Second Haplotype					
Family	A	B	C4A	C4B	BF	DR[c]	A	B	C4A	C4B[a]	BF	DR[c]
ADL	1	8	QO	1	S	3	28	40	3	1	S	4
	1	8	QO	1	S	3	3	14	2	1/2	S	3
	28	40	3	1	S	4	3	14	2	1/2	S	3
FEN	3	35	3/2	1	F	1	—	14	2	1/2	S	1
MIT	1	37	3	1	F	1	33	14	2	1/2	S	1
	29	44	QO	1	S	3	33	14	2	1/2	S	1
LAW	2	44	3	1	S	5	28	14	2	1/2	S	5
MCK	1	17	6	1	S	7	28	14	2	1/2	S	5
SHA	11	8	QO[b]	1	S	ND	28	14	3[b]	1/2	S	ND
JAH	1	8	QO	1	S	3	28	44	3	QO	F	4
HEA	1	8	QO[b]	1	S	3	31	40	3[b]	1	S	4
MCD[d]	1	8	QO	1	S	3	2	15	3	1	F	6
	1	8	QO	1	S	3	2	44	3	1	S	5
COO	2	60	QO	2	S	6	1	17	6	1	S	6
	2	60	QO	2	S	6	26	44	3	1	S	7
DAW	2	51	QO	1	S	5	2	14	3	1	F	6
HOD	11	51	QO[b]	1	S	6	26	44	3[b]	1	S	4
FIN[d]	2	17	6	1	S	7	24	7	3	1	S	2

[a]C4B1/2 duplication is usually, but not always proven.
[b]Allotype inferred, but not proven.
[c]ND indicates DR allele was not determined.
[d]H[c]D, FIN IgA <0.6 g/l.

Within the group of 14-S-2-1/2 supratypes, identity of MHC (major histocompatibility) alleles was maximal for the loci in the HLA-B and complotype region, with some variation in HLA-A and DR alleles. Similarly, recombinant versions of the more common 8/3-S-0-1 supratype were also identified with variation occurring at HLA-A or DR loci. In this supratype, there is evidence for deletion within the C4A locus

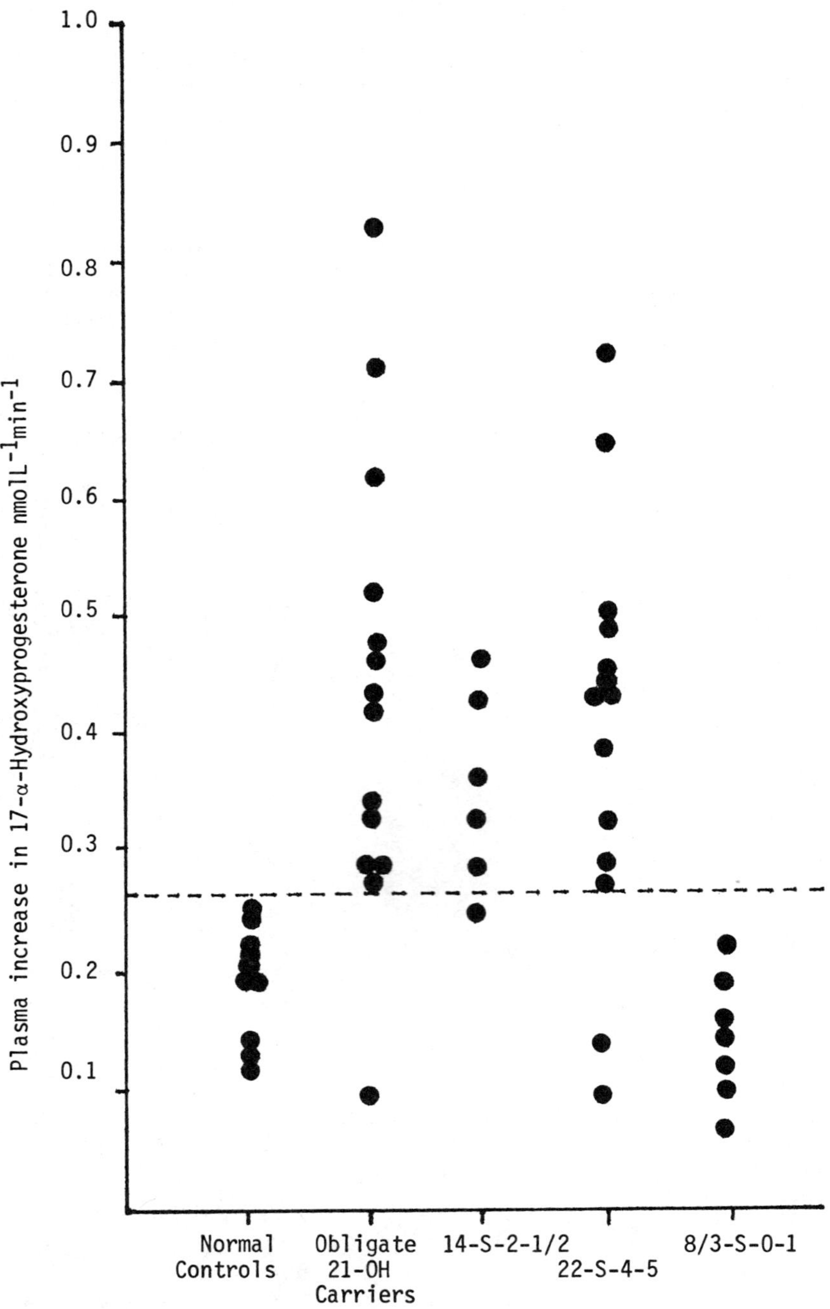

1.0
0.9
0.8
0.7
0.6
0.5
0.4
0.3
0.2
0.1
Plasma increase in 17-α-Hydroxyprogesterone nmolL-1min-1
Normal Controls
Obligate 21-OH Carriers
14-S-2-1/2
22-S-4-5
8/3-S-0-1

TABLE 2. MHC Supratypes Associated with CAH[a]

B	C4A	C4B	Bf	C2	DR	21-OH	N	
14	2	1/2	S	1		S	17[4]	Duplication not always detected. Bw65 where typed. Often DR1.
47	0	3	F	(1)		0	22[4,5]	Usually DR7, some DR4 (one C2-2 DR7).
*55	4	5	S	2		0	2[3]	
*55	4	2	S	2		0	4[4]	*Possibly identical.
	4	2	S	2		0	5[4,5]	Usually B35, DR4.
5	3	1	S	1		0	4[4,5]	Usually B51 where typed.
17	6	1	S	1	7	0	2[4,5]	B57 where typed.
7	3	1	S	1	2	0	3[4,5]	
44	3	0	F	1		0	2[5]	
44	3	0	S	1	4	0	3[3,5]	
60	3	0	S	1	1	0	2[5]	
35	3	1	S	1		S	2[5]	
62	3	1	S	1	4	0	2[4,5]	

[a]Identified from the literature.[3–5] Only supratypes occurring repeatedly are shown. The order of loci is arbitrary. N denotes the number of times the supratype was present in a total of 70 unrelated patients with different clinical forms of CAH.

leading to the C4AQ0 (null) phenotype. Other common supratypes with known functional defects were identified among the IgA deficient genotypes. HLA-B17, BfS, C4A6, C4B1, DR7 was present as 3/30 supratypes. When encoded on this supratype, the C4A locus gene product is hemolytically nonfunctional. One supratype with a duplicated C4A locus gene was also identified, which is HLA-A3, B35, C4A3/2, C4B1, BfF, DR1. Of the 30 haplotypes in TABLE 1, two-thirds possessed C4 null alleles, duplication events, or known functional defects involving one or other of the C4 loci. Furthermore, each individual carried at least one such unusual supratype. Of the remaining haplotypes, 4/10 were HLA-B44, BfS, C4A3, C4B1.

IgA Deficiency and Congenital Adrenal Hyperplasia (CAH) due to 21-Hydroxylase Deficiency

TABLE 2 lists those supratypes identified from the literature as being associated with CAH due to 21-hydroxylase deficiency. The 14-S-2-1/2 supratype found in patients with IgA deficiency is apparently identical to that associated with the cryptic and attenuated forms of 21-hydroxylase deficiency.[4] To determine whether deficiency of 21-hydroxylase function was found in IgA deficient individuals with this supratype, ACTH stimulation studies were carried out on members of two of the pedigrees described below. FIGURE 1 shows the rate of increase in 17-α-hydroxyprogesterone

FIGURE 1. Results of ACTH stimulation studies in five groups of subjects. (i) Normal controls comprise HLA-typed healthy normals and obligate normal sibs of affected children, (ii) reference positive controls are parents of affected children, (iii) members of pedigrees FEN and LAW carrying supratype 14-S-2-1/2, (iv) individuals carrying supratype 22-S-4-5 known to be associated with classical CAH,[3] and (v) individuals carrying supratype 8/3-S-0-1 negatively associated with CAH. Values are plotted as the rate of increase in plasma 17-α-hydroxyprogesterone measured 30 minutes after intravenous injection of ACTH. The 95% reference limit is shown as a dashed line.

following ACTH injections in five groups of individuals: (i) random healthy individuals and some obligate normal controls (HLA nonidentical sibs of CAH affected children), (ii) parents of children with CAH (obligate heterozygotes for 21-hydroxylase deficiency), (iii) members of pedigrees LAW (FIGURE 2) and FEN (FIGURE 3), who carry the 14-S-2-1/2 supratype, (iv) healthy normal subjects carrying the HLA-B22, BfS, C4A4, C4B5 (22-S-4-5) supratype associated with classical CAH[3], and, (v) healthy normal subjects carrying the supratype 8/3-S-0-1, which is negatively associated with CAH.[4,5] The results demonstrate biochemical evidence of heterozygous 21-hydroxylase deficiency in obligate carriers and those individuals

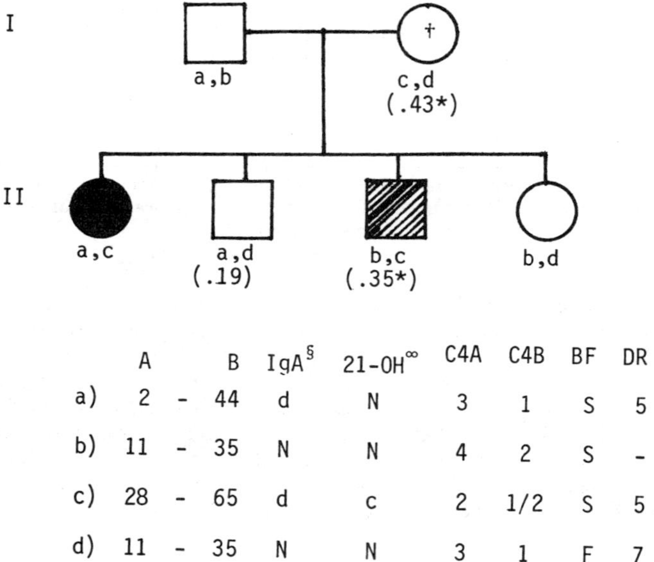

	A		B	IgA§	21-OH∞	C4A	C4B	BF	DR
a)	2	-	44	d	N	3	1	S	5
b)	11	-	35	N	N	4	2	S	-
c)	28	-	65	d	c	2	1/2	S	5
d)	11	-	35	N	N	3	1	F	7

FIGURE 2. Family LAW. Pedigree showing segregation of IgA deficiency and putative attenuated allele for 21-OH gene with supratype 14-S-2-1/2. Bw65 is a serological subspecificity of B14. Rate of increase in 17-OHP (nmol/1/min) after intravenous stimulation with ACTH is shown in brackets. * indicates biochemical 21-OH deficient carrier status. (∞) N represents normal gene and (∞) c represents cryptic or attenuated allele associated with late-onset 21-OH deficiency. Solid circle or square indicates IgA less than 0.3g/1. † indicates IgA not tested. Hatched circle or square indicates a IgA level lower than the normal range (0.65–3.4g/1). (§) N represents the normal IgA allele and (§) d represents the putative deficiency allele.

carrying the 14-S-2-1/2 or 22-S-4-5 supratypes. The 22-S-4-5 supratype is associated with classical CAH, but is not associated with IgA deficiency. By contrast, individuals carrying supratype 8/3-S-0-1 have normal ACTH responses despite the association of this supratype with IgA deficiency.

IgA Deficiency Segregates with MHC Haplotypes

The presence of IgA deficiency phenotypes in relation to the segregation of MHC haplotypes is shown in four pedigrees in FIGURES 2–5. The data is consistent with a

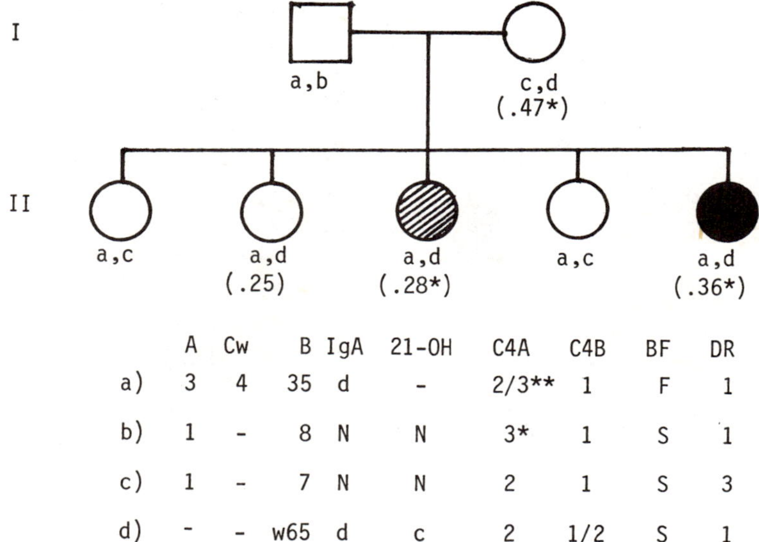

	A	Cw	B	IgA	21-OH	C4A	C4B	BF	DR
a)	3	4	35	d	-	2/3**	1	F	1
b)	1	-	8	N	N	3*	1	S	1
c)	1	-	7	N	N	2	1	S	3
d)	-	-	w65	d	c	2	1/2	S	1

FIGURE 3. Family FEN. Pedigree showing segregation of IgA deficiency and a putative cryptic or attenuated allele for the 21-OH gene with supratype 14-S-2-1/2. (α) HLA-A1, 2, 3, 9, 10, and 11 are excluded, with most probably A33 or A28 from population linkage disequilibrium. In d), duplication at C4B could not be proven, but it could be C4B2. The rate of increase in 17-OHP after intravenous stimulation with ACTH is shown in brackets. * indicates biochemical carrier status. ** C4A3 cannot be definitely assigned; C4A2 is also possible. For description of other symbols, see FIGURE 2.

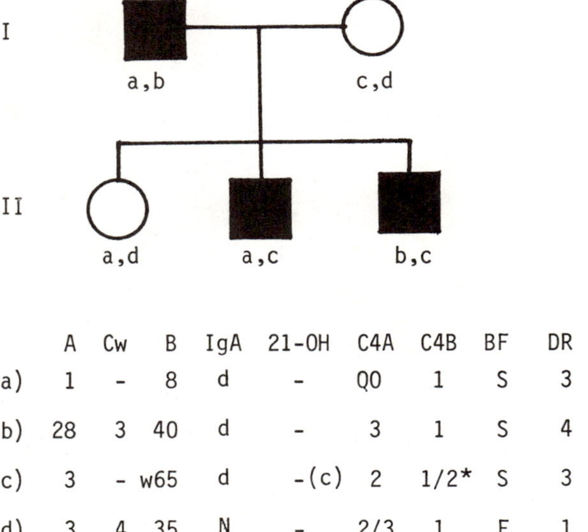

	A	Cw	B	IgA	21-OH	C4A	C4B	BF	DR
a)	1	-	8	d	-	QO	1	S	3
b)	28	3	40	d	-	3	1	S	4
c)	3	-	w65	d	-(c)	2	1/2*	S	3
d)	3	4	35	N	-	2/3	1	F	1

FIGURE 4. Family ADL. Pedigree showing segregation of IgA deficiency as a recessive gene. The 8/3-S-0-1 supratype (a) and the 14-S-2-1/2 supratype (c), as well as a third supratype (b), carry putative IgA deficiency determinants. * indicates that the C4B duplication could not be proven, and that C4B2 alone is possible. For description of other symbols, see FIGURE 2.

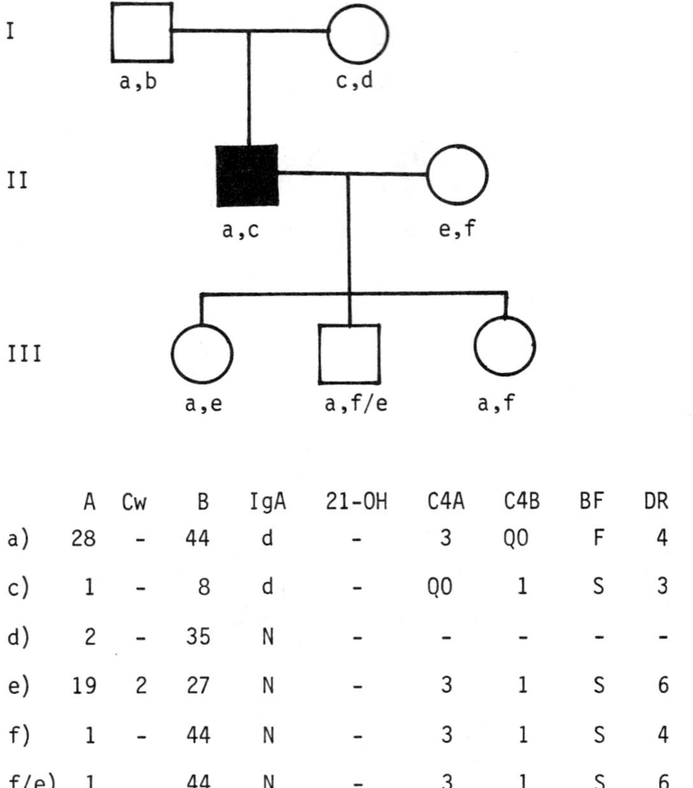

FIGURE 5. Family JAH. Pedigree showing segregation of IgA deficiency with 8/3-S-0-1 supratype. For description of other symbols, see FIGURE 2.

TABLE 3. Concentrations IgA (g/1) in Members of Families in FIGURES 2 to 6

Family	[a]I1	I2	II1	II2	II3	II4	II5	III1	III2	III3
LAW	2.1	1.8	<.3	1.7	0.5	1.1	—	—	—	—
FEN	2.3	1.0	0.9	0.8	0.6	0.8	<.3	—	—	—
ADL	<.3	1.6	1.7	<.3	<.3	—	—	—	—	—
JAH	NT	1.3	<.3	1.5	—	—	—	1.3	0.8	0.7
MIT	<.3	1.9	<.3	1.2	2.9	—	—	—	—	—

[a]Roman letters denote generation. Arabic letters denote members of a given generation from left to right. NT = not tested.

single MHC-linked locus with a recessive mode of inheritance (TABLE 3). By this model, pedigree FEN (FIGURE 3) and pedigree MIT (FIGURE 6) both contain individuals with normal serum levels of IgA, but, in whom, IgA deficiency would be predicted. These results may reflect differential complementation of variant mutant alleles involved in the regulation of serum IgA levels.

DISCUSSION

Eighteen IgA deficient subjects from 13 unrelated families have been MHC genotyped. Of the 30 MHC haplotypes associated with IgA deficiency, 20 are

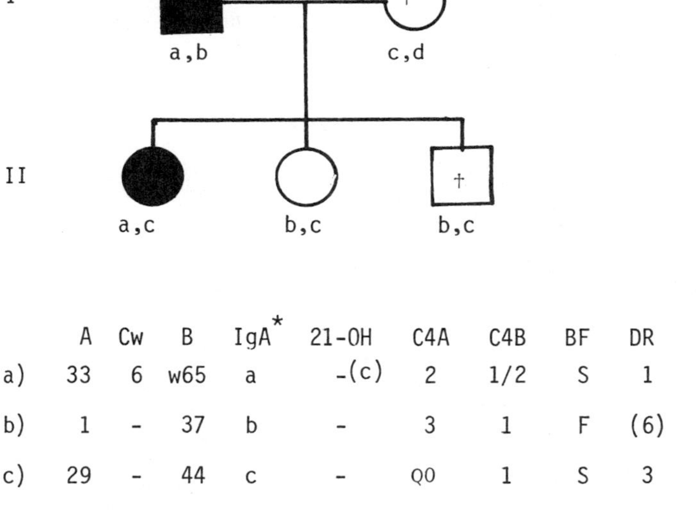

	A	Cw	B	IgA*	21-OH	C4A	C4B	BF	DR
a)	33	6	w65	a	-(c)	2	1/2	S	1
b)	1	-	37	b	-	3	1	F	(6)
c)	29	-	44	c	-	QO	1	S	3
d)	24	-	35	N	-	4	1	S	5

FIGURE 6. Family MIT. Pedigree of IgA deficient family. * a, b and c represent different putative IgA deficiency alleles. For description of other symbols, see FIGURE 2.

associated with null alleles, duplication events, or functionally abnormal gene products at one or other of the C4 loci. The supratype 14-S-2-1/2 is a marker for a cryptic or attenuated form of 21-hydroxylase deficiency, as well as for IgA deficiency. These two disorders are not coassociated on other MHC supratypes such as the classical CAH supratype 22-S-4-5 or the IgA deficiency-related supratype 8/3-S-0-1. Both IgA deficiency and CAH show associations with "recombinant" or truncated forms of supratypes recognized by their preservation in the region between HLA-B and complement. Analysis of several pedigrees informative for the segregation of IgA deficiency is consistent with linkage of this phenotype to MHC haplotypes. The data suggests the involvement of a single locus influencing serum IgA levels, which is best explained by a recessive inheritance with multiple alleles. Patients with a putative IgA deficiency genotype, but normal phenotype, may be compound heterozygotes for alleles showing a hierarchy of dominance. Such a model has been proposed for clinically heterogeneous congenital adrenal hyperplasia. With the recent identification of a probable duplicated structural gene for a cytochrome P450 21-hydroxylase (encoded within the duplicated C4 loci), the mechanism underlying the association of clinically heterogenous CAH with MHC supratypes can now be examined directly. Preliminary data suggest a deletion of one of the genomic fragments hybridizing to a bovine P450 21-hydroxylase cDNA in patients with CAH and the Bw47, BfF, C4AQ0, C4B3 supratype.[9] It is likely that CAH associated with different supratypes will demonstrate other abnormalities when the cloned genomic genes from these individuals are examined in detail. On the 14-S-2-1/2 supratype, structural modifications involving particularly the C4 loci and the cytochrome P450 21-hydroxylase locus appear to affect IgA levels. These data strongly suggest the existence of an important locus involved in the regulation of serum IgA levels and located close to or within the human class III MHC gene cluster.

[**Note Added in Proof:** Haplotypes a, b, and c of family FEN (FIGURE 3) most likely include C4 null alleles at C4B, C4A, and C4A, respectively. Haplotype b of family ADL (FIGURE 4) may also include C4BQ0. (See WILTON *et al.* 1985. Immunogenetics **21:** 333 for further information.)]

REFERENCES

1. COBAIN, T. J., M. A. H. FRENCH, F. T. CHRISTIANSEN & R. L. DAWKINS. 1983. Tissue Antigens **22:** 151–154.
2. HAMMARSTROM, L. & C. I. E. SMITH. 1983. Tissue Antigens **21:** 75–79.
3. MCCLUSKEY, J., P. H. KAY, M. S. STUCKEY, F. T. CHRISTIANSEN, R. L. DAWKINS & G. WILSON. 1983. Lancet **I:** 765–766.
4. O'NEILL, G. J., B. DUPONT, M. S. POLLACK, L. S. LEVINE & M. I. NEW. 1982. Clin. Immunol. Immunopathol. **23:** 312–322.
5. FLEISCHNICK, E., Z. L. AWDEH, D. RAUM, J. GRANADOS, S. M. ALOSCO, J. F. CRIGLER, P. S. GERALD, C. M. GILES, E. J. YUNIS & C. A. ALPER. 1983. Lancet **i:** 152–156.
6. CARROLL, M. C., R. C. CAMPBELL, D. R. BENTLEY & R. R. PORTER. 1984. Nature **307:** 237–240.
7. OEN, K., R. E. PETTY & M. C. SCHROEDER. 1982. Tissue Antigens **19:** 174–182.
8. PALSDOTTIR, A., S. J. CROSS, J. H. EDWARDS & M. C. CARROLL. 1983. Nature **306:** 615–616.
9. WHITE, P. C., M. I. NEW & B. DUPONT. 1984. (Abstract) Third H–2, HLA Cloning Meeting.

Neonatal Screening Program for Congenital Adrenal Hyperplasia in a Homogeneous Caucasian Population

E. CACCIARI, A. BALSAMO, A. CASSIO, S. PIAZZI,[a]
F. BERNARDI, S. SALARDI, A. CICOGNANI,
P. PIRAZZOLI, F. ZAPPULLA, M. CAPELLI,[a]
M. PAOLINI,[a] AND C.I. CORDARO

2nd Pediatric Clinic
University of Bologna
and
[a]Central Laboratory
S. Orsola Hospital of Bologna
40138 Bologna, Italy

The considerable variation in the reported incidence of congenital adrenal hyperplasia (CAH)[1–7] may be explained partially by the lack of a valid screening method. To evaluate the true prevalence of CAH, we examined all the newborns in the Emilia Romagna region during a period of approximately three years. Emilia Romagna is situated in northern Italy and is a homogeneous sample of a Caucasian population.

For the screening program, we took advantage of the specimens collected on filter paper used for neonatal screening of hypothyroidism and phenylketonuria. For the 17-OH progesterone assay, the microfilter paper method of Pang *et al.*[8–11] was used, although it was somewhat modified. After the statistical analysis of the results and the clinical examination of the subjects during the first period of the screening procedure, we established 20 pg/disc as the recall value. When values were higher than this, they were confirmed by a second microfilter paper assay and then the serum 17-OH progesterone concentration was determined.

Overall, 73,000 newborn babies were examined. One hundred and thirty-two (0.18%) had a value above the threshold of 20 pg/disc and were retested. Among those recalled, seven demonstrated a pathological value, with a prevalence of CAH of one case out of 10,428.

In TABLE 1, is represented the frequency of the various forms of 21-hydroxylase deficiency and the confidence limits at 95%, as well as the heterozygote and gene frequency.[12,13] Day of sampling, gestational age, and birthweight were all considered in the analysis of results. With regard to gestational age, infants were divided into two groups: either "term" (those delivered between 37 weeks and 42 weeks of gestation) or "preterm" (those delivered before 37 weeks of gestation).

17-OH progesterone values in relation to the day of sampling is reported in TABLE 2. No statistically significant differences were found between the various days. What might be found surprising is the low mean value of 17-OH progesterone given by the infants examined on the second day of life. This finding might be ascribed to the fact that the sampling was usually done before leaving the obstetrics ward, and it is possible that the infants examined and considered fit to be moved on days 2 and 3 of life were the most mature. Therefore, they exhibited lower 17-OH progesterone values.

TABLE 3 represents the number and percentage of the recalled according to the gestational age. The percentage of the recalled among the preterm infants was

TABLE 1. Frequency of the Various Forms of 21-Hydroxylase Deficiency

Prevalence of CAH	1:10,428
salt-losing forms	1:18,250
simple virilizing forms	1:73,000
nonclassical forms	1:36,500
95% confidence limits	
lower	1:21,978
upper	1:5062
Heterozygote frequency	1:51
Gene frequency	0.010

TABLE 2. Relation Between 17-OH-P Values and Day of Sampling

Day	Cases	Value pg/disc (mean ± SD)
2	219 (0.3%)	5.62 ± 1.01
3	1460 (2.0%)	6.03 ± 2.10
4	11,534 (15.8%)	6.42 ± 2.43
5	34,456 (47.2%)	6.31 ± 2.58
6	14,819 (20.3%)	6.98 ± 3.02
7	2555 (3.5%)	8.84 ± 8.39
>7	7957 (10.9%)	7.83 ± 7.74

significantly higher than that of the term infants. This justifies the setting of a higher recall threshold for the preterm infants. Furthermore, it must be pointed out that in spite of the fact that the number of recalled turned out to be high in the premature infants, the 17-OH progesterone values, even in these subjects, became rapidly normal. This could be seen in the first follow-up, which was usually made within the third week of life.

Considering the 17-OH progesterone values according to gestational age (term and preterm) and birthweight, FIGURE 1 shows a significant difference among the term infants between the groups having a birthweight of less than 2500 grams and the groups with a birthweight of more than 3500 grams. Another significant difference was evident between the term and preterm infants weighing 2500 to 3000 grams, while no such difference was present between the groups having a birthweight of 2000 to 2500 grams. From these data, it would seem that the gestational age has a greater influence on adrenal enzymatic maturity in infants having a birthweight greater than 2500 grams, whereas birthweight seems to be more important in infants less than 2500 grams.

As regards to the seven infants affected by 21-hydroxylase deficiency (four females and three males), they included four salt-wasting forms (TABLE 4, cases no. 1, 4, 5, and 7), one simple virilizing form (TABLE 4, case no. 2), and two patients (TABLE 4, cases

TABLE 3. Number and Percentage of Retests according to Gestational Age

	Patients Tested (%)	Retested (%)
Term infants	66,503 (91.1%)	53 (0.08)[a]
Preterm infants	6497 (8.9%)	79 (1.22)[a]

[a] $\chi^2 = 418$ ($p = 0.0000001$).

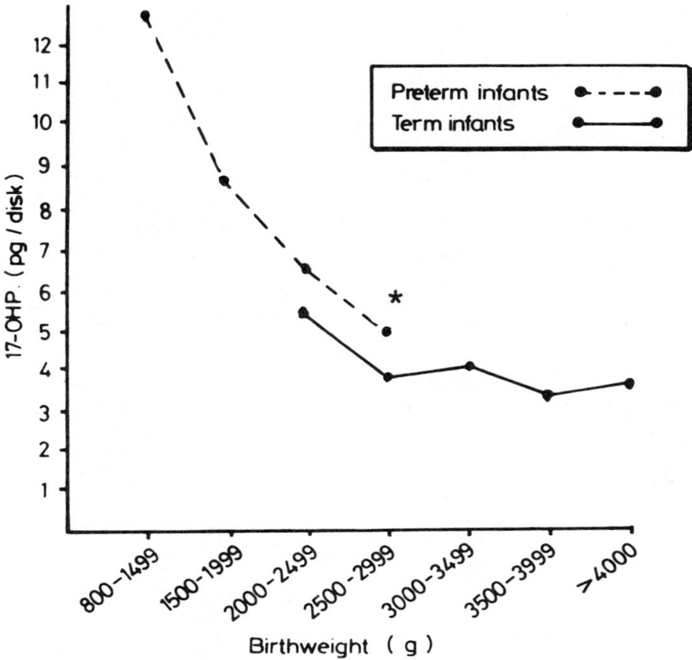

FIGURE 1. 17-Hydroxyprogesterone values (pg/disc) according to gestational age (term and preterm) and birthweight. * indicates $p < 0.05$ between term and preterm children in the group with birthweight 2500–3000 g.

TABLE 4. Clinical and Hormonal Features of the Seven Infants with CAH

Case n°	Age (days)	Gestation	Birthweight	17-OH-P: 1st Spot (pg/disc)	17-OH-P: Serum (ng/dl)	External Genitalia
1 (M)	4	Term	4000	>250		Mild macrogenitosomia
	24				>5000	
2 (F)	5	Term	3300	28[a]		Ambiguous
	150				5000	
3 (F)	5	Term	3870	47		Normal
	30				1400	
4 (F)	4	Term	3000	>250		Ambiguous
	14				>5000	
5 (F)	3	Term	3200	>250		Ambiguous
	7				>5000	
6 (M)	2	Preterm	1350	150		Normal
	2				560	
7 (M)	3	Term	3250	>250		Mild macrogenitosomia
	16				>5000	

[a]The blood of this subject collected on filter paper was stored in unfit conditions. This could explain the not particularly elevated value of 17-OH progesterone.

no. 3 and 6) with the late-onset or cryptic form of CAH having been based on a high 17-OH progesterone value with no clinical sign.[14–17] Six of them were born at term, and one was preterm.

In case no. 1 and in case no. 4, the screening enabled us to make a diagnosis, and, thus, immediately begin therapy. In the other infants, with the exception of the two having the nontypical form, the result of the screening did nothing but confirm the diagnostic suspicion of the physician. In the last six months, because of technical reasons, the CAH screening was restricted to the risk subjects only, and only after an intense educational campaign, especially in the obstetric and newborn environment. Following this method, two more cases of CAH have been diagnosed in this period. This is practically just what we expected. Naturally, though, the nontypical forms may have slipped our detection.

In conclusion, we can say that: (1) neonatal screening for CAH by means of 17-OH progesterone assay on filter paper discs is a reliable method for diagnosing 21-hydroxylase deficiency, and it is convenient to carry it out in the same laboratory that screens for phenylketonuria and hypothyroidism; (2) the maturity of the neonates and the particular day of sampling affect the value of 17-OH progesterone, but not the validity of the screening; and (3) although these data alone do not justify wide scale or even national scale screening for CAH, further experience is needed before definite conclusions may be drawn.

REFERENCES

1. HIRSCHFIELD, A. J. & J. K. FLESHMAN. 1969. An unusually high incidence of salt-losing congenital adrenal hyperplasia in the Alaskan Eskimo. J. Pediatr. **75:** 492–494.
2. QAZI, Q. H. & W. M. THOMPSON. 1972. Incidence of salt-losing form for congenital virilizing hyperplasia. Arch. Dis. Child. **47:** 302–307.
3. HUBBLE, D. 1966. Congenital adrenal hyperplasia. *In* Basic Concepts of Inborn Errors and Defects of Steroid Biosynthesis. K. S. Holts & D. N. Raine, Eds.: 168–75. Proc. 3rd Symp. Society for the Study of Inborn Errors of Metabolism. Livingstone, Edinburgh.
4. ROSENBLOOM, A. L. & D. W. SMITH. 1966. Congenital adrenal hyperplasia (letter). Lancet **i:** 660.
5. PRADER, A. 1958. Die haeufigkeit des kongenitalen adrenogenitalen Syndroms. Helv. Paediatr. Acta **13:** 426–431.
6. CHILDS, B., M. M. GRUMBACH & J. J. VAN WYK. 1956. Virilizing adrenal hyperplasia. A genetical and hormonal study. J. Clin. Invest. **35:** 213–221.
7. PANG, S., W. MURPHY & L. S. LEVINE. 1981. A pilot newborn screening for congenital adrenal hyperplasia (CAH) due to 21-hydroxylase deficiency at New York Hospital and Alaska. 1st Joint Meeting LWPES-ESPE. Geneva.
8. PANG, S., J. HOTCHKISS, A. L. DRASH, L. S. LEVINE & M. I. NEW. 1977. Microfilter paper method for 17-α-hydroxyprogesterone radioimmunoassay: Its application for rapid screening for congenital adrenal hyperplasia. J. Clin. Endocrinol. Metab. **45:** 1003–1008.
9. CACCIARI, E., A. BALSAMO, A. CASSIO, S. PIAZZI, F. BERNARDI, S. SALARDI, A. CICOGNANI, P. PIRAZZOLI, F. ZAPPULLA, M. CAPELLI & M. PAOLINI. 1982. Neonatal screening for congenital adrenal hyperplasia using a microfilter paper method for 17-α-hydroxyprogesterone radioimmunoassay. Experience gained from the study of 22,233 cases. Horm. Res. **16:** 4–9.
10. PIAZZI, S., M. CAPELLI, M. PAOLINI, D. PERUGINI, G. GROSSI, A. BALSAMO, P. SALOMONI, A. CASSIO, G. BUGIARDINI & E. CACCIARI. 1982. Neonatal screening for 21-hydroxylase deficiency: A microfilter paper method for 17-α-hydroxyprogesterone assay. J. Endocrinol. Invest. **5:** 87–90.
11. CACCIARI, E., A. BALSAMO, A. CASSIO, S. PIAZZI, F. BERNARDI, S. SALARDI, A.

CICOGNANI, P. PIRAZZOLI, F. ZAPPULLA, M. CAPELLI & M. PAOLINI. 1983. Neonatal screening for congenital adrenal hyperplasia. Arch. Dis. Child. **58:** 803–806.

12. DIEM, K. & C. LENTNER. 1972. Scientific Tables. 7th edition. p. 188. Ciba Geigy Limited. Switzerland.
13. FRISTROM, J. W. & P. T. SPIETH. 1980. Principles of Genetics. p. 533. Chiron Press. New York.
14. ZACHMANN, M. & A. PRADER. 1978. Unusual heterozygotes of congenital adrenal hyperplasia due to 21-hydroxylase deficiency. Acta Endocrinol. **87:** 557–565.
15. BLANKSTEIN, J., C. FAIMAN, F. I. REYES, M. L. SHROEDER & J. S. D. WINTER. 1980. Adult-onset familial adrenal 21-hydroxylase deficiency. Am. J. Med. **68:** 441–448.
16. NEW, M. I., F. LORENZEN, S. PANG, P. GUNCZLER, B. DUPONT & L. S. LEVINE. 1979. "Acquired" adrenal hyperplasia with 21-hydroxylase deficiency is not the same genetic disorder as congenital adrenal hyperplasia. J. Clin. Endocrinol. Metab. **48:** 356–369.
17. LEVINE, L. S., M. ZACHMANN & M. I. NEW. 1978. Genetic mapping of the 21-hydroxylase deficiency gene within the HLA linkage group. N. Eng. J. Med. **299:** 911–915.

Newborn Screening for Congenital Adrenal Hyperplasia with Special Reference to Screening in Alaska[a]

SONGYA PANG,[b] DAVID A. SPENCE,[c] AND MARIA I. NEW[b]

[b]*Division of Pediatric Endocrinology*
Department of Pediatrics
Cornell University Medical College
New York, New York 10021

[c]*The Section of Family Health*
Department of Health and Social Services
State of Alaska
Juneau, Alaska 99811

INTRODUCTION

Background of Newborn Screening Program

Over the last four decades, several factors have combined to effect changes in the direction of detecting genetic disorders. The concept of newborn screening began in 1962 when a simple and inexpensive screening test was developed by Guthrie and Susi.[1] With the advent of Guthrie's test in the mid-1960s, a new concept arose in genetic screening in pediatrics. The purpose of genetic screening came to be to discover disorders in early infancy prior to the onset of illness, rather than the identification and treatment of patients in later life. Subsequently, recent advances in laboratory technology, such as enzyme assay and radioimmunoassay, led to the current mandate of newborn screening programs for metabolic and congenital disorders.

Since technological advances have made an increasing number of tests available which may be employed in the early detection of health problems, interest in screening of specific disorders has increased. Thus, in 1974, William Frankenburg provided carefully thought-out guidelines for the screening of pediatric disease.[2] Subsequently, the National Academy of Sciences proposed the criteria for selection of diseases and tests for newborn screening programs (TABLE 1).[3] The first three criteria pertain to the suitability of the disease for screening. The next five criteria deal with the practical aspects of the screening program, including the availability of an early and accurate screening test and a feedback system to deliver proper health care subsequent to the screening test. The cost-benefit analysis is an important and yet difficult aspect of newborn screening, which must be based on both humanistic and economic factors.

[a]This work was supported in part by grant no. 6–274 from the March of Dimes Birth Defects Foundation, USPHS NIH awards no. HD–00072 and HD–15084, and grant no. RR–47 and RR–05396 from the Division of Research Resources, NIH. Additional support was granted by the Calista Corp., Anchorage, Alaska.

Screening Program for Congenital Adrenal Hyperplasia

The adrenal synthesizes three main classes of hormones, namely glucocorticoids and sex steroids, both under ACTH control, and mineralocorticoids, under the regulation of the renin-angiotensin system. In steroid biosynthesis, 21-hydroxylase activity converts progesterone (P) to deoxycorticosterone (DOC), an aldosterone precursor, and 17-hydroxyprogesterone (17-OHP) to deoxycortisol, a cortisol precursor. Thus, in congenital adrenal hyperplasia (CAH) due to 21-hydroxylase deficiency, cortisol deficiency with or without aldosterone deficiency occurs. This results in excessive ACTH secretion, which stimulates excessive androgen production by the adrenals, which is responsible for virilization of external genitalia in female newborns. CAH is transmitted by an autosomal recessive gene.[4] In addition, in newborns with CAH, salt-losing crises in the first few months of life as a result of aldosterone deficiency may be fatal.[5,6] Female newborns with CAH may be so virilized as to be given an erroneous sex assignment, thus, causing the trauma of sex reassignment or, in those retaining the male sex assignment, hysterectomy and sterility later in life.[5,7] Delayed diagnosis may also result in several other complications, including acceleration of skeletal maturation with ultimate short stature,[5] premature development of secondary sex characteristics in male children,[8] and further virilization in female children.[5] These complications can be prevented or quickly corrected if detected by newborn screening and early diagnosis.

TABLE 1. Criteria for the Selection of Diseases for Screening[a]

1. Importance of the problem when diagnosis is delayed:
   a. Mortality
   b. Morbidity
      —ability of person to learn
      —ability of person to earn a living
      —cause of other prolonged serious handicap
      —transmission of disease to others
      —transmission of certain genetic conditions to offspring
2. The condition occurs in the population with high frequency.
3. The condition is amenable to treatment.
4. A simple and inexpensive screening method is available.
5. The test is sensitive and specific, and quality control is maintained.
6. A mechanism exists to collect samples and deliver them to the laboratory.
7. A reliable means for reporting results exists.
8. Resources for treatment and counseling are available.
9. The cost of screening, diagnosis, and treatment during the asymptomatic phase should be outweighed by the savings in:
   a. human misery
   b. fiscal expenditure

[a]Based on W.K. Frankenberg[2] and the National Academy of Sciences.[3]

The incidence of CAH in the Caucasian population based on recent case survey is either more common or equally as common as phenylketonuria (PKU), a disorder in which newborn screening is mandated;[9–13] CAH is certainly more common than all disorders mandated for newborn screening, with the exception of congenital hypothyroidism (TABLE 2). The highest incidence of CAH by case survey has been reported in the Alaskan Eskimo, indicating that the newborns in this geographical area are at increased risk for CAH.[14] A reliable and valid screening test for CAH using a heel

TABLE 2. Comparison of Incidences

Disorders Being Screened (1981 New York State, USA)		CAH by Case Survey (1966–1980 Reports)	
Congenital Hypothyroidism	1:4000	Birmingham, England (ref. 9)	1:7244
Phenylketonuria	1:17,000	Wisconsin, USA (ref. 10)	1:15,000
Galactosemia	1:74,000	Munchen, Germany (ref. 11)	1:9831
Histidinemia	1:104,000	Switzerland (ref. 13)	1:13,924
Homocystinuria	1:240,000	Tyrol, Austria (ref. 12)	1:8991
Maple Syrup Urine Disease	1:340,000	Average Total:	1:10,998
Adenosine Deaminase Deficiency	1:750,000		
		Alaska Native (ref. 14)	1:1480
		Alaska Yupik (ref. 14)	1:490

prick capillary blood specimen impregnated on filter paper became available, in 1977, in our laboratory.[15] Thus, CAH in both the Caucasian population and the Alaskan Eskimo population meets all of the criteria proposed by Frankenburg and the National Academy of Science for the institution of a screening program for a disorder.[2,3] Newborn screening for CAH in the Caucasian population should prove to be beneficial, while such a program in the Alaska region is even more highly desirable. Thus, the technological advance prompted development of a regional or pilot CAH screening program in various nations, such as Italy,[16,17] France,[18] and Japan,[19,20] and in Alaska in the USA.[21]

Pilot Newborn Screening Program for CAH in Alaska

At the time of the filter paper sample collection for PKU screening, an additional ½ inch circle specimen was obtained for CAH screening. The blood samples completely saturated both sides of the filter paper. The specimens were identified by number and date of specimen. Samples were not flagged for newborns at risk for CAH.

A schematic timetable of the pilot newborn screening program for CAH is depicted in FIGURE 1. Since the majority of newborn specimens were obtained on the third day of life, the time schedule described in the program applied to the majority of newborns. The time required to transport the specimens from Alaska to the New York Hospital Laboratory was three to four days.

All newborns for whom the 17-OHP level in the filter paper specimen was >57 pg/disc (the lowest level observed in 20 affected newborns with CAH not via screening program) (FIGURE 1) were recommended for evaluation for suspected CAH, and a second specimen was requested. The subjects with 17-OHP levels <50 pg/disc (highest level found in control non-CAH newborns) were presumed to be not affected by CAH. The subjects with levels of 51–56 pg/disc were recalled to obtain a second specimen. All newborns in whom the 17-OHP in the second specimen was again >51 pg/disc were recommended for medical evaluation for suspected CAH. We recommend that the reference data must be established in each screening laboratory based upon the laboratory technique employed.

Laboratory Method

The laboratory method for 17-OHP radioimmunoassay (RIA) has been previously described in detail.[15] Each specimen was tested in duplicate. Filter paper samples were punched out with a 3 mm paper punch (blood volume 3.56 μl/disc), placed in extraction tubes, and eluted with buffer solution before extraction with diethyl ether. To achieve the sensitivity of the assay, two 3 mm discs per sample were used. The lowest concentration of 17-OHP detectable at a 99% confidence limit was 5 pg/tube. The high specificity of antiserum for 17-OHP (Sandoz, NJ) has previously been described.[15] The inter- and intra-assay coefficient variations were less than 11%.[21] In other newborn screening programs for CAH, either RIA using radiolabeled iodine[16–19] or enzyme-linked immunosolvent assay (ELISA)[20] was utilized for the determination of 17-OHP. The reports indicate that the sensitivity, specificity, and speed of these assays meet the criteria of a screening test, as described by Rosenberg and Scriver,[22] when the extract of the blood spot by an organic solvent was used in the assay.

Results of CAH Screening in Alaska

A histogram of 17-OHP levels measured in 53,357 consecutive specimens in a 65-month period (12/78–4/84) is depicted in FIGURE 2. 17-OHP levels in the majority of the specimens (99.9%) were below 50 pg per disc. Thus, these newborns were presumed not to be affected by CAH. Sixteen had 17-OHP levels between 51–56 pg/disc. Of these, fifteen were proven to be normal by a second specimen, and one was not available for follow-up. Twenty-two newborns had a 17-OHP level >57 pg/disc, which is the suspected level for CAH.

Clinical Information and Follow-up in Newborns with 17-OHP Levels >57 pg/disc

Follow-up information was available in 18 of 22 newborns with 17-OHP levels >57 pg/disc (TABLE 3).

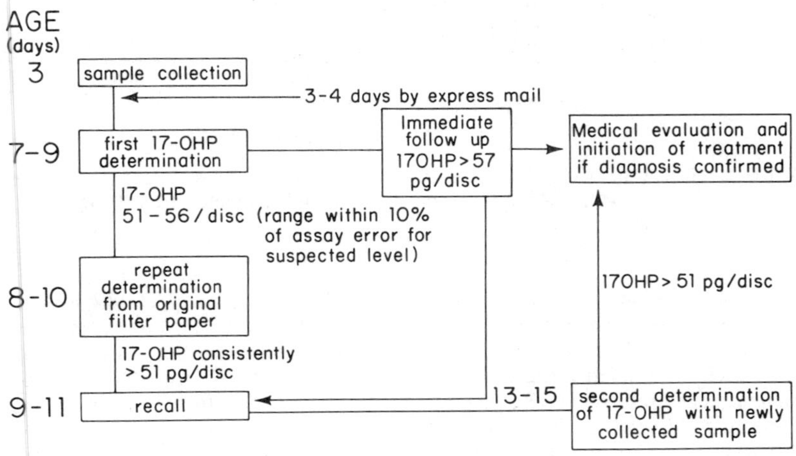

FIGURE 1. Schematic protocol for Alaskan newborn screening and follow-up program for congenital adrenal hyperplasia, in which the reference laboratory is New York. (From Pang *et al.*,[21] by permission of The Endocrine Society.)

The three female infants (one Caucasian and two Yupik Eskimos) and the one male infant (a Yupik Eskimo) with the highest 17-OHP levels (96, 120, 150, and 176 pg/disc) in this screening were subsequently proven to have CAH by confirmatory diagnostic tests. The onset of hyperkalemia without clinical symptoms of adrenal crisis occurred in all four newborns between the 6th and 14th day of life, and this occurred while undergoing medical evaluation which confirmed that thay had the salt-losing form of CAH. Treatment was instituted shortly thereafter, preventing the development of adrenal crisis. All female infants had ambiguous genitalia and were found to be genetic females on confirmation. In addition, one Yupik Eskimo newborn with ambiguous genitalia was clinically suspected and was thus treated for CAH at birth,

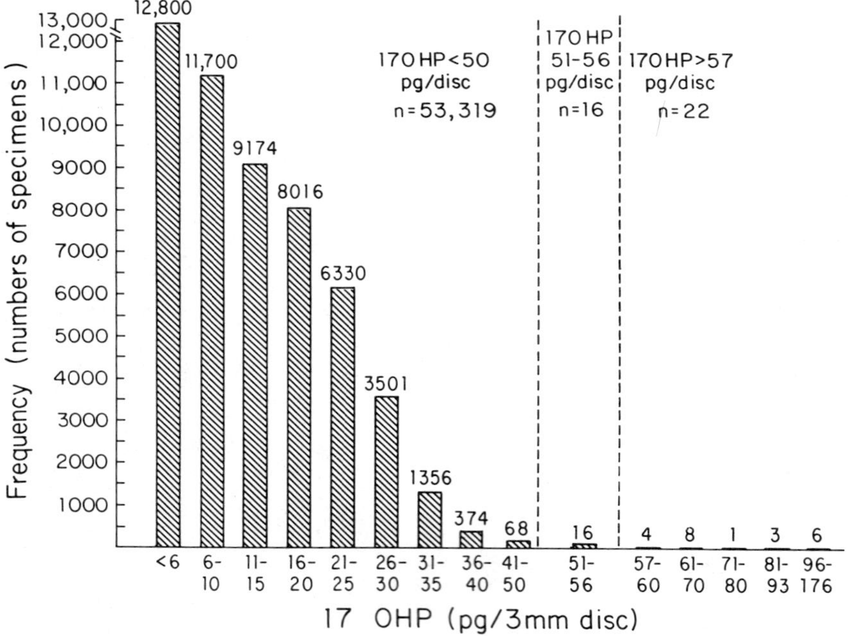

FIGURE 2. Histogram of 17-OHP levels in 3 mm disc filter paper specimens in 53,357 consecutive specimens in the Alaskan newborn population. (Adapted from Pang *et al.*,[21] by permission of The Endocrine Society.)

prior to confirmation of the diagnosis. The 17-OHP level obtained at 48 hours of age, while on intensive glucocorticoid treatment, was 19 pg/disc. However, the 17-OHP level rose to 250 pg/disc following discontinuation of treatment for 12 hours at an age of five weeks, confirming the clinical diagnosis of CAH. There was no history of consanguinity or interracial marriages in the parents of the affected newborns.

The other 11 newborns with 17-OHP levels >57 pg/disc were premature newborns with varying low birthweights (936–2386 gm). Repeat specimens in 9 of these 11 newborns were all <40 pg/disc. These nine premature infants were considered to be unaffected for CAH. The two remaining premature infants expired (nos. 8 and 12). One of these was a female (gestational age, 30 weeks; birthweight, 936 gm) with

TABLE 3. Eighteen of 22 Newborns with 17-OHP Level >57 pg/disc[a]

Newborn Infant No.	17-OHP level (pg/disc)	Gestation (BW, gms)	Age (Days)	Clinical Problem	Final Diagnosis		
1	58	Premature (<2380)	14	No	Non-CAH	Male	
2	60	Premature (<2380)	1	No	Non-CAH	Male	
3	60	Premature (<2380)	20	No	Non-CAH	Female	
4	61	Full-term (>2700)	1	No	Non-CAH	Female	
5	63	Premature	22	No	Non-CAH	Male	
6	63	Full-term	7	No	Non-CAH	Male	
7	63	Premature	6	No	Non-CAH	Male	
8	68	Extreme premature (<966)	5	High K^+, RDS	?	Male	Expired
9	76	Extreme premature (<966)	7	No	Non-CAH	Male	
10	88	Full-term	3	No	Non-CAH	Female	
11	89	Extreme premature	5	No	Non-CAH	Male	
12	93	Extreme premature	36	RDS	Non-CAH	Female	Expired
13	96	Extreme premature	10	No	Non-CAH	Female	
14	96	Full-term	1	Ambiguous genitalia	CAH, SL	Female	Caucasian
15	120	Full-term	3	Ambiguous genitalia	CAH, SL	Female	Yupik Eskimo
16	150	Full-term	2	Not suspected	CAH, SL	Male	Yupik Eskimo
17	157	Extreme premature	7	RDS	Non-CAH	Male	Indian/Caucasian
18	176	Full-term	6	Not suspected	CAH, SL w/ambig. genitalia	Female	Yupik Eskimo

[a]RDS = Respiratory distress syndrome; SL = salt-losing form of CAH.

neither ambiguous genitalia nor clinical symptoms of CAH. The other was one of twin male premature Aleut newborns, (gestational age, 28 weeks; birthweight, 966 gm), both of whom developed severe hyperkalemia and expired. A filter paper specimen was obtained in only one of these twins (no. 8). Autopsy findings showed massive intracranial hemorrhage and grossly and microscopically hyperplastic adrenal glands that were indistinguishable from the hyperplastic adrenal glands observed in CAH. Although the diagnosis of CAH could not be excluded, the male infant was not included as an affected newborn detected by screening because the diagnosis was not confirmed biochemically. Three newborns with 17-OHP levels >57 pg/disc were full-term newborns, and repeat specimen levels in these full-term newborns were <30 pg/disc, indicating these full-term newborns were not affected with CAH.

Frequency of CAH Gene in the Alaskan Newborn Population after Screening

Four of 2737 Yupik Eskimos, including one newborn who was clinically suspected, and one of 33,518 Caucasian newborns, were affected with CAH (TABLE 4). Thus, in the Yupik Eskimo, the observed incidence of the salt-losing form of CAH for homozygous affected cases, heterozygote frequency, and gene frequency were calculated to be 1:684, 1:13, and 0.038, respectively. The lower and upper 95th percentile confidence limits of these estimated incidence rates indicated that the incidence of CAH may be as high as 1:270 live births and as low as 1:2500 live births for homozygous affected cases, and as high as 1:8.8 persons and as low as 1:25 persons for heterozygotes. The 95th percentile confidence limits of the gene frequency are 0.060–0.02.

In the heterogeneous Caucasian population, the observed incidence of the salt-losing form of CAH based on the Alaska screening is estimated to be 1:33,518 for homozygous patients, 1:93 for the heterozygote frequency, and 0.005 for the gene frequency. The lower and upper 95th percentile confidence limits of these estimated incidence rates suggest salt-wasting CAH may be as common as 1:6000 live births and as rare as 1:1,325,000 live births for homozygous patients. Such wide confidence limits for the disease frequency in Caucasians indicate that a greater sample size is required for reliable estimates of disease frequency.

Reliability of CAH Screening Test

The false-positive rate of our screening test using a cutoff level of 57 pg/disc was 0.02%, and the recall rate using the cutoff level of 50 pg/disc was 0.02%, assuming all cases not available for follow-up were not affected with CAH. There were no cases diagnosed to have CAH whose screening tests were falsely low, with the exception of one newborn whose filter paper specimen was obtained while on intensive treatment. Thus, in this initial screening study, we did not determine any false-negative tests.

Costs

The direct cost for each screening test for CAH was $1.48, based on the scale of the screening program for Alaska and New York Hospital, which consisted of about 230 samples per week. This included wages for laboratory personnel ($250/week) and all expenses for reagents and supplies ($92/week).

TABLE 4. Frequency of CAH Gene Based on Newborn Screening and Case Surveys in Alaska

			Homozygous Affected Population		Heterozygous Population	
	Cases	Sample	Estimated Incidence	95th Percent Conf. Limit of Est. Rate[k]	Estimated Incidence	95th Percent Conf. Limit of Est. Rate[k]
Yupik Eskimo (salt-wasting only)				Lower–Upper		Lower–Upper
Alaska screening[a]	4	2737	1:684	1:270–1:2500	1:13.6	1:8.8–1:25.3
Case survey[b]	14	6860	1:490	1:292–1:896	1:11.6	1:9.0–1:15.5
Caucasian						
Alaska screening (salt-wasting only)	1	33,518	1:33,518	1:6000–1:1,325,000	1:92	1:39–1:565
Case survey[c–g] (salt-wasting only)	106	2,343,895	1:22,112	1:18,454–1:26,730	1:74.8	1:68–1:82
Case survey[c–e,g–j] (all 21-hydroxylase deficient)	215	3,619,891	1:16,836	1:14,798–1:19,239	1:65.4	1:61–1:69.8

[a] Pang *et al.*, 1984.
[b] Hirschfeld and Fleshman, 1969.
[c] Muller *et al.*, 1979.
[d] Werder *et al.*, 1980.
[e] Rosenbloom and Smith, 1966.
[f] Qazi and Thompson, 1972.
[g] Mauthe *et al.*, 1980.
[h] Childs *et al.*, 1956.
[i] Prader, 1958.
[j] Wilkins, 1962.
[k] Confidence limit for the expectation of Poisson variable for the estimated rate.

DISCUSSION

Our pilot screening program for CAH in the Alaskan newborn population demonstrates the feasibility of an effective newborn screening program for CAH. The direct benefit of CAH screening is the avoidance of adrenal crises, shock and its sequelae, and death. In this disorder, contrary to some other genetic disorders, death is avoidable through treatment, and affected patients have normal to above normal intelligence[23] and can become productive citizens. Salt-wasting crises in most cases of newborns with CAH occur after the seventh day of life.[6,14] In most instances, the result of screening was available between the seventh to the tenth day of life despite the great distance between New York and Alaska, which resulted in the prevention of adrenal crises in the affected newborns. Follow-up time for a second specimen collection in the early stage of screening was considerably delayed, but this also was shortened to not exceed 24 to 48 hours. However, a regional screening program would permit earlier diagnosis and follow-up even more efficiently.

In addition to the problem of adrenal crisis, severe virilization of external genitalia in female newborns may lead to an incorrect sex assignment for life[5,7] or a sex reassignment in later life and its attendant psychological damage,[24,25] including suicide.[24] Although ambiguity of the external genitalia should alert the physician to the possible diagnosis of CAH, it should be noted that three of the five deaths in Hirschfeld and Fleshman's report in Alaska occurred in female newborns who were not recognized to be affected with CAH, despite ambiguity of the genitalia.[14] While education of health professionals regarding CAH screening of female newborns with ambiguous genitalia is of paramount importance in all populations, in Alaska many infants are born in rural areas, where one cannot realistically rely on the selective screening for CAH by the health professional in the female newborn population. The recent experience of this clinic of cases referred with ambiguous genitalia and unrecognized CAH emphasizes the need for screening in urban, as well as rural areas, even in cases of females with ambiguous genitalia. Early recognition is important for the non-salt-waster, as well as for the salt-waster. CAH in males without salt-wasting symptoms often goes unrecognized because they do not have ambiguous genitalia. However, they do suffer the emotional problems associated with precocious puberty[26] or virilization,[5] and with short stature secondary to prematurely advanced bone age.[27] It is difficult to assign a cost to the treatment of emotional disorders stemming from these complications.

Frequency of CAH

The incidence of CAH based on case survey has been reported by several investigators and has varied from 1:490 to 1:67,000 live births.[4,9–14,29–32]

Recent reports based on case surveys indicate that in the European Caucasian population, CAH occurs 1:7500 to 1:13,900 live births.[11–13] Incidence of CAH in the heterogeneous USA population was reported to be 1:15,000 live births, based on a somewhat incomplete case survey in 1966.[10] Thus, the incidence of CAH between European and North American Caucasian populations may not greatly differ. In a literature survey of case reports of CAH since 1956, which include both salt-losing and simple virilizing types, 215 cases were reported in a population of 3.6 million newborns, giving an incidence of 1:16,836 live births.[4,9–11,13,30]

The screening of the Caucasian population of Alaska did not yield a valid estimate of incidence in the Alaskan Caucasian population because of the wide range of confidence limits of the estimated incidence resulting from the finding of only one affected case. Thus, a larger sample size for screening is required to predict the true incidence rate of CAH in the Alaskan Caucasian population. However, in the homogeneous Caucasian population of the provinces of Emilio Romagna and Lazio of Italy, recent screening programs for CAH have revealed an incidence of 1:5600 live births.[16,17]

Our preliminary Alaska screening suggests that salt-losing CAH in the Yupik Eskimo population is similar (1:684) to that of the case assessment method (1:490), which emphasizes the need for a screening program for this high risk group of newborns for CAH. More importantly, preliminary screening in both the Caucasian and Alaskan populations have revealed affected newborns who otherwise would not have been detected until the onset of adrenal crisis.

Analysis of the Reliability of the CAH Screening Test

Based on the Alaskan screening program (TABLE 5), the false-positive and recall rates of the microfilter paper technique for 17-OHP RIA for the diagnosis of CAH

TABLE 5. Screening Test

	CAH One time 17-OHP	Congenital Hypothyroidism: T4 plus TSH (Northwest [ref. 28], New England [ref. 34], Quebec [ref. 33])	Congenital Hypothyroidism: T4 alone (Toronto [ref. 35])
False-positive rate	0.02%	0.03–0.12%	1.87%
Recall rate	0.02%	0.05–0.8%	2%
False-negative rate	?	?	?

were comparable to those of the screening test for hypothyroidism (0.03–0.12% and 0.05–0.8%, respectively) utilizing T4 followed by TSH measurements in three regional screening laboratories.[28,33,34] The 17-OHP RIA false-positive and recall rates were considerably lower than the rates for the hypothyroidism screening test using T4 measurements alone (1.87% and 2%, respectively)[35] in the screening program in Alaska. Thus, a single measurement of 17-OHP alone for CAH screening is as effective as the T4/TSH combined measurement for the hypothyroid screening test. These data indicate that the signal to noise ratio of our screening test is high, thus fulfilling the criteria for screening tests proposed by Rosenberg and Scriver.[22] Similar results of false-positive or recall rates were also reported from the Italian CAH screening program.[17]

False-positive 17-OHP levels were found largely in premature neonates in the Alaskan newborn screening program (TABLE 3). This observation was further substantiated by the follow-up clinical data of newborns with false-positive results in other screening programs for CAH.[16–20] The preliminary reference data of 17-OHP in premature newborns indicate that the higher 17-OHP levels occur in the more severe and younger premature newborns (FIGURE 3). Thus, in cases suspected of CAH as a result of the screening test, false-positive results due to prematurity must be ruled out prior to a diagnostic workup for CAH.

Cost of CAH Screening Test

The reported direct cost for each screening test for CAH was \$1.48, based on the very small scale of the screening program in Alaska, which consisted of about 230 samples per week for the period of screening. This included wages for laboratory personnel (\$250/week) and all expenses for reagents and supplies (\$92/week).[21]

Direct cost of a microfilter paper analysis for screening for CAH is comparable to the cost of thyroid screening.[28] In CAH, in addition to the benefits described in the introduction to this chapter, the loss of a productive citizen must be reckoned in the cost-benefit ratio. The economic cost would be less in a larger regional screening program, thus improving the cost-benefit ratio of a screening program for CAH.

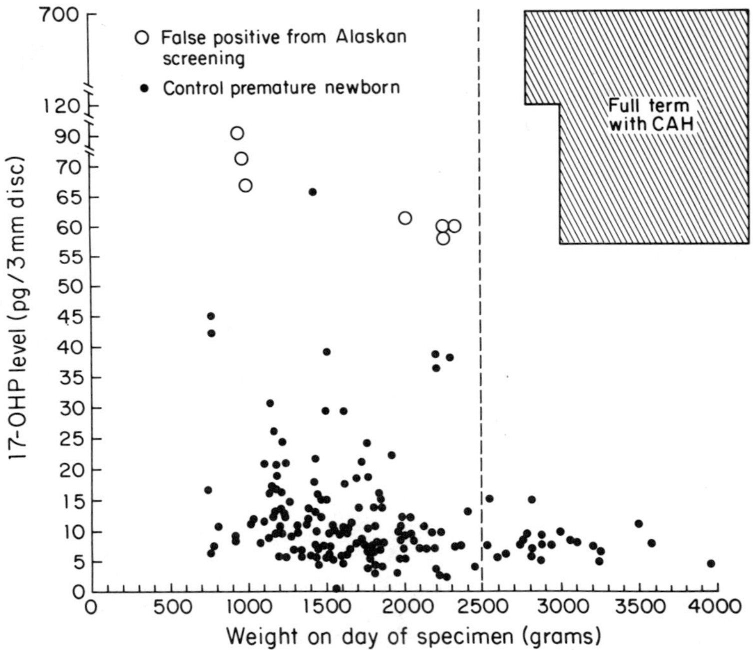

FIGURE 3. Reference data of 17-OHP concentration in capillary whole blood sample impregnated on filter paper specimen in premature newborns. The values are plotted according to weight of premature infants. Full-term newborns diagnosed with CAH at birth all weighed >2500 grams.[21] Note one false-positive result occurred in a premature infant weighing 1465 grams.

Screening males only would reduce the cost by half, but we are not convinced that females with ambiguous genitalia are detected routinely as patients with CAH. The risk of missing the diagnosis in these females is significant as judged by the number of referrals to us of patients raised as males who are genetic females virilized by CAH. This is a particularly tragic case of sex misassignment, since these genetic females would have been capable of normal life as a female, including reproduction, had they been properly diagnosed and treated from birth.

In summary, this study demonstrates the feasibility of a reliable newborn screening program for CAH utilizing the microfilter paper test for 17-OHP determination. CAH in Alaska meets all of the criteria proposed by Frankenberg for the institution of a screening program for a disorder—importance of early diagnosis; prevalence of the disorder; amenability to early treatment; and, availability of an inexpensive and reliable method for screening.[2] Finally, our pilot screening program emphasizes the necessity of regional screening programs for CAH that would allow the diagnosis before the seventh day of life. Thus, most affected newborns can be safely diagnosed before the onset of adrenal crisis.

REFERENCES

1. GUTHRIE, R. & A. SUSI. 1963. Pediatrics **32:** 338.
2. FRANKENBURG, W. K. 1974. Pediatrics **54:** 612–616.
3. NATIONAL ACADEMY OF SCIENCES. Division of Medical Science, Assembly of Life Sciences. 1975. Genetics screening, programs, principles, and research. Committee for the study of inborn errors of metabolism.
4. CHILDS, B., M. M. GRUMBACH & J. VAN WYK. 1956. J. Clin. Invest. **35:** 213.
5. NEW, M. I. & L. S. LEVINE. 1973. Congenital adrenal hyperplasia. *In* Advances in Human Genetics. H. Harris & K. Hirschhorn, Eds.: 251–326. Plenum Press. New York.
6. KOWARSKI, A. A. 1977. *In* Congenital Adrenal Hyperplasia. P. A. Lee, L. P. Plotnick, A. A. Kowarski & C. J. Migeon, Eds.: 113. University Park Press. Baltimore.
7. PERIS, L. A. 1960. Obstet. Gynecol. **16:** 156.
8. PENNY, R., N. OLAMBIWONNU & D. FRASIER. 1973. J. Clin. Endocrinol. Metab. **36:** 920–924.
9. HUBBLE, D. 1966. *In* Basic Concepts of Inborn Errors and Defects of Steroid Biosynthesis. Proceedings of the Third Symposium of the Society for the Study of Inborn Errors of Metabolism. K. S. Holt & D. N. Raine, Eds.: 68. Livingstone. Edinburgh.
10. ROSENBLOOM, A. L. & D. W. SMITH. 1966. Lancet **1:** 660.
11. MAUTHE, I., H. LASPE & D. KNORR. 1977. Klin. Paediatr. **189:** 172.
12. MULLER, W., M. PRADER, J. KOFLER, J. GLATZL & W. GEIR. 1979. Paediatr. Paedol. **14:** 151.
13. WERDER, E. A., R. E. SIEBENMANN, G. KNORR-MURSET, A. ZIMMERMAN, P. C. SIZONENKO, P. THEINTZ, J. GIRARD, M. ZACHMANN & A. PRADER. 1980. Helv. Paediatr. Acta **35:** 5.
14. HIRSCHFELD, A. J. & J. K. FLESHMAN. 1969. J. Pediatr. **75:** 492.
15. PANG, S., J. HOTCHKISS, A. L. DRASH, L. S. LEVINE & M. I. NEW. 1977. J. Clin. Endocrinol. Metab. **45:** 1003.
16. NATOLI, G., L. MOSCHINI, P. ACCONCIA, G. ALBINO, P. COSTA & G. PANSA. 1981. *In* Second International Symposium on Recent Progress in Pediatric Endocrinology, Serono Symposium, Milan, October 22–23, 1981. Raven Press. New York.
17. CACCIARI, E., A. BALSAMO, A. CASSIO, S. PIAZZI, F. BERNARDI, S. SALARDI, A. CICOGNANI, P. PIRAZZILI, F. ZAPPULLA, M. CAPELLI & M. PAOLINI. 1982. Lancet **1:** 1069.
18. DORCHE, C., D. BOZON, M. DAVID & M. O. ROLLAND. 1982. Programme and Abstract. Proceedings of the International Meeting on Neonatal Screening, August 16–21, 1982. Tokyo. p. 109.
19. SHIMOZAWA, K., N. SAITO, N. SAKURADA, J. YATA, Y. HIKITA, A. TAKEHIRO, K. KAWANAMI, Y. IGARASHI, K. OKADA, Y. ITO & M. IRIE. 1982. Programme and Abstract. Proceedings of the International Meeting on Neonatal Screening, August 16–21, 1982. Tokyo. p. 109.
20. FUKUSHI, M., O. ARAI, Y. MIZUSHIMA, Y. SATO, H. HAYASHI, N. TAKASUGI, H. ARAKAWA, M. MAEDA & A. TSUJI. 1982. Programme and Abstract. Proceedings of the International Meeting on Neonatal Screening, August 16–21, 1982. Tokyo. p. 110.
21. PANG, S., W. MURPHEY, L. S. LEVINE, D. A. SPENCE, A. LEON, S. LAFRANCHI, A. S. SURVE & M. I. NEW. 1982. J. Clin. Endocrinol. Metab. **55:** 413.

22. ROSENBERG, L. E. & C. R. SCRIVER. 1969. *In* Disorders of Amino Acid Metabolism. P. K. Bondy & L. E. Rosenberg, Eds.: 484. W. B. Saunders Co. Philadelphia.
23. ERHARDT, A. A. & S. W. BAKER. 1977. *In* Congenital Adrenal Hyperplasia. P. A. Lee, L. P. Plotnick, A. A. Kowarski & C. J. Migeon, Eds.: 447. University Park Press. Baltimore.
24. JONES, H. W., JR. & W. W. SCOTT. 1958. Hermaphroditism, Genital Anomalies, and Related Endocrine Disorders, 1st Edition. p. 215. Williams and Wilkins Co. Baltimore.
25. MONEY, J. & J. DALERY. 1977. *In* Congenital Adrenal Hyperplasia. P. A. Lee, L. P. Plotnick, A. A. Kowarski & C. J. Migeon, Eds.: 433. University Park Press. Baltimore.
26. WOOD, J. H. & L. I. GARDNER. 1975. *In* Endocrine and Genetic Diseases of Childhood. L. I. Gardner, Ed.: 1284. W. B. Sanders Co. Philadelphia.
27. BONGIOVANNI, A. M. & A. W. ROOT. 1963. N. Engl. J. Med. 268: 1283.
28. LAFRANCHI, S. H., W. H. MURPHEY, T. P. FOLEY, JR., P. R. LARSEN & N. R. M. BUIST. 1979. Pediatrics **63:** 180.
29. PRADER, A., G. J. P. A. ANDERS & H. HABICH. 1962. Helv. Paediatr. Acta **17:** 271.
30. PRADER, A. 1958. Helv. Paediatr. Acta **13:** 426.
31. EDWARDS, R. W. H. 1975. *In* Biochemistry of Steroid Hormones. H. L. J. Makin, Ed.: 273. Blackwell Scientific Publications. Oxford.
32. QAZI, Q. H. & M. W. THOMPSON. 1972. Arch. Dis. Child. **47:** 302.
33. DUSSAULT, J. H., P. COULOMBE, C. LABERGE, J. LETARTE, H. GUYDA & K. KHOURY. 1975. J. Pediatr. **86:** 670.
34. MITCHELL, M. L., P. R. LARSEN, H. L. LEVY, A. J. E. BENNETT & M. A. MADOFF. 1978. J. Am. Med. Assoc. **239:** 2348.
35. WALFISH, P. G. 1976. Lancet **1:** 1208.

Neonatal Screening for Congenital Adrenal Hyperplasia in Japan

H. NARUSE,[a] E. SUZUKI,[a] M. IRIE,[b] A. TSUJI,[c]
N. TAKASUGI,[d] M. FUKUSHI,[d] N. MATSUURA,[e]
AND K. SHIMOZAWA[f]

[a]*National Center for Nervous, Mental, and Muscular Disorders*
Tokyo, 187 Japan

[b]*Department of Internal Medicine*
Toho University
Tokyo, Japan

[c]*School of Pharmaceutical Science*
Showa University
Tokyo, Japan

[d]*Sapporo City Institute of Public Health*
Sapporo, Japan

[e]*Department of Pediatrics*
Hokkaido University
Sapporo, Japan

[f]*Department of Pediatrics*
Tokyo Medical and Dental College
Tokyo, Japan

INTRODUCTION

When people consider the problem of whether or not neonatal screening for Congenital Adrenal Hyperplasia (CAH) due to 21-hydroxylase deficiency might satisfy the necessary requirements for mass screening, the following ten headings, which have been proposed by Wilson[1] and are used often as the framework for mass screening, could be utilized as criteria:

(1) The condition sought should be an important problem.
(2) There should be an accepted treatment for patients with recognized disease.
(3) Facilities for diagnosis and treatment should be available.
(4) The natural history of the condition to be sought should be adequately understood.
(5) There should be a recognizable latent or early symptomatic stage.
(6) There should be a suitable test or examination.
(7) The test or examination should be acceptable.
(8) There should be an agreed policy on whom to treat.
(9) Case-finding process should be a continuous process.
(10) The cost of early diagnosis and treatment should be balanced economically in relation to the total expenditure on medical care.

Except for the last item, a screening for CAH meets all the conditions. As far as the cost-benefit analysis for the disease is concerned, the situation is complicated in comparison with that for PKU or congenital hypothyroidism since a patient suffering from CAH may not be covered by the social welfare system during his lifetime. However, if people take into account the early death of the babies, the problem of sex assignment, or the psychophysical complications caused by the disease, it will be obvious that a more precise analysis concerning the benefits resulting from the screening will be necessary.

If we could obtain a certain positive result for the cost-benefit analysis, screening for CAH would meet all criteria. Of course, there may be further objections when a special chemical test is proposed since the diagnosis of many cases of CAH can be based on clinical observations. When screening for galactosemia was started many years ago, and also, when we began the neonatal screening for congenital hypothyroidism, there arose the same kind of objections from the specialists in these diseases. However, after worldwide results of screenings were obtained, such objections became invalid since the diseases were detected in many patients by screening rather than by clinical diagnosis.

The result of large scale research on the screening for CAH could be an important factor to settle the theoretical discussion. If many cases detected through screening had already been diagnosed by clinical findings without a report from screening, objection to the screening would be valid. On the other hand, if many cases detected were not diagnosed well by clinical observations, such objections could not be accepted. Based on these considerations, various research on screening was initiated in Japan.

TABLE 1. Neonatal Screening in Japan

(1) Nationwide multiple screening for PKU, congenital hypothyroidism, galactosemia, maple syrup urine disease, homocystinuria, and histidinemia is performed by federal and local governments.
(2) The screening is performed on a voluntary basis without a law.
(3) The number of tested newborns in the fiscal year, 1982, was 1,488,664 (which was 97.8% of the total of newborns), and all babies were tested at least for six diseases.
(4) Fifty-three screening laboratories are working and all of them are covered by the "Interlaboratory Quality Control System," which is supported by the federal and local governments.

METHOD OF SCREENING

As reported by S. Pang *et al.* for the early detection of the disease, the use of dried blood taken for PKU or hypothyroid screenings would be the most efficient way.[2,3] However, in the most developed countries and in some developing countries, multiple newborn screening is already being undertaken.[4] Therefore, if screening for CAH were done with a dried blood sample, it should be integrated with screening for other genetic diseases or hypothyroidism, and it should be performed in a regional screening laboratory.

For example, in Japan, nationwide screening for PKU, congenital hypothyroidism, galactosemia, maple syrup urine disease, homocystinuria, and histidinemia has been established. Last year, more than 98% of the total number of the newborn were successfully screened for these diseases. Now, fifty-three screening laboratories under the direction of local governments are performing multiple screenings, which are covered by a national quality control system (TABLE 1).[5]

TABLE 2. Required Conditions of a Test of Regional Screening

(1) The test could be integrated as a part of the multiple neonatal screening when a dried blood sample is used.
(2) The test should be sensitive and specific so that the number of false-negative cases would be extremely low and the rate of the false-positives could be reduced.
(3) The test should be performed by a small number of technicians.
(4) Facilities for making the diagnosis should be available in each region.

Therefore, if medical people wish to add another screening test to their established screening program, the conditions described in TABLE 2 would be required. As in the case of screening for CAH, several problems, as shown in TABLE 3, could be expected. The first problem has almost been overcome in Japan. While screening for congenital hypothyroidism was under development several years ago, a project team for establishing nonisotopic techniques was organized, which succeeded in obtaining a good enzyme immunoassay(EIA) method (H. Naruse *et al.*, 1982).[6] A. Tsuji, who was a member of the project team, and his colleagues were able to develop a sensitive EIA method for 17-α-hydroxyprogesterone(17-OHP) assay.[7] An outline of the method is shown in TABLE 4.

Standard curves and interassay variations during four months are shown in FIGURE 1. The present method has a good sensitivity and a good reproducibility, and will be usable as a tool for mass screening. The procedure of the EIA method in NCNMMD is described in FIGURE 2. There may be another procedure, though, which might produce superior results. 17-OHP could be extracted by either buffer or ether. In a routine screening, the direct method using buffer extraction will be used first, and, then, the samples which show higher values of 17-OHP than an established cutoff point will be tested again by the ether extraction method.

Since there was a possibility that EIA for 17-OHP could produce many false-positive results when samples from premature babies were tested, we determined 17-OHP in about 600 babies who were admitted into NICU. The blood samples were collected during the first day of admission; then every several days, follow-up samples were collected until the baby was discharged from the hospital. FIGURE 3 shows the distribution patterns of 17-OHP in the premature babies. Many of the first samples

TABLE 3. Problems Anticipated in CAH Screening in Japan

(1) RIA method could obstruct integration of CAH screening with other genetic screening.

Countermeasure: Enzyme immunoassay could be useful.

(2) The number of specialists for CAH will be less than that for congenital hypothyroidism and there will be some delay in the final diagnosis as well as in the initiation of treatment.

Countermeasures: (a) Reduce false-positive cases (see item (3)); (b) Distinguish severe cases needing treatment urgently.

(3) Cross-reaction of some steroids to anti-17-OHP antiserum could increase a number of false-positive cases.

Countermeasures: (a) Develop HPLC or GC methods to confirm the increase of 17-OHP using dried blood spots; (b) Develop a better antibody.

TABLE 4. Enzyme Immunoassay for 17-α-OH-Progesterone

(1) A sensitive enzyme immunoassay for measuring 17-α-hydroxyprogesterone was developed by A. Tsuji *et al.*
(2) A competitive one using a double antibody coated bead.
(3) Conjugate: 17-OHP-horseradish peroxidase (HRP) was prepared from 17-OHP-3-carboxymethyl oxime and HRP.
(4) Anti-17-OHP antiserum: Prepared by 17-OHP-3-CMO-BSA conjugate injection.
(5) Sensitivity: A detection limit was 1 pg/tube.
(6) Stability of the reagent: More than one year.

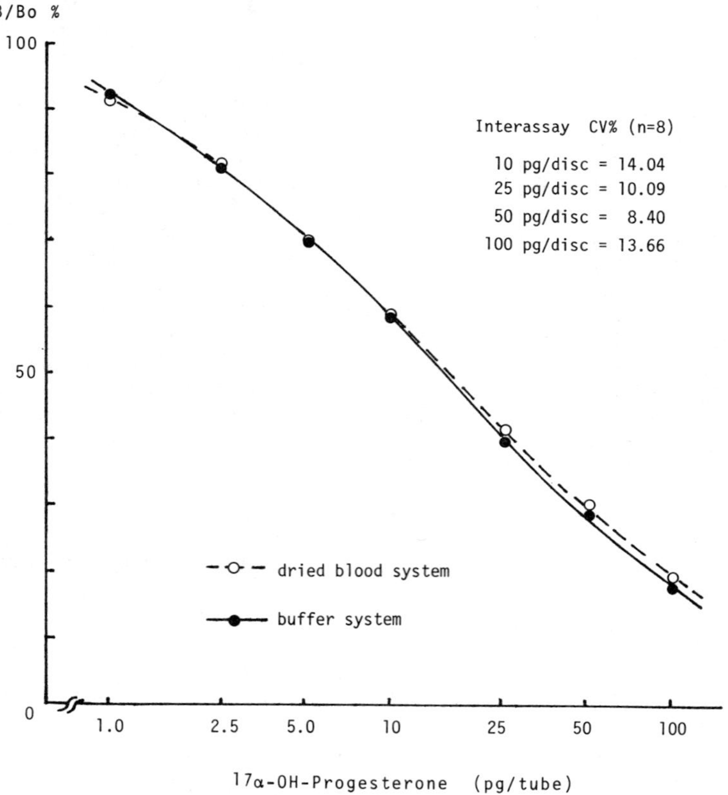

FIGURE 1. Standard curves of 17-OHP by EIA method and its interassay variations.

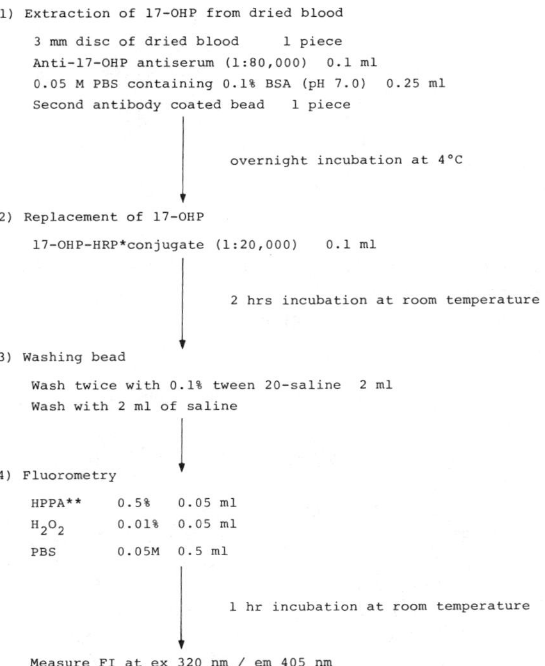

FIGURE 2. Procedures of EIA method for 17-OHP assay. *stands for horseradish peroxidase. **stands for 3-(p-hydroxyphenyl)propionic acid.

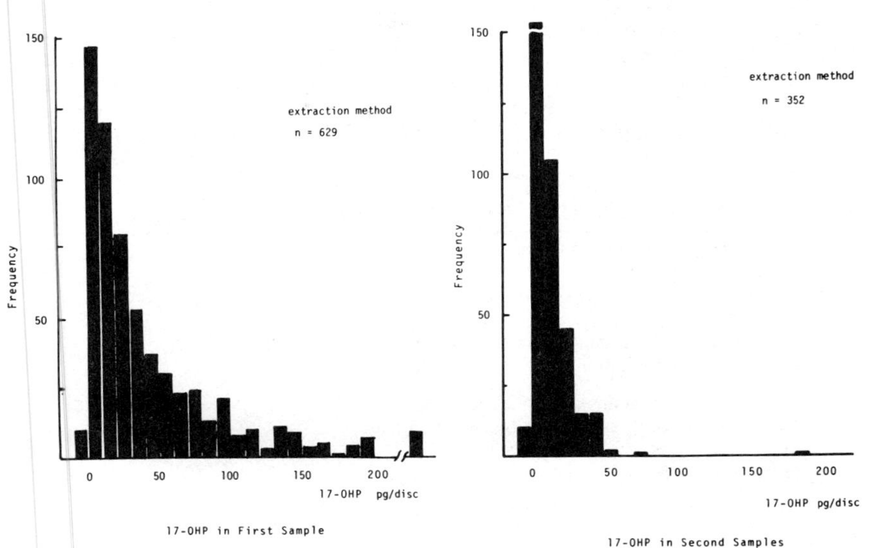

FIGURE 3. Distribution pattern of 17-OHP in premature babies.

TABLE 5. Result of Regional Screening for CAH in Japan (No. 1)

Area: Sapporo
(CAH screening was added to the program of established regional screening)
Method: EIA method developed by Tsuji *et al.*
Number of tested newborn: 42,176 (April 1982–March 1984)
Number requiring second blood samples: 547 (1.3%)
Number of referred babies: 39 (0.09%)
Number of confirmed cases of 21-OHD: 3 (1/14,000)

showed very high values of 17-OHP even with the ether extraction method, but the second samples showed a limited number of abnormal values. Thus, this method could be made available even for premature babies.

RESULTS OF THE SCREENING

From the fiscal year, 1982, the EIA method has been adopted in a screening laboratory in Sapporo. The result of the screening in Sapporo is shown in TABLE 5. About 42,000 babies were tested, and three cases in females were confirmed in the Department of Pediatrics, Hokkaido University. Two of them were the salt-losing form and one was the simple virilizing form. When people apply a new method to regional screening, the number of requested samples and of referred babies are very important. Actually, in this study, a cutoff point was set at a low level for safety. However, since details of the screening result were recorded, a number of referred babies can be calculated at different cutoff points. Thus, a number of referrals in this study were compared to that of a screening for congenital hypothyroidism (TABLE 6).

When a cutoff point was set at 50 pg/tube, the number of referrals was ten and the number of patients was three. On the other hand, in the case of neonatal screening for congenital hypothyroidism based on TSH assay in NCNMMD, the number of referrals was seventeen and that of the confirmed patients was ten. The result may suggest that the present method is working fairly well. However, in order to reduce a number of referrals, a better quality of antibody and a method which can determine 17-OHP in a dried blood exactly by HPLC or gas chromatography will be necessary.

Another group has performed regional screening based on Pang's method[2] and the

TABLE 6. Number of Referrals of CAH Screening and CH Screening

CAH screening in Sapporo city
Number tested: 42,176
Number requiring second blood samples (calculated) cutoff point 50 pg/tube: 70 (0.17%)[a]
Number of referrals (calculated) cutoff point 50 pg/tube: 10 (0.02%)[b]
Number of confirmed patients: 3

Neonatal thyroid screening in NCNMMD Tokyo
Number tested (TSH assay): 50,000 (1983 fiscal year)
Number of referrals: 17
Number of confirmed cases: 10

Ratio of Patient/Referral CAH:30% CH:58%

[a]Half of the 70 cases were babies with low birthweights.
[b]Two cases out of the ten were babies with low birthweights.

TABLE 7. Result of Regional Screening for CAH in Japan (No. 2)[a]

Area: West region of Shizuoka prefecture
(The remainder of dried blood after established screening tests was used for the research.)
Method: RIA method based on Pang's report (1977) and 17-OHP was extracted by assay buffer. No ether extraction was performed.
Number tested: 34,314 (May 1981–November 1983)
Recall: Newborn in highest 1% was examined carefully and 208 cases were referred to Prof. Igarashi, Dept. of Pediatr., Hamamatsu Medical College.
Number of confirmed cases of 21-OHD: 3 cases (1/11,400)

[a]Reference: SHIMOZAWA, K. et al. A neonatal mass screening study for Congenital Adrenal Hyperplasia in Japan. Acta Endocrinologica. In press.

result is shown in TABLE 7. The report will appear in *Acta Endocrinologica (Copenhagen)* soon.[8] The group has tested about 34,000 samples, and has detected three cases. An outline of the detected cases is shown in TABLE 8. Case 1 was not diagnosed until the report from the screening laboratory reached the patient's doctor. Case 2 was not diagnosed properly and had sex assignment trouble before the screening result arrived.

DISCUSSION

The results of screening in Japan are very limited and it is too early to conclude the incidence of CAH(21-hydroxylase deficiency) in Japan. However, based on these two studies, a tentative value of one in 10,000 to 20,000 might be usable. If the estimation proves to be correct, the incidence is far greater than that of PKU in Japanese, which is about one in 60,000 newborns. Thus, the incidence of CAH must not be neglected in

TABLE 8. Clinical and Laboratory Data of Three Patients with 21-OHD (in West Shizuoka Area)[a]

	Case 1	Case 2	Case 3
Diagnosed date (days)	31	26	28
Genetic sex	Male	Female	Female
Legal sex	Male	Male	Female
Family history	+	−	−
Genital ambiguity	−	+++	+++
Skin hyperpigmentation	±	−	±
Poor weight gain	+	+++	±
Hyponatremia (mEq/l)	+ (120)	+ (118)	− (133)
Hyperpotassemia (mEq/l)	+ (5.9)	− (5.4)	− (5.2)
Hypochloremia (mEq/l)	+ (88)	± (94)	± (93)
	*1 *2	*1 *2	*1 *2
Disc-17-OHP (pg/disc)	99.0 → 988	127 → 322	53.1 → 294
Plasma			
17-OHP (ng/ml)	270	245*3	113
21-DOF (ng/ml)	151	—	46.4
Diagnosis	SL	SL	SV

[a]Abbreviations: *1: In the 1st sample screening; *2: In the 2nd sample; *3: Measured by HPLC–UV photospectrometry; SL: salt-losing form; and, SV: simple virilizing form.

Japan. However, even though six cases were detected through our study, it is too early to evaluate the worth of screening for CAH. However, it should not be overlooked that two of the detected cases had not been diagnosed before the reports from the screening laboratories had reached the individuals' doctors.

There was another fact, which was reported by S. Suwa *et al.,* based on a questionnaire sent to all of the specialists of CAH in Japan.[9] According to the report, during the last ten years at least 488 cases of CAH were diagnosed, and out of these, 44 cases had died. In addition, many of the female cases had serious problems of reassignment of sex. If they had been diagnosed by neonatal screening, these problems could have been prevented.

Based on our experiments and the results of the questionnaire performed by Suwa *et al.,* the authors favor inaugurating nationwide screening for CAH in Japan in the future. However, in order to have a good system of nationwide screening for CAH, it seems necessary to reduce the number of false-positive cases. This is especially so, since there are fewer CAH specialists than those for congenital hypothyroidism. Therefore, it will be very important to make an effort to reduce the number of referred babies. For this purpose, improvement of EIA or RIA for 17-OHP assay, as well as for the development of the chromatographic method to determine 17-OHP in dried blood, will be necessary. Already, though, in Japan, several groups are involved in research to overcome these problems.

REFERENCES

1. WILSON, J. M. G. 1968. Presymptomatic detection and early diagnosis. Pitman Medical Publishing Co., Ltd. London.
2. PANG, S., J. M. HOTCHIKISS, A. L. DRASH, L. S. LEVINE & M. I. NEW. 1977. Microfilter paper method for 17-α-hydroxyprogesterone radioimmunoassay: Its application for rapid screening for congenital adrenal hyperplasia. J. Clin. Endocrinol. Metab. **45**(5): 1003–1008.
3. PANG, S., W. MURPHEY, L. S. LEVINE, D. A. SPENCE, A. LEONE, S. LEFRANCHI, A. S. SURVE & M. I. NEW, 1983. Newborn screening for congenital adrenal hyperplasia in Alaska. *In* Neonatal Screening. H. Naruse & M. Irie, Eds.: 316–323. Excerpta Medica. Amsterdam.
4. NARUSE, H. & M. IRIE, Eds. 1983. Neonatal Screening. Excerpta Medica. Amsterdam.
5. NARUSE, H., E. SUZUKI, N. WATANABE & T. OURA. 1983. Quality control on screening for inborn errors of metabolism in Japan. *In* Neonatal Screening. H. NARUSE & M. IRIE, Eds.: 430–435. Excerpta Medica. Amsterdam.
6. NARUSE, H., M. IRIE & A. TSUJI, Eds. 1982. Neonatal Hypothyroid Screening by Enzyme Immunoassay. Okado Publishing Co., Ltd. Tokyo.
7. ARAKAWA, H., M. MAEDA, A. TSUJI, H. NARUSE, E. SUZUKI & A. KAMBEGAWA. 1982. Fluorescence enzyme immunoassay of 17-α-hydroxyprogesterone in dried blood samples on filter paper and its application to mass screening for congenital adrenal hyperplasia. Chem. Pharm. Bull. **31**(8): 2724–2733.
8. SHIMOZAWA, K., S. SAISHO, N. SAITO, N. SAKURADA, J. YATA, Y. IGARASHI, Y. HIKITA, M. IRIE, T. HYODO, H. UESHIBA & K. OKADA. 1984. A neonatal screening study for congenital adrenal hyperplasia in Japan. Acta Endocrinol. In press.
9. SUWA, S., Y. IGARASHI, K. KATO, T. KUSUMOTO, A. TANAE, K. NIIMI & J. YATA. 1981. A case survey study for congenital adrenal hyperplasia in Japan. I. Study for the incidence. Acta Pediatr. Jap. **85**(2): 204–210. (In Japanese.)

Pitfalls of Prenatal Diagnosis of 21-Hydroxylase Deficiency Congenital Adrenal Hyperplasia[a]

SONGYA PANG,[b] MARILYN S. POLLACK,[c] MAY LOO,[d]
ORVILLE GREEN,[e] ROBERT NUSSBAUM,[f]
GEORGE CLAYTON,[f] BO DUPONT,[g]
AND MARIA I. NEW[b]

[b]*Division of Pediatric Endocrinology*
The New York Hospital-Cornell Medical Center
525 East 68th Street
New York, New York 10021

[c]*Department of Microbiology, Immunology, and Pathology*
Baylor College of Medicine
1200 Moursund Avenue
Houston, Texas 77030

[d]*5755 Cattle Road*
San Jose, California 95123

[e]*Department of Pediatrics*
Children's Memorial Hospital
2300 Children's Plaza
Chicago, Illinois 60614

[f]*Department of Pediatrics*
Baylor College of Medicine
1200 Moursund Avenue
Houston, Texas

[g]*Tissue Typing Laboratory*
Memorial Sloan-Kettering Cancer Center
1275 York Avenue
New York, New York 10021

INTRODUCTION

Ethical and Medical Aspects of Prenatal Diagnosis of CAH

Over the last decade, there has been an increased demand for prenatal diagnosis of CAH from parents of patients with 21-hydroxylase deficiency congenital adrenal hyperplasia. The major reasons for seeking prenatal information of the fetus at risk for CAH in our experience have been: (1) to reduce anxiety by excluding a diagnosis of CAH, (2) to fully prepare before the birth of the infant and to insure immediate

[a]This study was supported by USPHS NIH grant award nos. HD–00072, CA–37888, AM–33988, and AM–33990, and by a grant (RR–47) from the Division of Research Resources, General Clinical Research Centers Program, NIH.

medical attention at birth for an affected fetus, and (3) to help parents who may seek prenatal information with the intention of aborting an affected fetus. (This rare occurrence usually only happens with parents who had a tragic experience with a previously affected offspring, such as death of a child from adrenal crisis.) The availability of accurate prenatal diagnosis of CAH in recent years[1–17] has allowed some parents to plan a future pregnancy at ease,[9,17] as well as in some cases to avoid abortion of an unaffected fetus.[17]

With the increasing demand for complete and accurate prenatal information of the fetus at risk for CAH, we investigated the accuracy of the prenatal diagnostic tests. Accurate prenatal diagnosis of CAH may ultimately lead to prenatal treatment to reduce or prevent further virilization of external genitalia in an affected female fetus, as well as to provide immediate medical treatment at birth for an affected fetus.

Advances in Prenatal Diagnosis of Congenital Adrenal Hyperplasia

The feasibility of prenatal diagnosis of congenital adrenal hyperplasia (CAH) due to 21-hydroxylase deficiency (21-OH def.) by measurement of amniotic fluid 17-hydroxyprogesterone (17-OHP) was first reported in 1975.[1] Since then, a number of investigators have documented the usefulness of amniotic fluid hormone tests for prenatal diagnosis of CAH.[2–13] To date, studies of amniotic fluid 17-OHP concentration in a total of 66 midgestational fetuses at risk for CAH have been reported; in each case, the prenatal prediction was reported to be in agreement with the postnatal diagnosis[1–13] (TABLE 1). Thus, hormonal testing during the second trimester has been claimed to be an accurate prenatal predictor of 21-OH deficiency.

Successful HLA typing of cultured amniotic cells in the fetus and in the index case and parents became an additional tool for prenatal diagnosis of 21-OH deficiency[5,8,9,13–17] following the discovery of the close genetic linkage between the HLA complex and genes for 21-OH deficiency on chromosome 6.[18–22] In the literature to date, HLA typing of amniotic cells has been performed in 26 cases at risk for CAH,[5,8,9,13–17] and in 17 of these 26 cases, simultaneous amniotic fluid 17-OHP concentration was determined.[8,9,13,14] In all except two cases,[9] prenatal fetal prediction based on the HLA typing of amniotic cells was reported to be in agreement with the postnatal diagnosis and with the hormonal prediction (TABLE 1).[8,9,13,14] In the two cases where hormonal and HLA prediction did not agree, it was postulated that the discrepancy was due to either a genetic recombination on chromosome 6 between the HLA-A and HLA-B loci, or due to the absence of expression of 21-OH deficiency at the time of amniocentesis, with the implication that at least in one case the fetus might have been affected with the late-onset form of 21-OH deficiency CAH.[9] However, the final diagnoses in these fetuses could not be established since both fetuses were aborted during the second trimester.

Pitfalls of Prenatal Diagnosis of CAH Based on Hormonal Measurement and HLA Typing of Amniotic Fluid Specimen in Fetuses at Risk for CAH (ref. 23)

In the present report, we decribe the nature of errors in the interpretation of hormonal and HLA typing studies in the prediction of fetal genotypes for 21-OH deficiency. This report is based upon prenatal studies conducted during the second trimester of 32 fetuses at risk for CAH. We correctly predicted the postnatal outcome in 29 of the 32 fetuses using both hormonal and HLA typing tests of fetal amniotic fluid specimens. However, in 3 of the 32 fetuses, prenatal diagnosis was incorrect.

TABLE 1. Prenatal Diagnosis of 21-Hydroxylase Deficiency CAH since 1975

Author(s) (ref.)	Yr	No. of at Risk Cases Studied	No. of Cases Reported to Be Affected	Hormone Study: 17-OHP	HLA[j]	Index Case	Fetal Outcome
Frasier *et al.*[1]	1975	1	1	300	—	SW[i]	SW
Nagamani *et al.*[3]	1978	1	1	1000	—	SW	SW
Milunsky *et al.*[2]	1979	1	1	520	—	SW	SW
[a]Hughes *et al.*[4]	1979	retrospective 4	1	1080[d]	—	SW	SW
			retrospective	396[d]	—	??[h]	not affected
				180[d]	—	??	not affected
				98[d]	—	??	not affected
Marcus *et al.*[5]	1979	1	1	1467[b]	affected	SW	SW
Pollack *et al.*[14]	1979	2	1	110[c]	het	SW	het unaffected
				1300[c]	affected	SW	SW
Couillin *et al.*[15]	1979	1	0	—[e]	het	SW	het unaffected
Pang *et al.*[6]	1980	8	3	1300[c]	affected	SW	SW
				1800[b]	affected	SW	SW
				1900	—	SW	SW
				100	—	SW	not affected
				140	—	SW	not affected
				150	—	SW	not affected
				170	—	SW	not affected
				110[c]	het	SW	het unaffected
Warsos *et al.*[7]	1980	2	2	2239 } twin	—	SW }	affected (spontaneous abortion)
				2747 } twin	—	SW }	
Roseman *et al.*[8]	1980	3	0	280	homo nl	SW	not affected
				160	het	SW	het unaffected
				360	homo nl	SW	not affected
[a]Forest *et al.*[9]	1981	17	3 or 5	227	—	SW	not affected
				97	—	SW	not affected
				1354	affected	SW	aborted female fetus w/ ambiguous genitalia
				274	—	SW	not affected
				158	—	SW	not affected

TABLE 1. *Continued*

Author(s) (ref.)	Yr	No. of at Risk Cases Studied	No. of Cases Reported to Be Affected	Hormone Study: 17-OHP	HLA[j]	Index Case	Fetal Outcome
				144	—	SW	not affected
				1332	affected	SW	affected (aborted)
				101	—	SW	normal
				130	—	SW	normal
				1069	affected	SW	affected female w/clitoral hypertrophy at birth
				90	het	SW	not affected
				90	homo nl	SW	not affected
				126	homo nl	SW	not affected
				108[g]	affected[g]	SW	aborted male[g] w/normal adrenals
				130[g]	affected[g]	late-onset	aborted female[g] w/normal genitalia & normal adrenals
				162	homo nl	SW	not affected
				104	homo nl or het	SW	not affected
Couillin *et al.*[17]	1981	10	3	—	het	SW	het unaffected
				—	homo nl	SW	not affected
				fetal plasma: 9748	affected	SW	affected (aborted)
				—	homo nl	SW	not affected
				—	het	SW	het unaffected
				—	het	??	het unaffected
				—	affected	SW	affected (aborted)
				—	homo nl	SW	not affected
				—	affected	??	pregnancy continued, male
				—[e]	het	SW	het unaffected

[a]Hughes *et al.*[10,11]	1982	12	4	1080[d]	—	SW	SW
	1984	13[f]	3[f]	1188	—	SW	SW
				1440	—	SW	SW
				2160	—	SW	SW
				540	—	??	not affected
				430	—	??	not affected
				400	—	??	not affected
				396[d]	—	??	not affected
				252	—	??	not affected
				180[d]	—	??	not affected
				180	—	??	not affected
				98[d]	—	??	not affected
Carson *et al.*[12]	1982	7	4	2540	—	??	affected
				2330	—	??	affected
				607 (3rd trimester)	—	??	affected
				450 (3rd trimester)	—	??	affected
				200 (3rd trimester)	—	??	not affected
				200 (3rd trimester)	—	??	not affected
				160 (3rd trimester)	—	??	not affected
Kuhnle *et al.*[13]	1984	1	1	1620	affected	SW	affected female abortus w/ambiguous genitalia

[a]17-OHP values were converted from nmol/L to ng/dl.
[b]Same cases as under Pang *et al.* 1980. J. Clin. Endocrinol. Metab.; and Marcus. 1979. Am. J. Med. Genet.
[c]Same cases as under Pang *et al.* 1980. J. Clin. Endocrinol. Metab.; and Pollack. 1979. Lancet.
[d]Same cases as under Hughes *et al.* 1979. Lancet; and Hughes *et al.* 1982. Prenatal Diagnosis.
[e]Same case as under Couillin *et al.* 1979. Lancet; and Couillin *et al.* 1981. Prenatal Diagnosis.
[f]Data not provided (ref. 11).
[g]Two cases where there was disagreement between hormonal and HLA prenatal prediction of the fetal outcome.
[h]??: Information not reported.
[i]SW: Salt-waster.
[j]het: Heterozygous for 21-OH deficiency because one HLA haplotype is shared with the index case; homo nl: Homozygous normal for 21-OH deficiency because HLA haplotype is unidentical to the index case; affected: 21-OH deficiency because HLA haplotype is identical to the index case.

Based on these three cases, we discuss the pitfalls of prenatal diagnosis of 21-OH deficiency CAH by hormonal measurement and HLA typing of amniotic fluid specimens during the second trimester of pregnancy in fetuses at risk for CAH.

Method of Prenatal Study and Postnatal Follow-up

Amniotic fluid specimens in 32 pregnancies where the fetus was at risk for 21-OH deficiency CAH (as indicated by a previously affected offspring) were obtained by amniocentesis during the second trimester (15–22 weeks gestation) for prenatal diagnosis of CAH in the fetus. These prenatal amniotic fluid specimens were assayed for 17-OHP and Δ^4-androstenedione (Δ^4-A) by specific radioimmunoassay, as previously described.[1] HLA typing of the cultured amniotic cells was done in 19 of the 32 amniotic fluid specimens and in their respective index cases (sibs affected with CAH), as well as in their parents, by a previously described microcytotoxicity method, supplemented in a few cases by quantitative absorption and inhibition techniques.[14,16,18,22]

The gestational age of each fetus was estimated by subtracting two weeks from the first day of the last menstrual period or by fetal examination by ultrasound scan.

At midgestation, those fetuses whose amniotic fluid 17-OHP and amniotic fluid Δ^4-A concentrations were unequivocally elevated were diagnosed to have classical CAH, and those fetuses whose amniotic fluid 17-OHP and amniotic fluid Δ^4-A concentrations were in the normal range were predicted to be unaffected by homozygous classical 21-OH deficiency.

Where HLA typing of the cultured amniotic cells demonstrated that the fetus was HLA-identical to the index case, the fetus was predicted to be affected with 21-OH deficiency CAH. The fetuses who shared one HLA haplotype with the index case were predicted to be heterozygotes (carriers) for 21-OH deficiency CAH, while fetuses who shared no HLA haplotype with the index case were predicted to be homozygous normal. Whenever possible, the results of both hormonal testing and HLA typing were used to predict the fetal outcome. The sex of the fetus was usually not known at the time of the prenatal prediction for CAH.

Thirty of the thirty-two fetuses at risk for CAH who underwent studies for prenatal diagnosis of CAH were born at term and have been followed to date (median age = 2.68 years, range = 0.33–5.5 years). In two of the thirty-two cases, the parents chose an elective abortion during the second trimester for socioeconomic reasons; both sets of parents had undergone genetic counseling, and were informed that CAH is a life-compatible disorder and that an affected child may lead a normal life with treatment.

RESULTS OF PRENATAL PREDICTION AND POSTNATAL OUTCOME OF 32 FETUSES AT RISK FOR CAH

Correct Prediction of Fetuses Affected with CAH Based on Hormonal (Cases 1–7) and HLA Studies (Cases 1–4, TABLE 2, FIGURE 1): Classical Salt-Wasting CAH

In seven fetuses, amniotic fluid 17-OHP concentrations (950–1900 ng/dl) and Δ^4-A concentrations (220–739 ng/dl) were unequivocally higher than those hormone levels in the control normal male and female fetuses (17-OHP: 60–360 ng/dl; Δ^4-A: 11–170 ng/dl)[6] (FIGURE 2). Thus, these fetuses were hormonally predicted during the second trimester to be affected with CAH. Of five of these fetuses who had HLA typing of amniotic cells (cases 1–5), four (cases 1–4) were HLA-identical to their respective index cases, while in one (case 5), interpretation of the HLA genotype for prediction of the CAH genotype was not possible due to the presence of identical HLA A, B, C and DR antigens in both parents and in the affected sibling (index case). Thus, with the exception of one case where HLA prediction was not possible, there was agreement between hormonal and HLA prediction in fetuses predicted to be affected with CAH.

Six of seven fetuses predicted to be affected with CAH were born at term. Four were genetic females with varying degrees of ambiguous genitalia including fusion of the labia majora and clitoral enlargement, and two were genetic males with normal

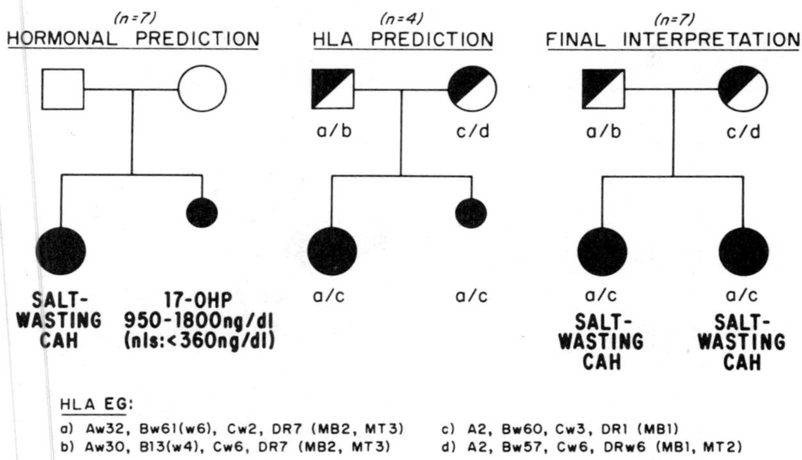

FIGURE 1. Prenatal diagnosis of salt-wasting classical 21-hydroxylase deficiency. Prenatal prediction of fetal outcome is shown based on the amniotic fluid hormone measurement and HLA typing of fetal amniocytes. The final diagnosis of the fetus is depicted under the final interpretation. All fetuses predicted to be affected with CAH based on the elevated amniotic fluid 17-OHP concentration and HLA identity to the index case were proven ultimately to be affected with the salt-wasting form of CAH; their index cases have salt-wasting classical 21-hydroxylase deficiency CAH. (From Pang *et al.*,[23] by permission of The Endocrine Society.)

external genitalia. All six newborns (cases 1–3, 5–7) were found to have elevated serum 17-OHP concentrations (4400–38,000 ng/dl; normal <500 ng/dl in the first three days of life) during the newborn period and developed electrolyte imbalance (low Na, high K) during early neonatal life (three days to two weeks). All affected sibs (index cases) of the six fetuses predicted to be affected with CAH and postnatally proven to have the salt-wasting form of CAH had been previously diagnosed to have salt-wasting CAH. Thus, in all six cases, amniotic fluid hormone measurements (17-OHP and Δ^4-A) and HLA genotypes of amniotic cells and of the respective index case accurately predicted an affected fetus with classical salt-wasting CAH (FIGURE 1).

One of the seven fetuses predicted to be affected with CAH (case 4) was electively aborted for socioeconomic reasons according to the parents' wishes. The 18-week

TABLE 2. Prenatal Fetal Prediction of CAH by Amniotic Fluid Hormonal Measurement and HLA Genotyping of Amniocytes and Final Diagnosis in the Fetus (Postnatally) and in the Index Case

Case Number	Amniotic Fluid Hormone Levels		Prenatal Hormonal Prediction	Prenatal HLA Haplotype[a]	Prenatal HLA Prediction	Final Postnatal Diagnosis of Fetus	Final Diagnosis of Index Case
	17-OHP[c]	Δ^4-A[c]					
	(ng/dl)	(ng/dl)					
1	1300	739	affected	a/c	affected	SW	SW
2	1900	330	affected	a/c	affected	SW	SW
3	950	515	affected	a/c	affected	SW	SW
4	1006	435	affected	a/c	affected	affected abortus	SW
5	1700	900	affected	—[d]	not possible	SW	SW
6	1800	220	affected	—	—	SW	SW
7	1200	—	affected	—	—	SW	SW
8	270	61	unaffected	a/c[e]	affected	nonclassical AH	nonclassical AH
9	158	97	unaffected	b/c[f]	het	SVAH[b] bxa/c = a/c	SVAH
10	110	38	unaffected	a/d	het	normal het	SW
11	150	20	unaffected	b/c	het	normal het	SW
12	140	—	unaffected	a/d	het	normal het	SW
13	230	48	unaffected	a/d	het	normal het	SW
14	180	—	unaffected	a/d	het	normal het	SW
15	210	112	unaffected	b/c	het	normal het	SW
16	200	36	unaffected	b/-	het or unaffected	normal het or homo normal	SW
17	140	47	unaffected	b/-	het or unaffected	normal het or homo normal	SW

18	220	64	unaffected	a/d or a/c[d]	het or affected	normal het	SW
19	160	39	unaffected	b/d	unaffected	homo normal	SW
20	300	53	unaffected	b/d	unaffected	homo normal	SW
21	170	57	unaffected	—	—	normal	SW
22	250	28	unaffected	—	—	normal	SW
23	220	42	unaffected	—	—	normal	SW
24	310	124	unaffected	—	—	normal	SW
25	98	34	unaffected	—	—	normal	SW
26	230	74	unaffected	—	—	normal	SW
27	120	48	unaffected	—	—	normal	stillborn, ambiguous genitalia, adrenal hyperplasia
28	90	63	unaffected	—	—	normal	SW
29	370	12	unaffected	—	—	normal	stillborn, ambiguous genitalia, adrenal hyperplasia
30	94	57	unaffected	—	—	normal	SW
31	235	29	unaffected	—	—	normal	SW
32	250	72	unaffected	a/c	affected	normal	normal

[a]a, c are HLA haplotypes linked to CAH gene inherited from the father and mother, respectively. b, d are HLA haplotypes not linked to CAH gene inherited from the father and mother, respectively.

[b]SVAH stands for simple virilizing classical CAH.

[c]Normal ranges: 17-OHP, 60-360 ng/dl; Δ^4-A, 11-170 ng/dl.

[d]Cases where accurate prenatal HLA prediction for CAH was not possible due to the presence of identical HLA-A, B, C, and DR haplotypes in a parent or between parents.

[e]A case where there was a disagreement between hormonal and HLA prediction of fetal outcome in a fetus affected with nonclassical, late-onset 21-hydroxylase defect.

[f]A case where recombination of paternal HLA A-B loci in the fetus was not detected because of the resulting HLA-B homozygosity. The detectable unaffected paternal HLA-A antigen reaction in the amniotic cells resulted in an inaccurate prediction of CAH genotype as heterozygote in the fetus.

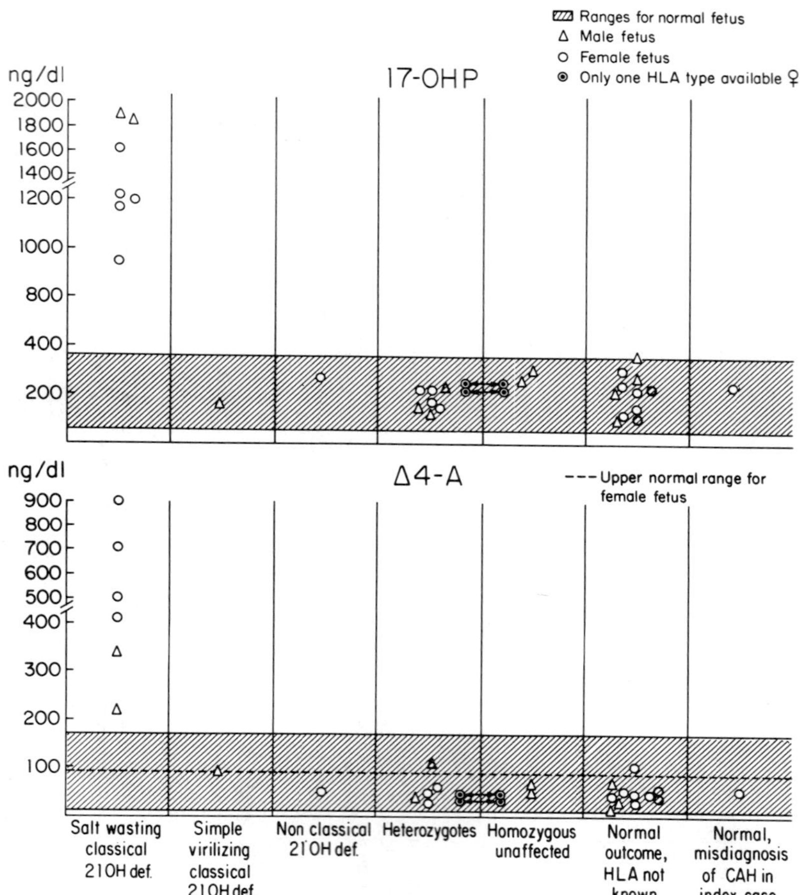

FIGURE 2. Amniotic fluid 17-OHP and Δ^4-A concentration in normal control fetuses and in 32 fetuses at risk for CAH. (From Pang *et al.*,[23] by permission of The Endocrine Society.)

abortus was found to be a genetic female with marked virilization of external genitalia including complete fusion of the labioscrotal folds and clitoral enlargement.

Correct Prediction of Fetuses Unaffected for CAH Based on Hormonal and HLA Studies (Cases 10–31, Table 2, Figure 2)

In 22 fetuses (cases 10–31), amniotic fluid 17-OHP (66–370 ng/dl) and Δ^4-A (12–124 ng/dl) were in the normal range (Figure 2). These fetuses were hormonally predicted to be unaffected with CAH. Typing of HLA-A and HLA-B antigens of amniotic cells in eleven of these fetuses indicated that: (1) Six fetuses were carriers for CAH (cases 10–15); (2) Two fetuses were either carriers for CAH or were unaffected with CAH based on only a positive paternal HLA haplotype antigen reaction (haplotype b/-) that was not linked to CAH (cases 16, 17); (3) Two fetuses were homozygous normal (cases 19, 20); and, (4) one fetus (case 18) was either a carrier for CAH (a/d) or homozygous affected with CAH (a/c) on the basis of homozygosity for HLA-A and B antigens in the mother (A29,Bw44(w4)/A29,Bw44(w4)).

All 22 fetuses (cases 10–31) hormonally predicted to be unaffected with CAH (including case 20, whose HLA genotyping predicted either a carrier or a homozygous affected fetus) were found at birth to be either normal female newborns (12 cases) or normal male newborns (10 cases) without any evidence of adrenal hormonal abnormalities (such as electrolyte imbalance or ambiguity of external genitalia in female newborns). Eight of these newborns (cases 11, 16, 18, 19, 22, 24, 25, and 30) were biochemically proven to be normal by the measurement of either serum 17-OHP (<110 ng/dl) or 17-OHP concentration in whole blood impregnated on filter paper (<10 pg/3mm disc).

The long-term follow-up of these 22 children (duration 0.3–5 yrs) indicated that they have all remained clinically asymptomatic to date. The index cases of these 22 children were previously diagnosed as having salt-wasting classical 21-OH deficiency.

Incorrect Prediction as a Normal Fetus Based on Hormonal Testing, but Correct Prediction of an Affected Fetus with CAH Based on HLA Studies (Case 8, TABLE 2, FIGURE 3): Nonclassical 21-Hydroxylase Deficiency

In a family in which the index case was previously diagnosed to have classical simple virilizing CAH, the fetus was predicted to be unaffected for CAH based on

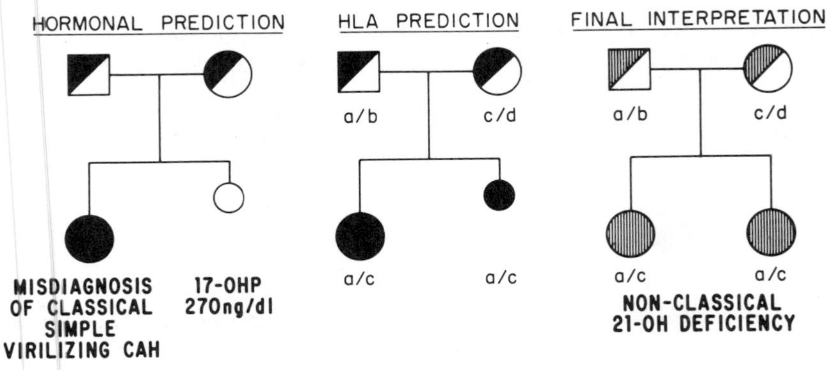

FIGURE 3. Pitfalls of prenatal diagnosis of nonclassical 21-hydroxylase deficiency. The fetus was prenatally predicted to be affected with 21-hydroxylase deficiency based on HLA typing and to be unaffected based on normal amniotic fluid 17-OHP concentration. Postnatally, the fetus presented with nonclassical 21-hydroxylase deficiency. The pitfalls in this case are: (1) index case was misdiagnosed as having classical simple virilizing CAH; and, (2) amniotic fluid 17-OHP concentration was normal in a fetus affected with nonclassical, late-onset 21-hydroxylase deficiency. HLA typing of amniotic cells is useful in prediction of fetuses affected with nonclassical CAH, while hormonal measurement in amniotic fluid is not useful in prenatal diagnosis of nonclassical, late-onset 21-hydroxylase deficiency. (From Pang *et al.*,[23] by permission of The Endocrine Society.)

normal amniotic fluid 17-OHP and Δ^4-A concentrations at 14 weeks gestation. However, the HLA genotype was identical to that of the index case, suggesting the fetus was affected. Thus, there was a disagreement between hormonal and HLA prediction of fetal outcome in this case. The fetus was born at term with birth weight 4.8 kg and birth length 51.3 cm, was reported at birth to have normal female external genitalia without any ambiguity, and had an uneventful nursery course. Thus, at birth, there was agreement between our prenatal hormonal prediction and the physical examination of the newborn. However, at the age of three weeks, the child was seen by a pediatrician who noted a 7.75 kg weight gain since birth. At this time, the child's labia majora were somewhat pigmented and the clitoris was measured to be 1 cm in length. A pediatric endocrinologist noted at this time questionable minimal posterior labia minoral fusion. At the age of one month, 17-OHP (as measured by a commercial laboratory) was elevated (11,800 ng/dl), supporting a diagnosis of nonclassical 21-OH deficiency CAH, and the child was started on hydrocortisone replacement therapy (2.5 mg tid). The index case had been diagnosed to have classical simple virilizing CAH based on an elevated 17-OHP level (2100 ng/dl) at three weeks of age, at which time she was evaluated for a slightly prominent clitoris, but without labial fusion.

The disagreement between prenatal hormonal and HLA prediction of fetal outcome, the absence of prenatal hormonal expression of 21-OH deficiency, and the postnatal presentation of the 21-OH deficiency in this case stimulated further investigation of the specific form of 21-OH deficiency in these sibs. An ACTH stimulation test was performed using cortrosyn (0.25 mg i.v. bolus) in the parents and in both affected sibs ten days after discontinuation of hydrocortisone treatment, at the age of 7 months in the younger sib and at the age of 22 months in the older sib. At the time of discontinuation of treatment, both sibs were not adequately suppressed, as evidenced by their adrenal hormone levels. Blood samples were obtained before and one hour after ACTH administration (25 U i.v. bolus). Baseline and ACTH-stimulated 17-OHP levels for the entire family are shown in FIGURE 4.

The nomogram for genotyping for 21-OH deficiency has been previously reported.[24] The hormonal results of both affected sibs were in the range of nonclassical, symptomatic (late-onset), or asymptomatic (cryptic) 21-OH deficiency, while hormonal results for the mother were clearly in the heterozygote range and for the father were at the overlapping range between heterozygotes and the general population.

The final diagnosis of nonclassical symptomatic (late-onset) adrenal hyperplasia in both sibs was, therefore, evident both by hormonal data and by the delayed postnatal clinical presentation of symptoms of androgen excess.

Incorrect Prediction as a Normal Fetus Based on Both Hormonal and HLA Studies (Case 9, TABLE 2, FIGURE 5): Classical Simple Virilizing CAH

In a case where the index case had been previously diagnosed to have classical simple virilizing CAH, the fetus was predicted to be unaffected with CAH based on normal amniotic fluid 17-OHP (270 ng/dl) and Δ^4-A (61 ng/dl) concentrations. HLA typing in the index case and in the parents indicated that the paternal HLA-A11, Cw-, B18, DR4 (haplotype a) and maternal HLA-A25, Cw-, B18, DR2 (HLA haplotype c) were linked to the gene for 21-OH deficiency in this family. HLA typing of cultured amniotic cells detected the expression of the maternal HLA-A25 and B18 antigens and the expression of the paternal HLA-A2 antigen. Thus, the fetus was predicted to be a carrier for CAH based on the maternal HLA-A and B antigen reactions and the paternal HLA-A antigen reaction of amniotic cells.

This fetus was born at full-term and was found to be a normal male with birth

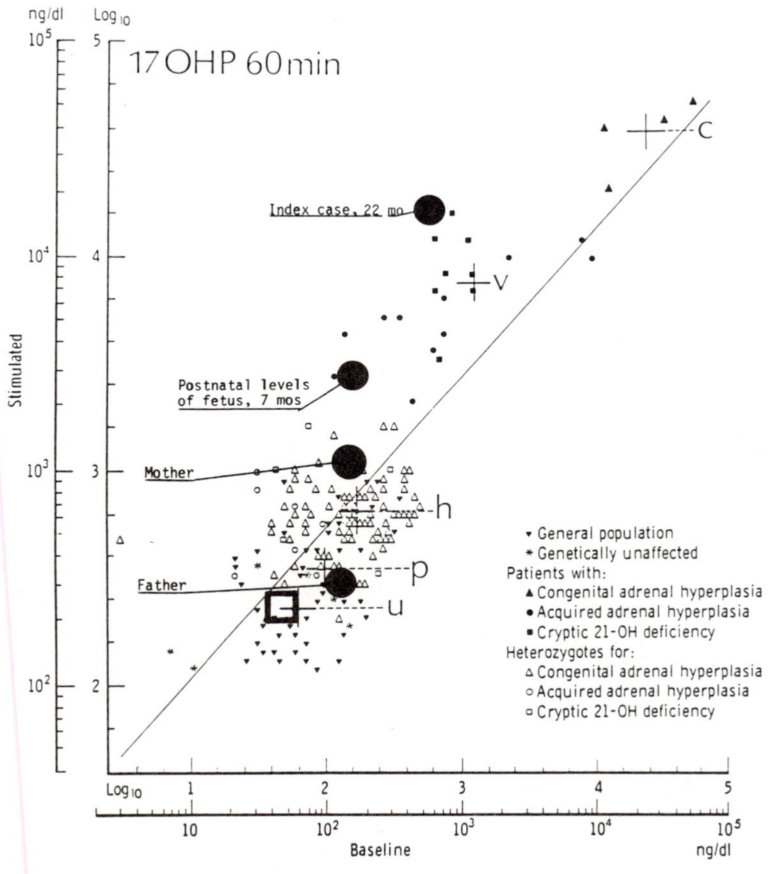

FIGURE 4. Baseline and ACTH-stimulated serum 17-OHP levels plotted on the nomogram. Case 8: The solid circles represent the hormonal responses in the index case, parents, and patient (case 8), in whom there was a disagreement between prenatal hormonal and HLA prediction of fetal outcome. The hormonal prediction was incorrect, while the HLA prediction was correct. The hormonal response to ACTH stimulation in both the patient (younger sib) and in the index case (older sib) indicates that these children are affected with nonclassical symptomatic 21-hydroxylase deficiency. The hormonal response in this mother and father are in the heterozygote range and in the overlapping range between the heterozygotes and the general population, respectively.

Case 32: The square represents the hormonal responses in the index case of the fetus (case 32) predicted by HLA typing to be affected with CAH and by hormonal studies to be unaffected. The hormonal response in this index case unequivocally excludes a diagnosis of CAH. (From Pang *et al.*,[23] by permission of The Endocrine Society.)

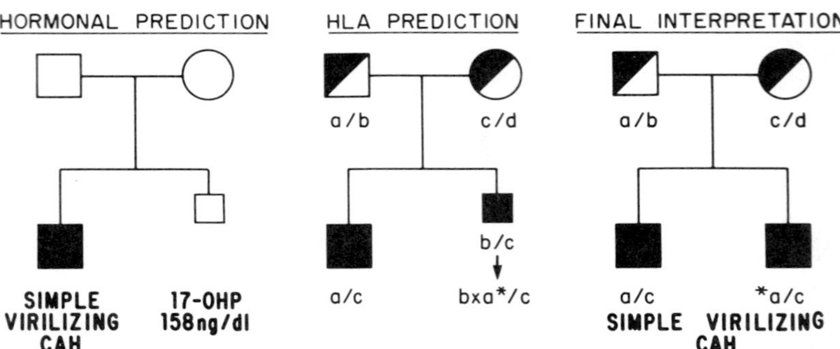

**HLA demonstrated an A/B paternal recombination but was B/DR identical to the index case:*

A2, Cw-, B18, DR4
A25, Cw-, B18, DR2

HLA:

a) A11, Cw-, B18, DR4
b) A2, Cw2, B27, DR5
c) A25, Cw-, B18, DR2
d) A25, Cw-, Bw38(39), DR7

FIGURE 5. Pitfalls in prenatal diagnosis of classical simple virilizing 21-hydroxylase deficiency. The fetus was prenatally predicted to be a carrier for 21-hydroxylase deficiency, as based on HLA typing and on a normal amniotic fluid 17-OHP concentration. Postnatally, the child was found to have a paternal HLA A/B recombinant haplotype; thereby, he was affected with simple virilizing classical CAH and was identical to his affected brother for the HLA-Cw, B, DR- haplotype of the father and the HLA-A, Cw, B, DR- haplotype of the mother. The pitfalls in this case are: (1) amniotic fluid 17-OHP measurement was normal in a fetus affected with simple virilizing classical CAH; and, (2) recombination of paternal HLA A-B antigen in the fetus was not detected because of the resulting HLA-B homozygosity. The detectable "unaffected" paternal HLA-A antigen on the amniotic cells resulted in an inaccurate prediction of CAH genotype as heterozygote in the fetus. (From Pang *et al.*,[23] by permission of The Endocrine Society.)

weight, 2.7 kg, and birth length, 53.3 cm. In this newborn, there was, clinically, no evidence of adrenal hormonal abnormalities, and no further biochemical or HLA studies were carried out in the immediate postnatal period. However, at age $2\frac{7}{12}$ years this male child was hormonally evaluated because of mild penile enlargement (length 3.8 cm). Testicular size was normal for a prepubertal boy, and there was no pubic hair growth or gynecomastia. The growth record of the child indicated growth acceleration from the 30th percentile at age $1\frac{2}{12}$ years to the 40th percentile at age $2\frac{7}{12}$ years. Bone age was 4–5 years at chronological age $2\frac{7}{12}$ years. Serum 17-OHP concentration at this time was highly elevated (10,086 ng/dl), resulting in a diagnosis of classical simple virilizing CAH, as in the index case. Thus, the hormonal and HLA testing failed to make the diagnosis of simple virilizing CAH prenatally.

Repeat HLA typing using peripheral blood lymphocytes in this child demonstrated a paternal HLA-A2 (haplotype b)/Cw-,B18,DR4 (haplotype a) recombination which could not have been detected prenatally in cultured amniotic cells because the HLA haplotype c (HLA-A25,Cw-,B18,DR2) inherited from the mother had the same HLA-B antigen. Thus, this younger sib and the index case were HLA-identical at the paternal HLA-Cw-, B18, and DR2 loci. Since HLA-A2 was, for retrospectively obvious reasons, the only paternal antigen clearly expressed in cultured amniotic cells, the rest of the fetal HLA haplotype was not correctly predicted, resulting in the prediction that the fetus was a carrier for CAH. In this case, where there was a recombination between the HLA-A and B loci, homozygous HLA-B18 antigen expression was the cause for incorrect prenatal prediction of CAH genotype based on the HLA typing interpretation which assumed that the father's alternate HLA-B27 allele had simply not been detected.

The final diagnosis of this younger sib is, therefore, classical simple virilizing CAH, and he shared the Cw-,B18,DR4 segment of his paternal haplotype and the maternal haplotype (HLA-A25,Cw-,B18,DR2) with his affected sib. Thus, the prenatal diagnosis was in error both by hormonal evaluation and by HLA genotyping of the fetus, the latter because of an undetected paternal A/B recombination.

Correct Prediction of a Normal Fetus Based on Hormonal Tests, but Incorrect Prediction as an Affected Fetus by HLA Typing because of Index Case Error (Case 32, Table 2, Figure 6)

In one case where the index case was previously diagnosed to have classical simple virilizing CAH, the fetus was predicted to be affected with CAH based on HLA identity of amniotic cells to the index case. Hormonally, however, the fetus was predicted to be unaffected with CAH, as based on normal amniotic fluid 17-OHP (250 ng/dl) and Δ^4-A concentrations (72 ng/dl).

The discrepancy between hormonal and HLA prediction of fetal outcome for CAH in this case stimulated reevaluation of the specific form of CAH in the index case. The index case is a male child who was evaluated initially for an enlarged phallus at age 3 months and was diagnosed to have classical simple virilizing CAH based on an elevated serum 17-OHP concentration (1200 ng/dl) from a commercial laboratory. The child was then treated with glucocorticoid hormone replacement therapy, which was discontinued at age 3 years. At age 4 years, the child had a normal penile size.

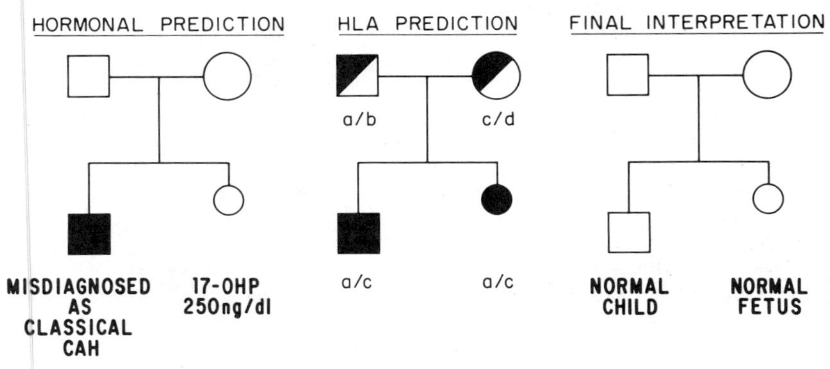

FIGURE 6. Misdiagnosis of CAH in index case led to an erroneous prenatal diagnosis of CAH (case 32). Hormonal data indicated that the index case did not have CAH (FIGURE 3). Amniotic fluid hormone levels accurately predicted a nonaffected fetus, while HLA typing of amniotic cells was irrelevant since the index case did not have CAH. (From Pang *et al.*,[23] by permission of The Endocrine Society.)

Reevaluation of the hormonal diagnosis of CAH was performed with ACTH stimulation (cortrosyn, 0.25 mg i.v. bolus). Blood samples were obtained at the baseline and one hour after ACTH administration. The baseline and ACTH-stimulated serum 17-OHP concentrations in this child (FIGURE 3) were in the overlapping range between the general population and the carriers for 21-OH deficiency, thus, excluding the diagnosis of classical or nonclassical 21-OH deficiency.[24] In addition, baseline and ACTH-stimulated serum deoxycorticosterone (9.7 → 91 ng/dl), corticosterone (0.5 → 4.3 μg/dl), cortisol (9.5 → 36 μg/dl), aldosterone (3 → 28 ng/dl), dehydroepiandrosterone (<10 → 40 ng/dl), Δ^4-A (<5 → <5 ng/dl), and testosterone (<5 → <5 ng/dl) levels were in the ranges of normal prepubertal males. This confirmed that this index case was a normal child.

Misdiagnosis of CAH in the index case, therefore, led to an erroneous prenatal diagnosis of 21-OH deficiency in the younger sib. The HLA prediction was erroneous as it was irrelevant, since the index case was not affected with CAH. The hormonal prediction of a fetus without 21-hydroxylase deficiency was correct.

DISCUSSION

Salt-Wasting Classical 21-OH Deficiency

Prenatal amniotic fluid hormone data presented in this study demonstrate that in classical salt-wasting 21-OH deficiency CAH, prenatal diagnosis of the homozygous affected fetus is reliable by measurement of amniotic fluid concentrations of 17-OHP and Δ^4-A during the second trimester. In all fetuses prenatally predicted to be affected with CAH and postnatally proven to have salt-wasting CAH, and in one affected abortus whose index case has salt-wasting CAH, amniotic fluid 17-OHP and Δ^4-A concentrations were unequivocally elevated. Thus, amniotic fluid hormone levels accurately predicted an affected fetus with salt-wasting classical 21-OH deficiency.

In the literature to date, all fetuses correctly predicted by amniotic fluid hormone measurement to be affected with CAH have been reported to be affected with the salt-wasting form of classical 21-OH deficiency and all their index cases have also had the salt-wasting form of CAH.[1–7,9,10–13] Thus, the hormonal data in fetuses affected with classical salt-wasting CAH unequivocally documented the presence of a significant 21-OH deficiency in the fetal adrenal during midgestation. This is consistent with the fact that in the classical salt-wasting form of 21-OH deficiency CAH, there is evidence at birth of prenatal and immediate postnatal 21-OH deficency, as manifested by virilization in affected females and symptoms of glucocorticoid and mineralocorticoid deficiency in affected males and affected females.

Thus, it is evident that amniotic fluid hormone measurement is useful in the prenatal diagnosis of salt-wasting classical 21-OH deficiency. We, therefore, propose that prenatal diagnosis of CAH based on hormone measurement alone is useful only in fetuses at risk for the salt-wasting form of 21-OH deficiency and in families where the index case has been unequivocally diagnosed as having salt-wasting 21-OH deficiency CAH.

Nonclassical Late-Onset 21-OH Deficiency and Classical Simple Virilizing CAH (CVAH)

In the nonclassical, late-onset symptomatic form of 21-OH deficiency, which is an allelic variant of classical CAH, there is usually no evidence of prenatal virilization, but symptoms of excess androgen production appear postnatally.[25,26] Thus, in female patients, the differentiation of classical CAH from the nonclassical, late-onset 21-OH deficiency is based on the clinical presence or absence of prenatal excess androgen effect on the external genitalia, as well as on the hormonal data using the reference nomogram for 17-OHP.[24] In male patients, the distinction between the two types of 21-OH deficiency is based on hormonal reference data since male newborns with classical CAH do not exhibit any clinical abnormalities of androgen excess.

Our data suggest that hormonal evidence of 21-OH deficiency may not be present prenatally when the fetus is affected with the nonclassical form of 21-OH deficiency. The normal concentration of amniotic fluid 17-OHP and Δ^4-A at midgestation in a fetus affected with nonclassical 21-OH deficiency (case 8) indeed supports this prediction. An error in prenatal diagnosis of CAH in this case occurred due to: (a) the misdiagnosis of the index case as having classical CVAH; and, (b) the normal amniotic fluid hormonal levels in the affected fetus. Thus, the index case must be correctly diagnosed as nonclassical 21-OH deficiency since amniotic fluid hormonal measurement of 17-OHP and Δ^4-A are not elevated and cannot predict a fetus affected with the nonclassical, late-onset form of 21-OH deficiency.

One would predict that a fetus affected with classical simple virilizing CAH (CVAH) should express hormonal evidence of 21-OH deficiency during midgestation. However, a male fetus (case 9) with the postnatal diagnosis of CVAH based on hormonal data, had a normal concentration of amniotic fluid 17-OHP and Δ^4-A. Since both this patient and the index case are male patients with clinical and biochemical manifestations of CVAH, the absence of hormonal expression of 21-OH deficiency in the fetus affected with classical CVAH during midgestation is a puzzling finding. Certainly, further prenatal hormonal data are needed in classical simple virilizing CVAH to validate this observation.

Based on the preliminary hormonal findings in classical simple virilizing CVAH, we suggest that amniotic fluid hormonal measurements may not be useful in predicting a fetus affected with classical CVAH. Thus, it is evident that the diagnosis of the specific form of 21-OH deficiency in the index case must be well established in order to employ amniotic fluid hormonal measurements for prenatal diagnosis of CVAH.

Misdiagnosis of Normal Index Case as Affected with 21-OH Deficiency CAH

When the index case is misdiagnosed as having 21-OH deficiency CAH, HLA typing of fetal amniotic cells is irrelevant in the prenatal diagnosis of CAH. However, amniotic hormonal measurements (17-OHP and Δ^4-A) would accurately reflect the absence of classical 21-OH deficiency in the fetus. Our experience points out the importance of conducting hormonal studies for prenatal diagnosis in such cases. Similarly, in cases when the index case is affected with either 11β-hydroxylase deficiency or 3β-hydroxysteroid dehydrogenase deficiency CAH, HLA typing of fetal amniocytes is not relevant since the genes for these enzyme deficiencies are not linked to the HLA complex.[27–29]

While hormonal studies of amniotic fluid are useful only in predicting a fetus affected with the salt-wasting classical form of CAH, correct HLA typing of the amniotic cells, index case, and parents is useful in prenatal diagnosis of both classical and nonclassical forms of 21-OH deficiency. Vastly improved amniotic cell HLA typing procedures now render the HLA results even more reliable.[30]

In conclusion, our experience in prenatal prediction of CAH in 32 prospective cases revealed that there was a strong prenatal hormonal expression of 21-OH deficiency in the salt-wasting form of classical CAH. However, there was no hormonal expression of 21-OH deficiency in a fetus with postnatal-onset nonclassical 21-OH deficiency or in a fetus with simple virilizing classical 21-OH deficiency CAH. Therefore, prenatal diagnosis may not be possible by hormonal measurement in these forms. We suggest that both hormonal studies and HLA typing of the amniotic fluid specimens, as well as of the index case and the parents, are essential to accurately predict the ultimate fetal outcome of phenotype and genotype of 21-OH deficiency CAH.

REFERENCES

1. FRASIER, S. D., I. H. THORNEYCROFT, B. A. WEISS & R. HORTON. 1975. Elevated amniotic fluid concentration of 17α-hydroxyprogesterone in congenital adrenal hyperplasia. J Pediatr. **86:** 310.
2. MILUNSKY, A. & D. TULCHINSKY. 1977. Prenatal diagnosis of congenital adrenal hyperplasia due to 21-hydroxylase deficiency. Pediatr. **59:** 768.
3. NAGAMANI, M., P. G. MCDONOUGH, J. O. ELLEGOOD & V. B. MAHESH. 1978. Maternal and amniotic fluid 17α-hydroxyprogesterone levels during pregnancy: Diagnosis of congenital adrenal hyperplasia *in utero*. Am. J. Obstet. Gynecol. **130:** 791.
4. HUGHES, I. A. & K. M. LAURENCE. 1979. Antenatal diagnosis of congenital adrenal hyperplasia. Lancet **2:** 7.
5. MARCUS, E. S., J. H. HOLCOMBE, D. TULCHINSKY, R. R. RICH & V. M. RICCARDI. 1979. Prenatal diagnosis of congenital adrenal hyperplasia. Am. J. Med. Genet. **4:** 210.
6. PANG, S., L. S. LEVINE, L. CEDERQVIST, M. FUENTES, V. M. RICCARDI, J. H. HALCOMBE, H. M. NITOWSKY, G. SACHS, C. E. ANDERSEN, M. A. DUCHON, R. OWENS, I. MERKATZ & M. I. NEW. 1980. Amniotic fluid concentrations of Δ^5 and Δ^4 steroids in fetuses with congenital adrenal hyperplasia due to 21-hydroxylase deficiency and in anencephalic fetuses. J. Clin. Endocrinol. Metab. **51:** 223.
7. WARSOS, S. L., J. W. LARSEN, S. G. KENT, K. N. ROSENBAUM, G. P. AUGUST, C. J. MIGEON & J. D. SCHULMAN. 1980. Prenatal diagnosis of congenital adrenal hyperplasia. Obstet. Gynec. **55:** 751.
8. ROSENMANN, A., Z. SCHUMERT, R. THEODOR, T. COHEN & C. BRAUTBAR. 1980. Amniotic 17-αhydroxyprogesterone and HLA typing for the prenatal diagnosis of 21-αhydroxylase deficiency congenital adrenal hyperplasia. Am. J. Med. Genet. **6:** 295.
9. FOREST, M. G., H. BETUEL, P. COULLIN, A. BOUE, M. DAVID, D. FLORET, R. FRANCOIS, P. GUIBAUD, H. PLAUCHU & R. RAPPAPORT. 1981. Antenatal diagnosis of congenital adrenal hyperplasia (CAH) due to 21-hydroxylase by steroid analysis in the amniotic fluid of midpregnancy: Comparison with HLA typing in 17 pregnancies at risk for CAH. Prenatal Diagnosis **1:** 197.
10. HUGHES, I. A. & K. M. LAURENCE. 1982. Prenatal diagnosis of congenital adrenal hyperplasia due to 21-hydroxylase deficiency by amniotic fluid steroid analysis. Prenatal Diagnosis **2:** 97.
11. HUGHES, I. A. 1984. Personal communication.
12. CARSON, D. J., A. OKUNO, P. A. LEE, G. STETTEN, S. M. DIDOLKAR & C. J. MIGEON. 1982. Amniotic fluid steroid levels: Fetuses with adrenal hyperplasia, 46, XXY fetuses, and normal fetuses. Am. J. Dis. Child. **136:** 218.
13. KUHNLE, U., B. BOHM, G. WOLFF, A. MAYEROVA, H. G. DORR, F. BIDLINGMAIER & D. KNORR. 1984. Virilization without adrenal hyperplasia in 21-hydroxylase deficiency during fetal life. J. Clin. Endocrinol. Metab. **58:** 574.
14. POLLACK, M. S., L. S. LEVINE, S. PANG, R. P. OWEN, H. M. NITOWSKY, D. MAURER, M. I. NEW, M. DUCHON, I. R. MERKATZ, G. SACHS & B. DUPONT. 1979. Prenatal diagnosis of congenital adrenal hyperplasia (21-hydroxylase deficiency) by HLA typing. Lancet **1:** 1107.
15. COUILLIN, P., H. NICOLAS, J. BOUE & A. BOUE. 1979. HLA typing of amniotic-fluid cells applied to prenatal diagnosis of congenital adrenal hyperplasia. Lancet **1:** 1076.
16. POLLACK, M. S., D. MAURER, L. S. LEVINE, M. I. NEW, S. PANG, M. A. DUCHON, R. P. OWENS, I. R. MERKATZ, H. M. NITOWSKY, G. SACHS & B. DUPONT. 1979. HLA typing of amniotic cells: The prenatal diagnosis of congenital adrenal hyperplasia (21-OH deficiency type). Transplant. Proc. **XI:** 1726.

17. COUILLIN, P., J. BOUE, H. NICOLAS, C. CHERUY & A. BOUE. 1981. Prenatal diagnosis of congenital adrenal hyperplasia (21-OH deficiency type) by HLA typing. Prenatal Diagnosis **1:** 25.
18. DUPONT, B., S. E. OBERFIELD, E. M. SMITHWICK, T. D. LEE & L. S. LEVINE. 1977. Close genetic linkage between HLA and congenital adrenal hyperplasia (21-hydroxylase deficiency). Lancet **2:** 1309.
19. PRICE, D. A., P. T. KLOUDA & R. HARRI. 1978. HLA and congenital adrenal hyperplasia linkage confirmed. Lancet **1:** 930.
20. WEITKAMP, L. R., M. BRYSON & G. E. BACON. 1978. HLA and congenital adrenal hyperplasia linkage confirmed. Lancet **1:** 931.
21. FU, S. M., R. STERN, H. G. KUNKEL, B. DUPONT, J. A. HANSEN, N. K. DAY, R. A. GOOD, C. JERSILD & M. FOTINO. 1975. Mixed lymphocyte culture determinants and C2 deficiency: LD–7a associated with C2 deficiency in four families. J. Exp. Med. **142:** 495.
22. LEVINE, L. S., M. ZACHMANN, M. I. NEW, A. PRADER, M. S. POLLACK, G. J. O'NEILL, S. Y. YANG, S. E. OBERFIELD & B. DUPONT. 1978. Genetic mapping of the 21-hydroxylase deficiency gene within the HLA linkage group. N. Engl. J. Med. **299:** 911.
23. PANG, S., M. S. POLLACK, M. LOO, O. GREEN, R. NUSSBAUM, G. CLAYTON, B. DUPONT & M. I. NEW. 1984. Pitfalls of prenatal diagnosis of 21-hydroxylase deficiency congenital adrenal hyperplasia. J. Clin. Endocrinol. Metab. **61:** 89.
24. NEW, M. I., F. LORENZEN, A. J. LERNER, B. KOHN, S. E. OBERFIELD, M. S. POLLACK, B. DUPONT, E. STONER, D. J. LEVY, S. PANG & L. S. LEVINE. 1983. Genotyping steroid 21-hydroxylase deficiency: Hormonal reference data. J. Clin. Endocrinol. Metab. **57:** 320.
25. KOHN, B., L. S. LEVINE, M. S. POLLACK, S. PANG, F. LORENZEN, D. J. LEVY, A. LERNER, G. F. RONDANINI, B. DUPONT & M. I. NEW. 1982. Late-onset steroid 21-hydroxylase deficiency. A variant of classical congenital adrenal hyperplasia. J. Clin. Endocrinol. Metab. **55:** 817.
26. CHROUSOS, G. P., D. L. LORIAUX, D. MANN & G. B. CUTLER, JR. 1982. Late-onset 21-hydroxylase deficiency is an allelic variant of congenital adrenal hyperplasia characterized by attenuated clinical expression and different HLA haplotype associations. Horm. Res. **16:** 193.
27. BRAUTBAR, C., A. ROSLER, H. LANDAU, I. COHEN, D. NELKEN, T. COHEN, C. LEVINE, J. SACK, A. BENDERLI, S. MOSES, E. LIEBERMAN, B. DUPONT, L. S. LEVINE & M. I. NEW. 1979. No linkage between HLA and congenital adrenal hyperplasia due to 11β-hydroxylase deficiency. N. Engl. J. Med. **300:** 205.
28. PANG, S., L. S. LEVINE, F. LORENZEN, D. CHOW, M. S. POLLACK, B. DUPONT, M. GENEL & M. I. NEW. 1980. Hormonal studies with obligate heterozygotes and siblings of patients with 11β-hydroxylase deficiency congenital adrenal hyperplasia. J. Clin. Endocrinol. Metab. **50:** 586.
29. PANG, S., L. S. LEVINE, E. STONER, J. M. OPITZ, M. S. POLLACK, B. DUPONT & M. I. NEW. 1983. Non-salt-losing congenital adrenal hyperplasia due to 3β-hydroxysteroid dehydrogenase deficiency with normal glomerulosa function. J. Clin. Endocrinol. Metab. **56:** 808.
30. MAURER, D. H. & M. S. POLLACK. 1985. The use of gamma interferon to increase HLA antigen expression on cultured amniotic cells used for the prenatal diagnosis of 21-hydroxylase deficiency. *In* Congenital Adrenal Hyperplasia, vol. 458. M. I. New, Ed.: 148–155. Ann. N.Y. Acad. Sci. New York.

Pitfalls in Prenatal Diagnosis of 21-Hydroxylase Deficiency by Amniotic Fluid Steroid Analysis?

A Six Years Experience in 102 Pregnancies at Risk[a]

MAGUELONE G. FOREST

Unité de Recherches Endocriniennes et Métaboliques chez l'Enfant
INSERM-U.34
Hôpital Debrousse
29 rue Soeur Bouvier
69322 Lyon Cedex 05 France

INTRODUCTION

Congenital adrenal hyperplasia (CAH) due to 21-hydroxylase (21-OH) deficiency results in virilized female fetuses from *in utero* exposure to high levels of adrenal androgens (testosterone particularly) who are associated or not with clinical symptoms of salt loss after birth. Several years ago, Frasier *et al.*[1] showed that 17α-hydroxyprogesterone (OHP), the immediate precursor above the enzyme block, was elevated in the amniotic fluid of a CAH affected fetus at 24–38 weeks gestation. Confirming their prediction that amniotic fluid steroid analysis could be used as "a valuable *in utero* predictor of CAH," several authors have reported the feasibility of the prenatal diagnosis of CAH by measuring OHP in amniotic fluids collected at midgestation.[2–10]

Despite some debate on whether the prenatal detection of CAH[11–13] is desirable, it is in accordance with the ninth guideline for ethical issues in prenatal diagnosis: "Prenatal diagnosis should not be denied to the woman who is at risk and desires the procedure, but has decided that she will not undergo an abortion no matter what the diagnosis."[14] There is, undoubtedly, an increasing demand for prenatal diagnosis of CAH. Most parents with a previously affected child have experienced considerable distress because of a sexual ambiguity in a CAH female or the delay in diagnosis and proper treatment in an affected male infant. Nowadays, parents wish to have prior knowledge of the outcome of a planned pregnancy in relation to CAH and sex of the fetus. Provision of such information is the purpose of the prenatal diagnosis of 21-OH deficiency. Therefore, because of its importance, it should not be fraught with false predictions.

The reports available in the literature have clearly demonstrated a success in the prediction made. However, they are all "pilot studies" of a small number of patients,[1–10] and the question whether or not prenatal diagnosis of CAH by steroid analysis in amniotic fluid presents limitations or pitfalls has not yet been fully answered. Also because control values differ significantly between laboratories, review of the literature is useless (TABLE 1) in this regard. The purpose of this report is to

[a]This work was supported by a grant from le Ministère des Universités, Mission de la Recherche, by l'Association Française pour le Dépistage et la Prévention des Maladies Métaboliques et des Handicaps de l'enfant, and INSERM.

answer the question by presenting our experience, which is based on both a large number (100) of prenatal diagnosis of CAH and a reasonable length of time (six years) of observation.

On the other hand, there is little or no[15] reported experience of attempted prenatal diagnosis in families in whom the 21-OH enzyme defect is expressed as the late-onset or attenuated form of the disease (peri- or postpubertal signs of virilization or hirsutism). Therefore, the value of prenatal diagnosis of the late-onset form of 21-OH deficiency has been questioned.[16,17] However, since the feasibility of this diagnosis is not yet clearly established nor contradicted, we shall also report our limited experience of two pregnancies at risk in the same family.

CLINICAL MATERIAL

Prenatal diagnosis of 21-OH deficiency by amniotic fluid steroid analysis was attempted over a six-year period of time in a total of 102 pregnancies at risk in 93 families (diagnosis being required nine times for two successive pregnancies in the same family). Patients were numbered by chronological order.

In the largest series ($n = 100$), the fetus was at risk for a severe congenital form of 21-OH deficiency. In most instances ($n = 93$), there was a high risk for CAH with one or several previously affected siblings. In the seven other pregnancies the risk for CAH was rather low, but the demand still pressing—in five, the CAH affected member of the family (index case) was either the mother (case no. 64) or a first cousin on one side (case nos. 32, 77, and 96) or both sides (case no. 36) of the family; in the last two, the mother previously had a CAH affected child, but either the father was different (case no. 83) or the HLA typing excluded the paternity of the index case (case no. 63). In two families, 21-OH deficiency was associated with another hereditary disease, familial hypercholesterolemia (family no. 59) and hemophilia A (family no. 80, in whom the parents considered only the latter risk to be unbearable although the fetus was revealed to have both diseases).

Only one family at risk for the late-onset form of 21-OH deficiency was studied. The previously affected girl developed hypertrophy of the clitoris and precocious pubarche at nine years of age. Basal levels of OHP were 3100 ng/dl rising to 6430 ng/dl 15 minutes after an i.v. injection of 0.25 mg of ACTH-$_{1\text{-}24}$ (Synacthen®) stimulation. The mother thus required a prenatal diagnosis for the next two pregnancies (case nos. 16 and 58) at three-year intervals.

All families received counseling prior to amniocentesis. Amniotic fluids were collected at midgestation (16–18 weeks) except twice (20th week). Steroid analysis was retrospective, i.e., performed after genetic counseling and parental decision based on HLA prediction only, in families 1–8, 10, and 16, as previously reported.[9] In all other instances, prediction based on steroid analysis was used for prenatal diagnosis whether or not HLA typing was available and/or concordant. Detailed pedigrees of families 1–8 and 10,[18] and families 11–18[9] have been reported, as well as individual OHP amniotic fluid concentrations in these 17 pregnancies.[9] In the present report intended to examine the reliability of steroid analysis in the prenatal diagnosis of 21-hydroxylase deficiency, only a few HLA typings useful for the discussion will be presented.

In a previous study,[19] we determined the normal ranges for 14 steroid hormone concentrations in 63 amniotic fluids collected between 14 and 20 weeks from mothers who had requested amniocentesis for increased risk of chromosome defect or neural tube defect and had produced normal infants. A further 77 normal amniotic fluids collected under the same conditions were analyzed to establish normal values in a larger population and to eliminate any technical variation over the six-year period of the study. Mean and ranges were strictly comparable between the two groups, and the pooled 140 specimens were used as controls in the present report.

TABLE 1. Reported Cases of Antenatal Diagnosis of CAH due to 21-OH Deficiency Made on Amniotic Fluid Concentrations of 17α-Hydroxyprogesterone (OHP): 19 Cases of Confirmed CAH out of 50 Pregnancies Studied

		CAH Affected Fetuses			Control Groups (M & F)			
Ref. No.	Cases of CAH/Total	Age (weeks)	Sex	OHP[a] (ng/dl)	Age (weeks)	(n)	OHP (ng/dl) mean ± SD	(range)
1	1[b]/1	24–38 (n = 6)	M	188 ± 80 (110–295)	24–38	(13)	71 ± 20	(40–100)
2	1[b]/1	16	F	525[c]	16–20	(24)		(20–150)
3	1/1	14	M	824	14–24	(48)	157 ± 58	(50–350)[c]
		22		2069				
		28		293	24–40	(22)	79 ± 32	(25–150)[c]
		34		368				
4	1[b]/5	17	F	1090	15–21	(36)	250 ± 75	(132–380)
5	1/1	14.5	M	1467[e]	Mid[g]	(5)	234 ± 69	
6	1/2	16th	F	1500[e]	Mid[g]	(43)		(0–400)
7	2/1[d]	17	F	2239	Mid[g]	(10)	229 ± 67	(145–348)
			F	2747				
8	3/8	14	F	1700	12–20	(46)[c]		(65–300)
		14.5	M	1900				
		17	F	1300				(20–150)
		Term		1700	Term	(10)[c]		
9	3[f]/17	17th	F	1244[b]	12–20	(75)	119 ± 43	(36–270)
		17th	M	1221[b]				
		16.5	F	982				
10	5/13	14 to 17	3F 2M	992 to 2016	14–20	(44)	267 ± 79	(102–410)

[a]Significantly higher than in age matched controls.
[b]Retrospective study.
[c]Read on a figure.
[d]Twin pregnancy.
[e]HLA typing gave concordant results.
[f]Cases 3, 7, and 11 of the present report.
[g]Midgestation.

METHODS

Amniotic fluid concentrations of OHP, Δ^4-androstenedione (Δ^4), and testosterone (T) were determined in the same laboratory by specific radioimmunoassay (RIA) techniques after ethyl ether extraction and after purification on celite microcolumn chromatography, as previously reported in detail.[19] OHP was always measured twice in two separate assays, as well as Δ^4 and T in the first 54 patients. Results were within the established interassay variation coefficients of the methods (9% for OHP, 8.4% for Δ^4, and 7% for T). In three instances (cases 15, 16, and 88), determination of OHP, Δ^4, and T was repeated on amniotic fluids collected three to four weeks later at the time of an abortion required by the parents. Various steroids were also measured by the same RIA techniques in blood specimens obtained by cardiac puncture in these three fetuses.

Fetal chromosome analyses were performed by karyotyping cultured amniotic cells by conventional methods. When performed, most HLA typing on either cultured amniotic cells or blood lymphocytes (of family members and of, occasionally, the fetus at risk when repeated after birth) were made in the laboratory of either H. Betuel (Lyon) or A. Boué (Paris), using techniques that they have reported in detail.[9,18,20,21]

RESULTS AND DISCUSSION

Methodological Limitations

Methodological limitations may theoretically arise from technical problems and/or difficulties in interpreting results.

Steroid Analysis

Technical Problems. There was no technical problem in measuring OHP, Δ^4, or T in amniotic fluid samples. Steroids were present in such concentrations that measurement was always made well above the sensitivity (5 to 10 pg/tube) of our methods[19] provided that a 2–3 ml sample was used to estimate T levels. It was only because of the importance of the prediction to be made that we repeated the measurement of OHP (the best indicator of 21-OH deficiency, see below) on all samples of this series. This was chiefly intended to exclude any human mistake. As mentioned above, results have always been concordant. The mean of those duplicated measures was, therefore, used as the final result.

Establishment of Normal Values. This was obviously the first step. Reference values reported in the literature cannot be used blindly as they differ considerably (TABLE 1). For instance, we have correctly and unambiguously predicted a CAH affected fetus in four families (nos. 51, 74, 102, and 103) based on OHP amniotic fluid levels of 436 to 499 ng/dl, which would have fallen in the upper normal range of other laboratories (TABLE 1).

Interpretation of Results. A problem could theoretically arise if there was an overlap between values observed in normal controls and CAH affected fetuses. Determination of the limits of normality (and not mean ± SD) was, therefore, of paramount importance for interpreting results. In normal pregnancies, we found values of amniotic fluid OHP concentrations significantly ($p < 0.001$) higher in male ($n = 76$) than in female ($n = 74$) fetuses, despite an overlap in individual values. The sex of the fetus was usually unknown (results of karyotyping being not yet available) when steroid analysis was performed. Therefore, amniotic fluid levels of OHP of both sexes have been combined and these pooled results were considered by us as the basis for normal controls. The mean value was 119 ng/dl. About 10% of normal controls showed OHP levels between 200 and 271 ng/dl. We decided to take the value of 300 ng/dl for the upper limit of normal for OHP since this number was 10% above the highest individual value observed in our population of 140 normal controls. This 10% represented the interassay coefficient variation or twice the intra-assay coefficient variation (5.6%) of the RIA method. The upper limit of normal was defined prior to the interpreted results in prenatal diagnosis. As yet, there was no problem in interpreting

TABLE 2. Steroid Concentrations (ng/dl) in Two Separate Fractions of Amniotic Fluid (AF) Collected during a Difficult Amniocentesis

	OHP	Δ^4	T
First fraction:AF contaminated with blood	343[a]	303[a]	12.5[a]
Second fraction: clear AF	199	57.5	5.3

[a]Values clearly above the upper limit of normal for a female fetus.

results since CAH affected fetuses had OHP values well above this limit. Clear-cut differences between CAH affected pregnancies and normal controls were also observed in all studies in the literature (TABLE 1).

Aspect of Amniotic Fluid Samples. In opposition to others,[10] we found no difference in steroid levels whether or not the amniotic fluids were centrifuged. However, it was of paramount importance that fluids should not be contaminated in the course of the amniocentesis. Indeed, the only pitfall was when the amniotic fluid was contaminated by placental or maternal blood. For example, in the course of a somewhat laborious amniocentesis, a mixture of blood and amniotic fluid was first obtained and, then, a clear amniotic fluid was produced. Fortunately, the obstetrician collected the two fractions separately. They were both analyzed for steroid content. As shown in TABLE 2, the prediction would have been a CAH affected fetus (and thus, erroneous) if based on the steroid analysis of the contaminated fraction. In this particular pregnancy (case no. 27), karyotype was 46, XX and HLA typing indicated a heterozygote. This example also illustrated that correct prenatal diagnosis by amniotic fluid steroid analysis relies on the expertise of the obstetrician performing the amniocentesis. With the exception of this case, all amniotic fluid specimens analyzed for prenatal diagnosis of 21-OH deficiency were apparently not contaminated.

Karyotype

From a large series in the literature,[22] the error rate for full chromosomal analysis is very low (0.07%). Accuracy is even greater in karyotyping of sex chromosomes, which are the only important ones to identify in the context of prenatal diagnosis of 21-OH deficiency. As pointed out earlier,[23] there is a possibility of error in diagnosis of sex due to contamination of amniotic cells by maternal cells. Thus, uncertainty exists than an XX karyotype may reflect a maternal rather than a fetal cell line. To the best of our knowledge, this risk has not been estimated. It has been evidenced once in our series of controls.

HLA Typing

The technical limitations of HLA typing have been previously discussed.[18] It includes the problems inherent to cell cultures—failure of fetal cell growth and limitations in the antigens available. Interpretive difficulties may arise from antigen cross-reactions, antigens poorly expressed in amniotic cells, HLA genotypes which are noninformative, and, although low, a risk of recombination. Also, the possibility of contamination by a maternal cell line may explain apparent multiple antigen cross-reactions. This likely happened once in our series when both maternal HLA haplotypes were apparently identified (case no. 38, see below for details).

Steroid Analysis and Prenatal Diagnosis of CAH

Based on our criteria for interpreting OHP values, results were classified according to the prediction made of either unaffected or CAH affected fetuses. Information of the outcome of pregnancy or abortion was subsequently obtained.

CAH Affected Fetuses

Twenty pregnancies were predicted to be affected. In all instances, OHP levels were well above the upper limits of normal (illustrated in FIGURE 1), with a wide range from 436 to 2145 ng/dl. In pregnancies proceeding to term, OHP values found at 16–17 weeks of gestation did not correlate with the severity of the disease in the newborn infant (documented in ref. 6). As previously reported,[7,8] Δ^4 was also abnormally high in CAH fetuses. However, Δ^4 values fell thrice in the normal range for fetal sex, and until the results of the karyotype were available, they did not help the diagnosis in six cases (FIGURE 1). Ranging from 131 to 1592 ng/dl, Δ^4 amniotic fluid levels did not always correlate with those of OHP. However, the lowest Δ^4 values were observed with the lowest OHP concentrations. Amniotic fluid concentrations of T were similar in female and male affected fetuses (FIGURE 1). Still, when the sex of the fetus was determined by karyotype, the findings of high levels of T in the amniotic fluid of a female fetus reinforced the diagnosis.

There was no false-positive prediction from steroid analysis. HLA typing was performed in 16 out of these 20 pregnancies. Results were inconclusive once, and

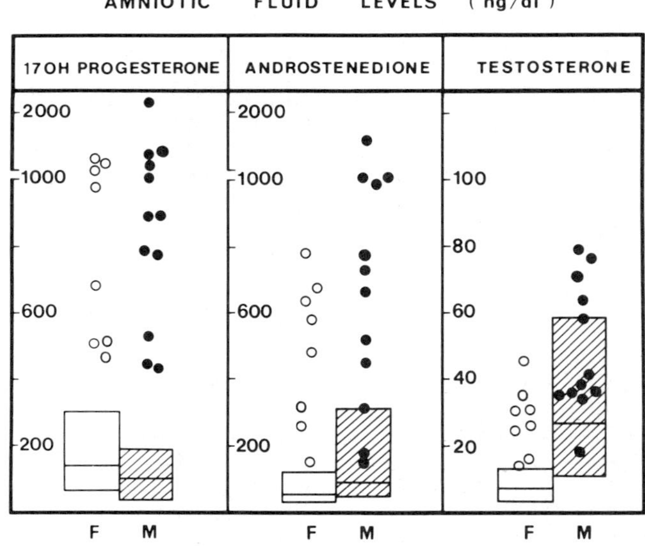

FIGURE 1. Amniotic fluid concentrations of 17α-hydroxyprogesterone, Δ^4-androstenedione, and testosterone in female (○) and male (●) affected CAH fetuses. Means and ranges observed in 140 normal pregnancies bearing either a male (M) or a female (F) fetus are represented by open and hatched columns, respectively.

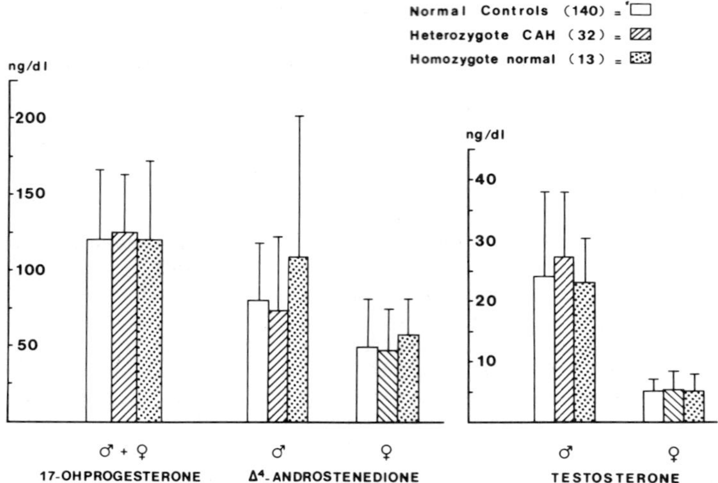

FIGURE 2. Mean (±1 SD) amniotic fluid (AF) concentrations of 17α-hydroxyprogesterone, Δ^4-androstenedione, and testosterone in amniotic fluids collected from normal controls ($n = 140$) and from 45 pregnancies at risk for 21-hydroxylase deficiency. The latter, predicted to be unaffected as based on AF steroid concentration, were grouped according to HLA typing in heterozygote or homozygote normal infants. In the three groups, values of Δ^4-androstenedione and testosterone were significantly higher for male than for female infants. There was, however, no difference between heterozygotes, homozygote normals, and controls, no matter the AF hormone measured in each fetal sex.

discordant twice, having one misdiagnosis (case no. 21) and one with a situation secondary to prenatal treatment (case no. 67). Both will be discussed below. The products of the five pregnancies predicted to be affected on the sole steroid analysis were full-term infants (two males and one female) with typical clinical and biological features of CAH. The last two females were aborted and presented an obvious virilization of the external genitalia.

*Unaffected Fetuses (*n *= 79)*

In the series of fetuses at risk for CAH, there was no false-negative prediction by steroid analysis. HLA typing was not available in 17 cases (index cases were deceased). It was obviously noninformative in four out of the seven cases at low risk described in the clinical material section, and results were unconclusive thrice (parental homozygous haplotypes). In five other instances, fetal HLA typing excluded the risk for CAH, but could not discriminate heterozygous carriers from homozygous normals (lack of expression of some HLA antigens). Classification was, however, possible thrice when repeating HLA typing on peripheral or cord blood lymphocytes after birth. In addition, HLA typing was at discordance with steroid analysis twice (cases nos. 15 and 38 presented below). Finally, this has permitted us to recognize 33 heterozygotes and 15 homozygous normal infants at the time of prenatal diagnosis. As illustrated in FIGURE 2 and extending our earlier results,[9] amniotic fluid steroid concentrations did not distinguish between these two groups. This is, though, of no consequence for prenatal diagnosis as heterozygous carriers can be detected postnatally.

Difficulties Encountered in Prenatal Diagnosis of CAH due to Discordance between Results of Amniotic Fluid Steroid Analysis and Fetal HLA Typing

This happened thrice during the first two years of our experience:

(1) *Family no. 15.* This case has been reported in detail,[9] and results are illustrated in FIGURE 3. HLA identity with the affected brother (index case) indicated an affected fetus, and based on this information, the parents chose to interupt the pregnancy. HLA typing was repeated on blood lymphocytes and confirmed the genotype. However, retrospective steroid analysis showed normal steroid levels in the amniotic fluid samples collected at 17 weeks for prenatal diagnosis (FIGURE 3) and at 21 weeks at the time of abortion (OHP = 105 ng/dl; Δ^4 = 105 ng/dl; T = 30 ng/dl). The abortus,

Family 15

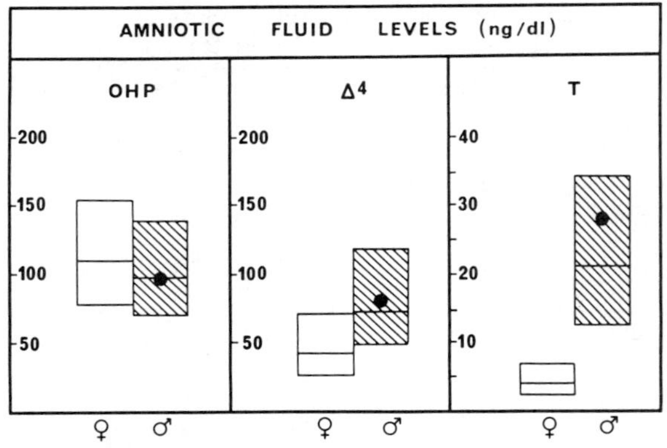

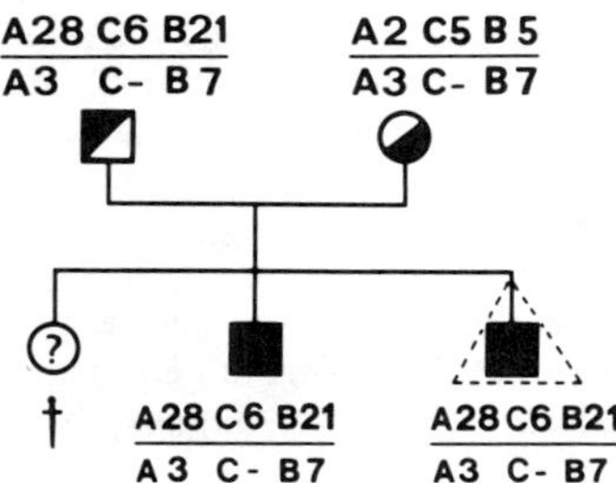

FIGURE 3. Amniotic fluid (AF) concentrations of 17α-hydroxyprogesterone (OHP), Δ^4-androstenedione (Δ^4), and testosterone (T) in a pregnancy at risk for 21-hydroxylase deficiency. Retrospective analysis showed that AF steroids were within the normal range (represented as mean and 95% confidence limits by open and hatched columns in normal females and males, respectively). HLA typing on amniotic cells (identical to the previously CAH affected brother) had predicted a CAH affected fetus (see text for details).

being male, had normal genitalia. Fetal plasma levels of steroids are given in TABLE 3. There is, however, a lack of normative data in the literature on normal fetal blood levels of various hormones except for T[24] or Δ^4.[25] Still, from these informations and the comparison of steroid levels observed in a CAH affected abortus (TABLE 3), we concluded that the fetus was biologically normal. The hypothesis of a recombination was made to explain the discrepancy, but could not be proven (TABLE 4).

(2) *Family no. 38.* A quite similar discrepancy was observed once again, two years later—normal amniotic fluid steroid levels and HLA identity with the index case. Results are illustrated in FIGURE 4. On the other hand, there was an apparently unexplained high positive reaction for HLA-A_2 antigen; HLA-B8 was not expressed. We made the hypothesis that the discrepancy could be explained by a contamination of amniotic cells by a maternal cell line. Also, because of our increasing experience and faith in steroid analysis, genetic counseling was to advise the mother to proceed with the pregnancy. In keeping with the hypothesis made, the full-term newborn was a girl with normal genitalia and normal plasma steroid levels. HLA typing performed on cord blood lymphocytes revealed a homozygous normal status (FIGURE 4). The girl is now three years old and is perfectly normal (TABLE 4).

TABLE 3. Plasma Concentration (ng/dl) of Steroid Hormones in Three Aborted Fetuses (Blood Obtained by Cardiac Puncture)

	Fetus 15 ♂	Fetus 16 ♀	Fetus 3 CAH ♀
17α-OH-Pregnenolone	1240	2407	1820
Progesterone	45,050	22,450	6205
17α-OH-Progesterone	490	1886	9748[a]
Δ^4-Androstenedione	645	123	2350[a]
Testosterone	36	—	320[a]
Estrone	196	86	165
17β-Estradiol	125	24	60

[a]Values abnormally high.

(3) *Family no. 21.* An opposite situation was encountered in this case (FIGURE 5)—amniotic fluid levels of both OHP and Δ^4 were unambiguously elevated, while those of T were high normal for a male fetus (as predicted by the karyotype), and HLA typing showed no interpreting difficulties in predicting a heterozygote. Pregnancy was continued to term. However, the newborn infant rapidly developed clinical symptoms of a severe salt loss. At three weeks of age, basal levels of OHP were 35,000 ng/dl and those of T = 337 ng/dl. HLA typing repeated on cord blood in two different laboratories confirmed the heterozygote genotype found in amniotic cells. Repeated again at six months of age on peripheral blood lymphocytes, HLA typing performed with a different set of HLA antibodies showed an identity with the index case. This case illustrates the difficulties of HLA typing, but emphasizes the accuracy and safety of amniotic fluid steroid analysis for prenatal diagnosis of CAH (TABLE 4).

Prenatal Diagnosis in Late-Onset 21-OH Deficiency

Results obtained in the only two pregnancies at risk studied (in the same family) are presented in FIGURE 6. The first prenatal diagnosis (pregnancy no. 16) was performed during the first year of our program and has been reported.[9,15] HLA typing

TABLE 4. Difficulties Encountered in the Prenatal Diagnosis of 21-Hydroxylase Deficiency (21-OH-D) in Six Pregnancies for Which Steroid Analysis and HLA Typing Gave Apparent Discordant Predictions

Risk for	Case No. (FIGURE No.)	Sex	Prediction of Steroid Analysis	Prediction of HLA Typing	Final Diagnosis
CAH	15 (3)	M	unaffected	21-OH-D	recombination ?
	21 (4)	M	21-OH-D	heterozygote	21-OH-D
	38 (5)	F	unaffected	21-OH-D	homozygote normal
	67 (8)	F	normal levels[a]	21-OH-D	21-OH-D
Late-Onset	16 (6)	F	unaffected	21-OH-D	recombination ?
	58 (6)	M	unaffected	21-OH-D/heterozygote ?	heterozygote

[a]Efficient prenatal treatment.

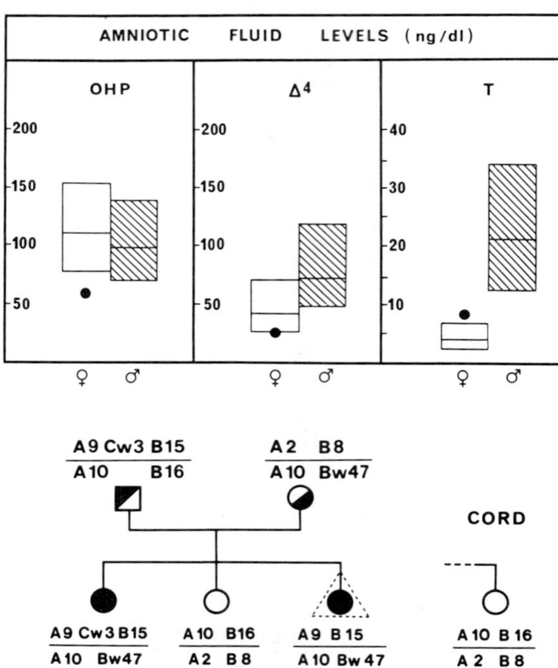

FIGURE 4. Steroid amniotic fluid (AF) concentrations in a male fetus at risk for 21-hydroxylase deficiency predicted a CAH affected, as based on the results of AF hormone levels well above the normal range (legends as in FIGURE 3). HLA typing made on amniotic cells predicted a heterozygote infant, but gave concordant results when determined on lymphocytes six months after birth (see text for details).

revealed identity with the affected sister (index case), whereas amniotic fluid steroid levels were normal. The latter were somewhat disregarded during counseling. The mother, unwilling to take any risk of having another affected girl, required an interruption of pregnancy. From all subsequent hormonal studies made on amniotic fluid and fetal blood collected (TABLE 3) at the time of abortion, and from the normality of external genitalia and of the adrenal grands, it was concluded that the fetus was biologically normal. Initially,[9] two hypotheses were made: (1) a genetic recombination centromeric to HLA-B, and, (2) the 21-OH deficiency was not expressed in the fetus.[9,17] They could not be demonstrated nor contradicted (TABLE 4).

The second prenatal diagnosis (case no. 58) was made three years later when the mother became pregnant again. Apparently, discordant results were again observed (FIGURE 6)—normal steroid levels and HLA identity with the index case. However, a weak positive reaction was observed with HLA-A_9 (part of the theoretically normal paternal haplotype) (FIGURE 6). Pregnancy was continued on term. After birth, the child did perfectly well and still does at two years of age. Unfortunately, it has not been possible yet for us to obtain repetition of HLA typing on peripheral blood lymphocytes.

Hormonal studies made after short ACTH testings are illustrated in FIGURE 7. The kinetics of OHP responses in patient 58 (aged 7 and 16 months) are presented. In all members of the family, results are presented as the OHP rate of increase at 30 min. In our series of normal controls, values were not different with age (and were, therefore, pooled in FIGURE 7) and were quite similar to those reported by Gutai *et al.*[26] From this study, it was concluded that the mother had a normal response, whereas the father and patient no. 58 showed typical heterozygote responses. Assuming that a hereditary defect in 21-OH enzyme can be detected biochemically before the appearance of clinical symptoms, it was inferred that patient no. 58 did not present 21-OH deficiency. Also, by taking into account the 3-allele genetic model of 21-OH deficiency[28,29] (i.e., normal 21-OH allele, the deficient allele 21-OH^{O} for the congenital or severe forms, and the "susceptibility" 21-OH^{S} allele for the late-onset or attentuated forms), the father was assumed to be 21-OH^{O}/21-OH, the mother 21-OH^{S}/21-OH, and the patient 21-OH^{O}/21-OH, rather than a compound heterozygote 21-OH^{S}/21-OH^{O} as should be the index case.

Nevertheless, steroid level analysis in the amniotic fluid obviously indicates the

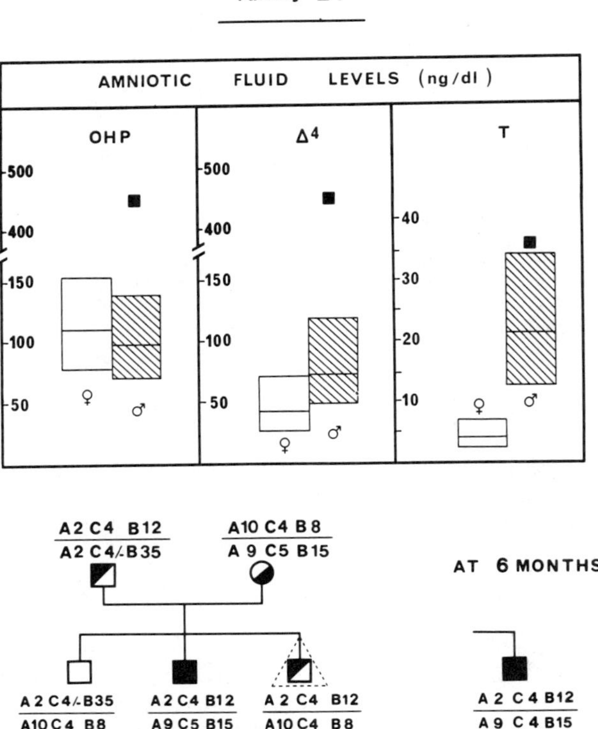

FIGURE 5. Steroid amniotic fluid (AF) concentrations in a female fetus at risk for 21-hydroxylase deficiency predicted an unaffected, as based on the results of AF levels (legends as in FIGURE 3). HLA typing on amniotic cells gave apparent discordant results, but revealed a homozygous normal infant when determined on lymphocytes at birth (see text for details).

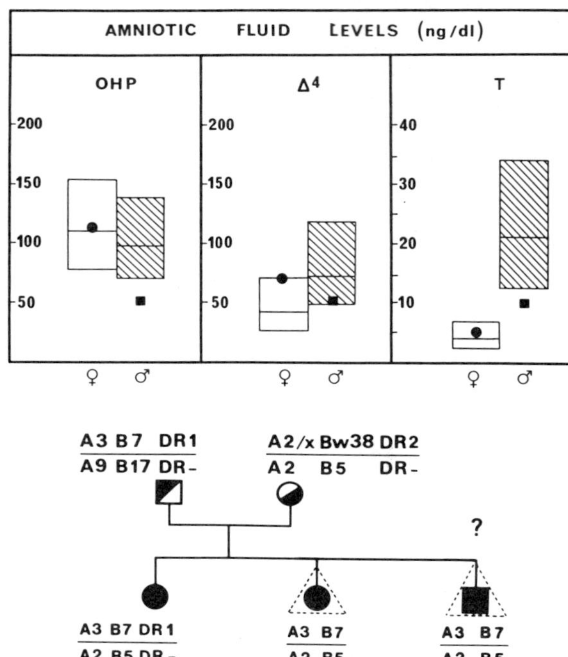

FIGURE 6. Steroid amniotic fluid concentrations and HLA typing on amniotic cells in two successive pregnancies (female ●, then male ■) of a mother who previously had a girl with the late-onset form of 21-hydroxylase deficiency (index case). The apparent discordance between results of steroid analysis and those of HLA typing is discussed in the RESULTS section.

disease status of a fetus. It is, therefore, logical to predict that the late-onset form is unlikely to be detectable *in utero* by steroid analysis. On the other hand, as most parent's first concern is the risk for a female fetus for being virilized *in utero,* demand for prenatal diagnosis would probably be infrequent in late-onset pregnancies. Pilot studies are still needed for a better understanding of this form of 21-OH deficiency.

Prenatal Diagnosis and Prenatal Treatment of CAH

David and Forest[29] were first in experiencing prenatal treatment of fetuses at risk for CAH by giving the mother oral dexamethasone (0.5 mg twice daily) from early pregnancy. The subject of a separate report,[29] prenatal treatment of CAH is only mentioned here because it is the unique situation in which discordant results between steroid analysis and HLA typing are to be expected (TABLE 4). Indeed, treatment is intended to prevent the overproduction of adrenal androgens in CAH affected fetuses. Efficient adrenal suppression must be reflected by normal amniotic fluid steroid levels. Such an example is illustrated in FIGURE 8. The female fetus was predicted to be affected on the results of the HLA typing, and the diagnosis was confirmed at birth. In this patient, prenatal treatment was successful in preventing virilization of the external genitalia. In this situation, prenatal diagnosis relies on HLA typing only, and its limitations and pitfalls are those of the techniques involved. Steroid analysis in the amniotic fluid collected in midgestation serves as a control of the efficacy of the treatment.

CONCLUDING REMARKS

According to our experience, there is practically no technical limitation in the methods used in prenatal diagnosis of 21-OH deficiency by steroid analysis. It appears rather easy to overcome interpreting difficulties by using a prior and rigorous delineation of the normal limits. Human mistakes are still to be avoided, and these are

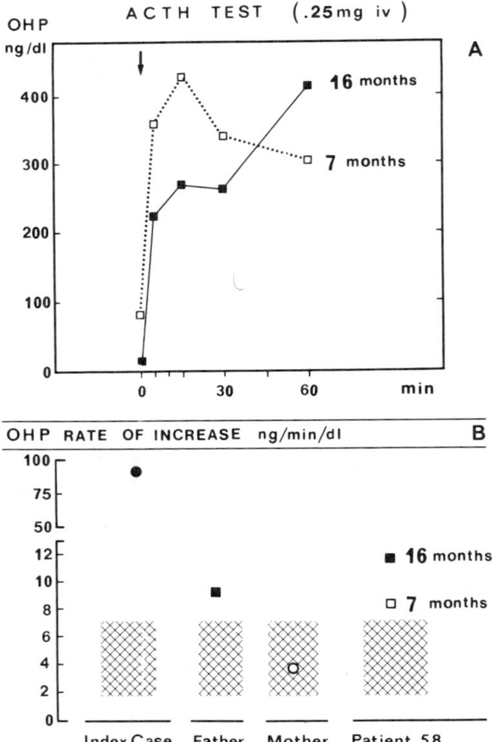

FIGURE 7. Short ACTH tests performed at 7 and 16 months of age in patient no. 58, whose results of prenatal diagnosis are illustrated in FIGURE 6. On the top panel, the kinetics of the 17-hydroxyprogesterone (OHP) responses are shown. The lower panel gives the results expressed as a rate of increase of OHP in ng/dl per minute during the same ACTH test performed in the previously affected sister (index case), father, mother, and patient no. 58, the third child of the family. The shaded areas represent the range of the responses observed in 41 normal controls.

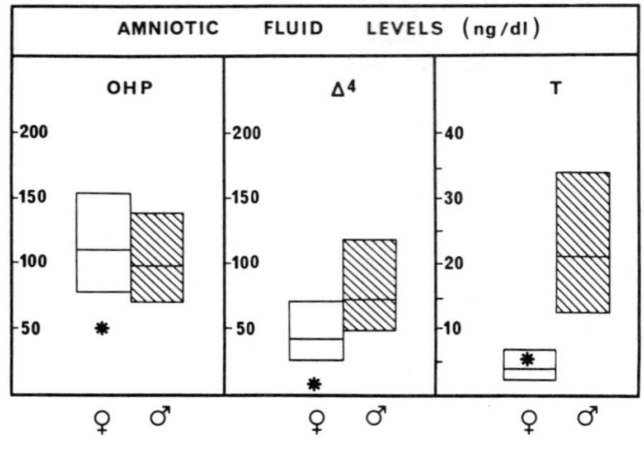

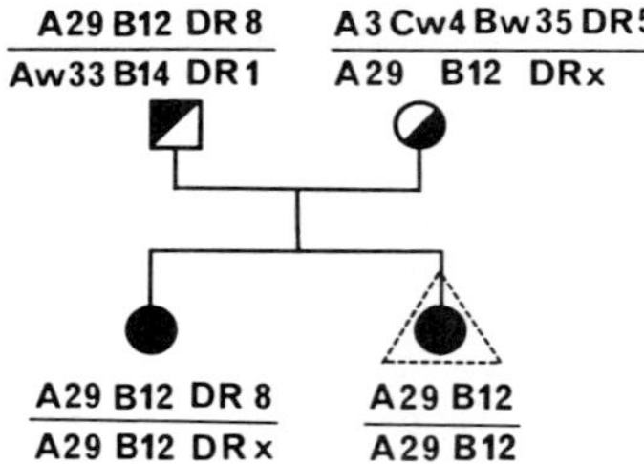

FIGURE 8. Amniotic fluid (AF) concentration of steroid hormones and HLA typing in a female fetus with CAH due to 21-hydroxylase deficiency treated *in utero.* Levels of OHP and Δ^4, lower than normal, accounted for efficient fetal adrenal suppression. Prediction of a CAH affected infant based on HLA typing was confirmed by hormonal studies after birth.

TABLE 5. Final Diagnosis in 100 Pregnancies at Risk for CAH Resulting from 21-Hydroxylase Deficiency

	Male	Female	Total
CAH affected	12	9	21
Heterozygote	13	21 (1[a])	34
Homozygote normal	8	9 (2[a])	17
Heterozygote/homozygote normal ?[b]	12	15	27
Recombination (?)	1	—	1
Total	46	54	100

[a]Classified only on HLA typing after birth.
[b]HLA not made.

minimized by the experienced and conscientious processing of the samples and by care in communicating not only numbers, but full and correct interpretation of the results.

We can conclude that we have not really experienced any pitfalls in the prenatal diagnosis of the congenital form of 21-OH deficiency based on steroid analysis. The question as to the risk for the fetus to have CAH has always been answered properly. The difficulties encountered in only 6 out of 102 prenatal diagnosis are accounted for by three different factors: (1) incorrect HLA typing prediction in the first three, (2) feasibility or not of the prenatal diagnosis of the late-onset form; and, (3) in the last case, it was artificially introduced by treating the fetus *in utero.*

Thus, the data presented here reemphasize[9] that measurement of OHP levels in the amniotic fluid is a simple, rapid, and, above all, highly reliable and accurate method for the prenatal diagnosis of CAH due to 21-OH deficiency. By contrast, its value in the late-onset or attenuated forms of the disease remains to be established. For epidemiologic purpose, compilation of the results obtained are presented in TABLE 5 and TABLE 6. Prenatal diagnosis of CAH due to 21-OH deficiency is no longer at the

TABLE 6. Consecutive Prenatal Diagnosis of 21-Hydroxylase Deficiency in Nine Families

Risk for	Case No.	First Diagnosis	Second Diagnosis
CAH[a]	4,54	homozygote normal male	heterozygote female
	5,22	heterozygote female	heterozygote male
	6,37	heterozygote female	heterozygote female
	20,89	CAH affected male[b]	heterozygote male
	23,60	homozygote normal female	heterozygote male
	25,86	CAH affected male	CAH affected female
	44,78	heterozygote female	CAH affected male
	50,85	unaffected male[b]	unaffected male[b]
Late-Onset	16,58	affected (?) female	heterozygote (?) male

[a]Congenital adrenal hyperplasia.
[b]HLA typing not done, index case deceased.

stage of attempts for research purpose. It is nowadays becoming a reliable diagnostic service. Constant quality control between the numerous partners involved—pediatricians (correct biological diagnosis and follow–up), obstetricians (expertise in amniotic puncture), biologists (high quality laboratory work), and genetic counselors (full and adequate information to the parents)—is, however, required to maintain safety and accuracy in this diagnosis.

ACKNOWLEDGMENTS

We thank the pediatricians, obstetricians, and geneticists who sent us amniotic fluid samples for steroid analysis. The HLA typings mentioned in this study were made in the laboratories of Drs. H. Bétuel (Lyon) and A. Boué (Paris). The excellent technical help by M.P. Monneret and the secretarial assistance by M. Montagnon are greatly appreciated.

REFERENCES

1. FRASIER, S. D., I. H. THORNEYCROFT, B. A. WEISS & R. HORTON. 1975. Elevated amniotic fluid concentration of 17α-hydroxyprogesterone in congenital adrenal hyperplasia. J. Pediatr. **86:** 310–312.
2. MILUNSKY, A. & D. TULCHINSKY. 1977. Prenatal diagnosis of congenital adrenal hyperplasia due to 21-hydroxylase deficiency. Pediatrics **59:** 768–770.
3. NAGAMANI, M., P. G. MCDONOUGH, J. O. ELLEGOOD & U. B. MAHESH. 1978. Maternal and amniotic fluid 17α-hydroxyprogesterone levels during pregnancy: Diagnosis of congenital adrenal hyperplasia *in utero*. Am. J. Obstet. Gynecol. **130:** 791–794.
4. HUGHES, I. A. & K. M. LAURENCE. 1979. Antenatal diagnosis of congenital adrenal hyperplasia. Lancet **ii:** 7–9.
5. MARCUS, E. S., J. H. HOLCOMBE, D. TULCHINSKY, R. R. RICH & V. M. RICCARDI. 1979. Prenatal diagnosis of congenital adrenal hyperplasia. Am. J. Med. Genet. **4:** 201–204.
6. POLLACK, M. S., L. S. LEVINE, S. PANG, R. P. OWENS, H. M. NITOWSKY, D. MAURER, M. I. NEW, M. DUCHON, I. MERKATZ, G. SACHS & B. DUPONT. 1979. Prenatal diagnosis of congenital adrenal hyperplasia (21-hydroxylase deficiency) by HLA typing. Lancet **i:** 1107–1108.
7. WARSOS, S. L., J. W. LARSEN, S. G. KENT, K. N. ROSENBAUM, G. P. AUGUST, C. J. MIGEON & J. P. SCHULMAN. 1980. Prenatal diagnosis of congenital adrenal hyperplasia. Obstet. Gynecol. **55:** 751–753.
8. PANG, S., L. S. LEVINE, L. L. CEDERQVIST, M. FUENTES, V. M. RICCARDI, J. H. HOLCOMBE, H. M. NITOWSKY, G. SACHS, C. E. ANDERSON, M. A. DUCHON, R. OWENS, I. MERKATZ & M. I. NEW. 1980. Amniotic fluid concentrations of Δ^5 and Δ^4 steroids in fetuses with congenital adrenal hyperplasia due to 21-hydroxylase deficiency and in anencephalic fetuses. J. Clin. Endocrinol. Metab. **51:** 223–229.
9. FOREST, M. G., H. BÉTUEL, P. COUILLIN, A. BOUÉ *et al.* 1981. Prenatal diagnosis of congenital adrenal hyperplasia (CAH) due to 21-hydroxylase deficiency by steroid analysis in the amniotic fluid of midpregnancy: Comparison with HLA typing in 17 pregnancies at risk for CAH. Prenatal Diagnosis **1:** 197–207.
10. HUGHES, I. A. & K. M. LAURENCE. 1982. Prenatal diagnosis of congenital adrenal hyperplasia due to 21-hydroxylase deficiency by amniotic fluid steroid analysis. Prenatal Diagnosis **2:** 97–102.
11. SIBERT, J. R. 1979. Antenatal detection of congenital adrenal hyperplasia. Lancet **ii:** 37.
12. STEFFES, M. M. & E. T. WONG. 1979. Prenatal diagnosis of congenital adrenal hyperplasia. Lancet **ii:** 303–304.
13. GREENBERG, F. 1979. Prenatal diagnosis of congenital adrenal hyperplasia. Lancet **ii:** 908.
14. POWLEDGE, T. M. & J. FLETCHER. 1979. Guidelines for the ethical, social, and legal issues in prenatal diagnosis. N. Engl. J. Med. **300:** 168–172.
15. FRANCOIS, R., H. BÉTUEL & M. G. FOREST. 1981. Misleading prediction from fetal HLA typing in the prenatal diagnosis of 21-hydroxylase deficiency adrenal hyperplasia (21-OH-DAH). Pediatr. Res. **15:** 79 (abstract 30).
16. VALENTIN-THON, E., E. KATTNER & E. PASSARGE. 1982. Prenatal diagnostic of 21-hydroxylase deficiency. Prenatal Diagnosis **2:** 237–238.
17. FOREST, M. G., H. BÉTUEL, P. COUILLIN and A. BOUÉ. 1982. Prenatal diagnosis of 21-hydroxylase deficiency. Prenatal Diagnosis **2:** 238–239.
18. COUILLIN, P., J. BOUÉ, H. NICOLAS, C. CHERUY & A. BOUÉ. 1981. Prenatal diagnosis of congenital adrenal hyperplasia (21-OH deficiency type) by HLA typing. Prenatal Diagnosis **1:** 25–33.
19. FOREST, M. G., E. DE PERETTI, A. LECOQ, E. CADILLON, M. T. ZABOT & J. M. THOULON. 1980. Concentration of 14 steroid hormones in human amniotic fluid of midpregnancy. J. Clin. Endocrinol. Metab. **51:** 816–822.
20. BÉTUEL, H., L. GEBUHRER, L. DESCOS, H. PERCEBOIS, Y. MINAIRE & J. BERTRAND. 1980. Adult celiac disease associated with HLA DR W 3 and DR W 7. Tissue Antigens **15:** 231–238.
21. COUILLIN, P., H. NICOLAS, M. C. GRISARD, J. BOUÉ & A. BOUÉ. 1980. Méthodologie pour le typage HLA des cellules foetales du liquide amniotique. Ann. Genet. **23:** 40–45.

22. GOLBUS, M. S., W. D. LOUGHMAN, C. J. EPSTEIN, G. HALBASCH, J. D. STEPHENS & B. D. HALL. 1979. Prenatal genetic diagnosis in 3000 amniocentesis. N. Engl. J. Med. **300:** 157–163.
23. JUDD, H. L., J. D. ROBINSON, P. E. YOUNG & O. W. JONES. 1976. Amniotic fluid testosterone levels in midpregnancy. Obstet. Gynecol. **48:** 690–695.
24. REYES, F. I., R. S. BORODITSKY, J. S. D. WINTER & C. FAIRMAN. 1974. Studies on human sexual development. II. Fetal and maternal serum gonadotropin and sex steroid concentration. J. Clin. Endocrinol. Metab. **38:** 612–617.
25. TAKAGI, S., T. YOSHIDA, K. TSUBATA, H. OZAKI, T. K. FUJII, Y. NOMURA & M. SAWADA. 1977. Sex differences in fetal gonadotropins and androgens. J. Steroid Biochem. **8:** 609–620.
26. GUTAI, J. P., A. A. KOWARSKI & C. J. MIGEON. 1977. The detection of the heterozygous carrier for congenital virilizing adrenal hyperplasia. J. Pediatr. **90:** 924–929.
27. MIGEON, C. J., Z. ROSENWAKS, P. A. LEE, M. D. URBAN & W. B. BIAS. 1980. The attenuated form of congenital adrenal hyperplasia as an allelic form of 21-hydroxylase deficiency. J. Clin. Endocrinol. Metab. **51:** 647–649.
28. LARON, Z., M. S. POLLACK, R. ZAMIR, A. RAITMAN, Z. DICKERMAN, L. S. LEVINE, F. LORENZEN, G. J. O'NEILL, S. PANG, H. I. NEW & B. DUPONT. 1980. Late-onset 21-hydroxylase deficiency and HLA in the Ashkenazi population: A new allele at the 21-hydroxylase locus. Hum. Immunol. **1:** 55–66.
29. DAVID, M. & M. G. FOREST. 1984. Prenatal treatment of congenital adrenal hyperplasia resulting from 21-hydroxylase deficiency. J. Pediatr. **105:** 799–803.

The Use of Gamma Interferon to Increase HLA Antigen Expression on Cultured Amniotic Cells Used for the Prenatal Diagnosis of 21-Hydroxylase Deficiency[a]

DAVID H. MAURER AND MARILYN S. POLLACK

Department of Microbiology and Immunology
Baylor College of Medicine
Houston, Texas 77030

INTRODUCTION

As previously stated, HLA typing can be used as the primary test for the prenatal diagnosis of the HLA-linked monogenetic disease congenital adrenal hyperplasia due to 21-hydroxylase deficiency.[1–5] It can also be used for the prenatal diagnosis of another HLA-linked disease, complement C4 deficiency,[6] and for the prenatal determination of paternity.[7,8] Most importantly, successful test results from prenatal HLA typing of amniotic fluid cells have so far proved accurate in all cases in which postnatal samples have been available. However, in a few instances, technical and theoretical difficulties have made HLA test results for amniotic fluid cells difficult to interpret.[9] These difficulties have resulted because B and C locus antigens are poorly expressed on most amniotic cell cultures[1–5] because the existence of positive associations between HLA-linked disease factors and specific HLA antigens[10,11] increases the chance that both parents in a pregnancy at risk for that disease will have the same HLA antigens on their HLA-linked disease haplotypes, and because one parent has been HLA-A,B homozygous.

In one particular case (Pang *et al.*),[12] weak HLA-B antigen expression resulted in failure to detect an HLA-A/B recombination in a fetus at risk for 21-OH deficiency. The parental 21-OH deficiency haplotypes both had HLA-B18, and the HLA-A/B recombination resulted in conversion to homozygosity for both B18 and the 21-OH deficiency alleles. In another case, the haplotype HLA-A2, Bw44, which has a frequency of 6% in North American Caucasians,[13] had homozygous expression in the mother of one 21-OH deficiency family, which precluded the use of HLA typing for prenatal diagnosis (FIGURE 1).

The increase in quantitative HLA-A, B, and C antigen expression and the induction of DR antigen expression that results from incubation of amniotic cells with gamma interferon, as described in this report, have already had an important effect on the certainty with which fetal cell HLA-A,B,C typing results can be obtained. In the future, this should allow determination of the fetal DR allotype whenever use of an additional polymorphic genetic marker might be necessary.

[a]This study was supported in part by grant nos. R01–AM33990, R01–AM33988, and R01–CA37888 from the National Institutes of Health.

MATERIALS AND METHODS

Cell Cultures

Primary amniotic fluid cell cultures were established in Chang's medium (Hana Media) from cells obtained by amniocentesis at 15–17 weeks of gestation by standard

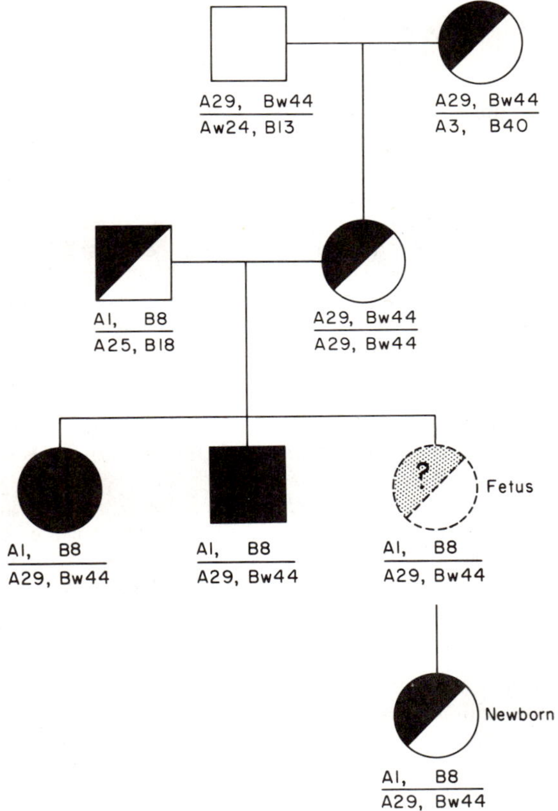

FIGURE 1. A family (Family K) referred for prenatal diagnosis of 21-hydroxylase deficiency by HLA typing for which the polymorphism of HLA-A and HLA-B was insufficient to allow diagnosis. In this case, the fetus carried the paternal deficiency haplotype, but the status of the maternal haplotype was unknown because the mother was homozygous for her HLA-A and HLA-B antigens.

techniques. Fetal fibroblast cultures were established from limb tissue fragments from first trimester abortuses and were cultured in alpha MEM (M.A. Bioproducts) supplemented with 10% heat inactivated fetal calf serum (GIBCO). In some cases, the fetal amniotic cells and fibroblasts studied in this report had been obtained months earlier for prenatal tests, and had been frozen, stored, and thawed by standard techniques prior to use. Single cell suspensions of the cultured cells were obtained by brief exposure of the monolayer cultures to Trypsin-Versene (M.A. Bioproducts).

HLA antigen expression was modified by incubating subconfluent cultures in the presence of 1000 units per ml of partially purified human gamma interferon (Meloy Laboratories) for various times, typically for two or five days. In some experiments, monoclonal antibodies to human gamma interferon (Meloy Laboratories) were added at the beginning of the induction period.

Detection of Cell Surface Antigens

Indirect immunofluorescence staining was performed using a panel of monoclonal antibodies including the DR framework specific monoclonal antibodies NEI–011 (New England Nuclear) and L243 (courtesy Dr. R. Levy), the DQw1, w3(DS) framework specific monoclonal antibody Leu–10 (Becton-Dickinson FACS Systems), a "control" antibody (Leu–5) specific for the sheep RBC receptor (E) on T lymphocytes (Becton-Dickinson), and the antibody 4E1 which detects a common determinant of the HLA-B antigen (courtesy Dr. B. Dupont). For each test, 0.25 million suspended fetal cells were incubated with the appropriate dilution of monoclonal antibody in phosphate buffered saline with 2% bovine serum albumin (to minimize nonspecific binding) and 0.1% sodium azide at 4°C for 20 minutes. After washing twice, fluorescein-conjugated goat anti-mouse $F(ab'_2)$ immunoglobulin (Cappel) was added for 20 minutes at 4°C. The cells were then washed twice and fixed in suspension in 1% paraformaldehyde in phosphate buffered saline. Stained cells were analyzed by flow cytometry (EPICS V, Coulter Electronics). The percent of cells bearing each marker was determined using immunoanalysis programs purchased from Coulter Electronics in comparison with cells stained only with the "control" antibody (Leu–5) and fluorescein-conjugated (second) antibody, or with the second antibody alone. Modal fluorescence intensities were determined by converting logarithmic channel number to signal voltage by the method of Muirhead *et al.*[14]

For the detection of HLA allospecificities on fibroblasts and amniotic cells, micromonolayers of cell plated in microtiter trays and preincubated with interferon were sequentially incubated with preselected HLA and DR typing sera and complement. Living cells and complement killed cells were simultaneously visualized by staining with fluorescein diacetate and ethidium bromide, respectively, and by observing the staining patterns with an inverted fluorescence microscope, as previously described.[1,2]

RESULTS

Tests of ten different amniotic cell cultures and three different fetal fibroblast cultures indicated that preincubation with gamma interferon increased the cell surface expression of HLA-A, B, and C locus antigens in all cases and induced the appearance of HLA-DR antigens detectable by cytofluorometric analysis. The role of interferon in these phenomena was confirmed by their abrogation by a monoclonal antibody to interferon. A second set of class II molecules, the DQ(DS) molecules, were not detected.

Typical flow cytofluorometric analyses illustrating these results are shown in FIGURE 2. Increases in modal fluorescence intensity for HLA-B expression (antibody

4E1) ranged from 1.8 to 3.5-fold. Since all cells were already expressing HLA-B molecules, there was little or no change in the percent positive cells. For DR molecules, the percent positive cells increased from 0 to 28–58%, depending on the monoclonal antibody used (TABLE 1). It appears likely from inspection of the cytofluorograph profiles that all the cells became DR positive, with the difference from 100% reflecting the large overlap of the weakly staining cells with the control (nonstained) curve.

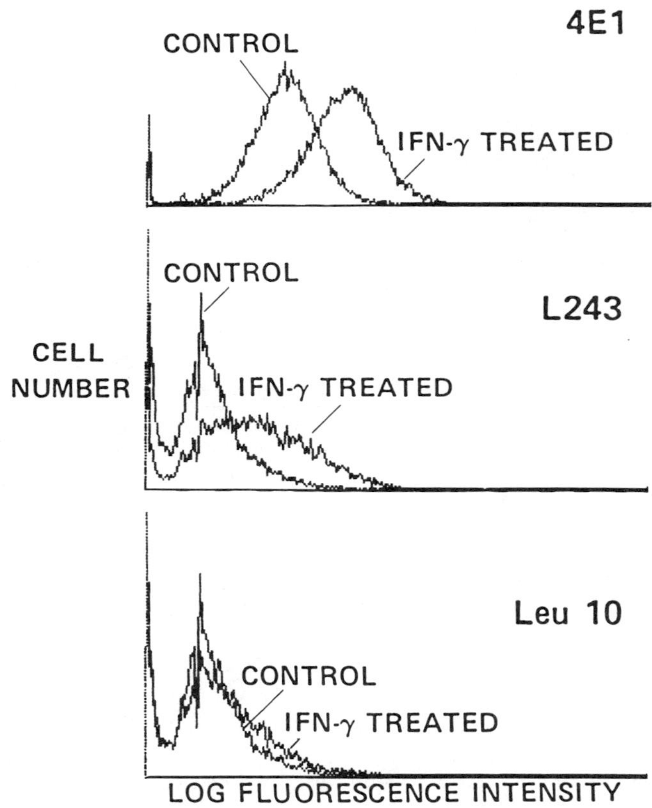

FIGURE 2. Cytofluorometric analysis results for untreated and gamma interferon (IfN-γ) fetal cells after staining with A: a monoclonal antibody specific for HLA-B (4E1); with B: a monoclonal antibody specific for HLA-DR (L-243); and with C: a monoclonal antibody specific for DQ (MB, DC) (Leu-10), as described in MATERIALS AND METHODS. Prior to gamma interferon treatment, cells stained for either DR or DQ were indistinguishable from nonstained control cells (not shown), and cells stained only for DQ remained indistinguishable from control cells even after five days of incubation with gamma interferon.

Expression of DQ(DS) molecules was not detected for any of the specimens tested (FIGURE 2, TABLE 1), although the antibody used was clearly able to detect these molecules on control B-cells (data not shown).

The increase in HLA-B antigen expression demonstrated by cytofluorographic techniques was accompanied by a dramatic improvement in the ease with which all

HLA class I antigens could be detected by the microcytotoxicity test described earlier. Some examples are shown in TABLE 2. In example A, the increased expression of HLA-B antigens allowed definite assignment of the unaffected HLA-B14 maternal haplotype to the fetal cells, and, in example B, the definite detection of the HLA-C locus antigen Cw2 without detection of HLA-Cw1 allowed assignment of the unaffected paternal HLA haplotype to the fetal cells. All cells tested showed stronger expression of appropriate HLA-A, B, and C alloantigens, and no significant nonspecific reactions were noted with the negative control alloantisera.

DISCUSSION

Gamma interferon has previously been noted by other groups to increase HLA-A, B, and C antigen expression on a variety of cell types and to induce DR antigen expression on at least some types of cells that usually fail to express DR.[15–18] The current report also indicates that incubation with gamma interferon can and should be used as a routine procedure to aid in the assignment of specific HLA alloantigens to cultured amniotic cells used for the prenatal determination of 21-hydroxylase deficiency status. Since all HLA-A, B, and C locus antigens can be detected with this method, problems relating to potentially HLA homozygous fetal cells or A/B

TABLE 1. Induction of DR Expression by Interferon[a]

Monoclonal Antibody Identification	Specificity	Percent Positive Cells (Above Background) Without Interferon	With Interferon
4E1	HLA-B	89%	89%
L243	HLA-DR	0%	58%
NEI-011	HLA-DR	0%	29%
LEU-10	HLA-DQ (DS,MB)	0%	0%
LEU-5	E-Receptor (control)	0%	0%

[a]MJ fetal fibroblasts were treated with 1000 units per ml of gamma interferon for five days prior to staining with monoclonal antibodies and were analyzed with the Epics V cytofluorograph as described in the section entitled MATERIALS AND METHODS.

recombinations can be avoided. In particular, for example, if both parents are heterozygous, but both have the antigens, A2 and Bw44, detection of only A2 and Bw44 on gamma interferon treated fetal cells could now be interpreted confidently to indicate that the fetus is homozygous for those determinants.

In relation to DR antigen induction, additional studies in progress suggest that serological detection of DR allotypes can be accomplished. This would be useful in the few cases (e.g., FIGURE 1) for which the polymorphism of HLA-A, B, and C locus antigens is not sufficient to act as a marker for the 21-OH deficiency status. Although serologically detected DQ(DS, MB) antigens offer far less polymorphism (only DWQw1, DQw2, and DQw3 (MB1, 2, 3) determinants can be serologically detected, and these are in strong linkage disequilibrium with specific DR antigens), it is intriguing that the DQ Class II molecules do not appear to be induced on amniotic cells (or fibroblasts) in parallel with the induction of DR Class I antigens. Additional studies of this and other phenomena related to the differential regulation of expression of different Class II molecules on different cell types are described elsewhere (Maurer *et al.*[19]).

In addition to their role in improving the prenatal diagnosis of 21-OH deficiency, it

TABLE 2. Examples of Interferon-Increased Expression of HLA Antigens on Cultured Amniotic Cells

A. *Family A with 21-OH Deficiency*

	HLA Genotypes
Patient	A29, B7 (w6), Cw7 / A29, Bw39(w6), Cw7
Father	A29, B7(s6), Cw7 / A3, (B7(w6), Cw7)
Mother	A29, Bw39(w6), Cw7 / A2, B14(w6), Cw8

	Serum	Specificity	Reaction of Untreated Cells	Reaction of Interferon-Treated Cells
Fetal Cells[a]	Lee	A2	+++	+++
	Schwarz	A2	+++	+++
	King	A3	+++	+++
	CT529	A3	+++	+++
	BRA173	B14	−	+++
	NJ2407	B14	+	+++
	CC336.1	Cw7	−	+++
	Clos	B7	−	−

B. *Family B with 21-OH Deficiency*

	HLA Genotypes
Patient	A−, Bw56(w6), Cw1 / A2, Bw51(w4), Cw2
Father	A−, Bw56(w6), Cw1 / A28, Bw51(w4), Cw2
Mother	A2, Bw51(w4), Cw2 / A28, Bw44(w4), Cw7

	Serum	Specificity	Reaction of Untreated Cells	Reaction of Interferon-Treated Cells
Fetal Cells[b]	Welch	A28	+	++
	Mor	A28	+	++
	Haselow	Bw51(+w52)	++	+++
	Bechard	Bw51	+++	+++
	Hinz	Cw2	−	+++
	Beyer	Cw2	−	+++
	Werner	Bw56(+w55)	−	−
	Haas	Cw1	−	−
	CC336.1	Cw7	−	−

[a]The fetal cells have the unaffected maternal haplotype.
[b]The fetal cells carry the unaffected paternal haplotype.

is important to note that the gamma interferon treated fetal cells have improved HLA typing results for cells tested in relation to the prenatal determination of paternity or the prenatal determination of bone marrow transplantation options for leukemia patient siblings or potentially immunodeficient fetuses. These additional indications for prenatal HLA typing, together with the possibility that HLA typing may someday be used routinely to document the fetal origin of cells used for other genetic tests, demonstrate how useful this modification in the amniotic cell HLA typing procedure may prove to be.

ACKNOWLEDGMENTS

The authors wish to thank Drs. Maria New and Lenore S. Levine for referring many of the families whose members' cells were studied in the present report, Drs. R. Levy and B. Dupont for providing two of the monoclonal antibodies, Dr. R. Rich for the use of the flow cytometer, and Brenda Connelly for typing and editing the manuscript.

REFERENCES

1. Pollack, M. S., D. Maurer, L. S. Levine, M. I. New, S. Pang, M. Duchon, R. P. Owens, I. R. Merkatz, H. M. Nitowsky, G. Sachs & B. Dupont. 1979. Prenatal diagnosis of congenital adrenal hyperplasia (21-hydroxylase deficiency) by HLA typing. Lancet **i:** 1107–1108.
2. Pollack, M. S., D. Maurer, L. S. Levine, M. I. New, S. Pang, M. Duchon, R. P. Owens, I. R. Merkatz, H. M. Nitowsky, G. Sachs & B. Dupont. 1979. HLA typing of amniotic cells: The prenatal diagnosis of congenital adrenal hyperplasia (21-OH deficiency type). Transplant. Proc. **11:** 1726–1728.
3. Couillin, P., H. Nicolas, J. Boué & A. Boué. 1979. HLA typing of amniotic fluid cells applied to prenatal diagnosis of congenital adrenal hyperplasia. Lancet **i:** 1076.
4. Coullin, P., H. Nicolas, M. C. Grisard, J. Boué & A. Boué. 1980. Methodologie pour le typage HLA des cellules foetales du liquid amniotique. Ann. Genet. In press.
5. Couillin, P., J. Boué, H. Nicolas, C. Cheruy & A. Boué. 1980. Antenatal diagnosis of congenital adrenal hyperplasia (21-OH deficiency type) by HLA typing. Prenatal Diagnosis **1:** 25–33.
6. Pollack, M. S., H. D. Ochs & B. Dupont. HLA typing of cultured amniotic cells for the prenatal diagnosis of a complement C4 deficiency. Clin. Genet. **18:** 197–200.
7. Pollack, M. S., I. A. Schafer, D. Barford & B. Dupont. 1980. HLA typing for the prenatal identification of paternity. J. Am. Med. Assoc. **244:** 1954–1956.
8. Boué, J., P. Couillin & F. Yvert. 1980. Utilisation des marqueurs chromosomiques et des antigenes HLA pour le controle prenatal du procreateur dans une insemination artificielle pour indication genetique. Ann. Genet. **23:** 79–82.
9. Pollack, M. S., S. D. Heagney, D. Braun, Jr. & G. J. O'Neill. 1981. Technical and theoretical considerations in the HLA typing of amniotic fluid cells for prenatal diagnosis and paternity testing. Prenatal Diagnosis **1:** 183–195.
10. Pollack, M. S., L. S. Levine, M. Zachmann, A. Prader, M. I. New, S. Oberfield & B. Dupont. 1979. Possible genetic linkage disequilibrium between HLA and the 21-hydroxylase deficiency gene (congenital adrenal hyperplasia). Transplant. Proc. **11:** 1315–1316.
11. Dupont, B., R. Virdis, A. J. Lerner, C. Nelson, M. S. Pollack & M. I. New. 1984. Distinct HLA-B antigen associations for the salt-wasting and simple virilizing forms of congenital adrenal hyperplasia due to 21-hydroxylase deficiency. *In* Histocompatibility Testing 1984. E. D. Albert, M. P. Baur & W. R. Mayr, Eds.: 660. Springer-Verlag. Heidelberg.

12. PANG, S., M. S. POLLACK, M. LOO, O. GREEN, R. NUSSBAUM, G. CLAYTON, B. DUPONT & M. I. NEW. 1985. Pitfalls of prenatal diagnosis of 21-hydroxylase deficiency congenital adrenal hyperplasia. J. Clin. Endocrinol. Metab. **61:** 89–97.
13. BAUR, M. P. & J. A. DANILOVS. 1980. Reference tables of two and three-locus haplotype frequencies for HLA-A, B, C, DR, Bf, and GLO. *In* Histocompatibility Testing 1980. P. I. Terasaki, Ed.: 994–1210. UCLA Tissue Typing Laboratory. Los Angeles.
14. MUIRHEAD, K. A., T. C. SCHMITT & A. R. MUIRHEAD. 1983. Determination of linear fluorescence intensities from fluorocytometric data accumulated with logarithmic amplifiers. Cytometry **3:** 251–256.
15. KILLANDER, D., P. LINDAHL, L. LUNDIN, P. LEARY & I. GRESSER. 1976. Relationship between the enhanced expression of histocompatibility antigens on interferon treated L-1210 cells and their position in the cell cycle. Eur. J. Immunol. **6:** 56–59.
16. LINDAHL, P., I. GRESSER, P. LEARY & M. TOVEY. 1976. Interferon treatment of mice: Enhanced expression of histocompatibility antigens on lymphoid cells. Proc. Natl. Acad. Sci. USA **73:** 1284–1287.
17. HERON, I., M. HOKLAND & K. BERG. 1978. Enhanced expression of B - microglobulin and HLA antigens on human lymphoid cells by interferon. Proc. Natl. Acad. Sci. USA **75:** 6215–6219.
18. POBER, J. S., M. A. GIMBRONE, JR., R. A. COTRAN, C. S. REISS, S. J. BURAKOFF, W. FIERS & K. A. AULTS. 1983. Ia Expression by vascular endothelium is inducible by activated T cells and by human interferon. J. Exp. Med. **157:** 1339–1353.
19. MAURER, D., W. E. COLLINS, J. H. HANKE, M. VAN, R. R. RICH & M. S. POLLACK. 1985. Class II positive human dermal fibroblasts restimulate cloned allospecific T-cells, but fail to stimulate primary allogenic lymphoproliferation. Hum. Immunol. In press.

Prenatal Therapy in Congenital Adrenal Hyperplasia

Attempted Prevention of Abnormal External Genital Masculinization by Pharmacologic Suppression of the Fetal Adrenal Gland *in Utero*

GEORGE P. CHROUSOS,[a] MARK I. EVANS,[b] D. LYNN LORIAUX,[a] JAMES McCLUSKEY,[c] JOHN C. FLETCHER,[d] AND JOSEPH D. SCHULMAN[e]

[a]*Developmental Endocrinology Branch*
National Institute of Child Health and Human Development, NIH
Bethesda, Maryland 20205

[b]*Division of Reproductive Genetics*
Department of Obstetrics & Gynecology
Wayne State University, Hutzel Hospital
Detroit, Michigan 48201

[c]*Laboratory of Immunology*
National Institute of Allergy and Infectious Diseases, NIH
Bethesda, Maryland 20205

[d]*Assistant for Bioethics*
Warren G. Magnuson Clinical Center, NIH
Bethesda, Maryland 20205

[e]*Fairfax-Northern Virginia Genetics*
and
In Vitro Fertilization Institute
Fairfax, Virginia 22031

INTRODUCTION

In severe congenital adrenal hyperplasia (CAH), the adrenal androgens secreted by the fetal adrenal gland are thought to masculinize the external genitalia of female fetuses between 10 and 16 weeks gestation.[1–7] This is the period when the testes normally masculinize the bipotential genital tubercle and adjacent regions in male fetuses.[8,9] Thus, suppression of the fetal pituitary-adrenal axis with a glucocorticoid compound during gestational weeks 10–16 should lead to prevention of ambiguous external genitalia in female fetuses with the severe form of CAH.

However, prenatal diagnosis of CAH due to 21-hydroxylase deficiency (which is based on detection of elevated amniotic fluid 17-hydroxyprogesterone and adrenal androgen concentrations, or is performed by linkage analysis and HLA typing of cultured amniotic fluid cells)[10–13] cannot, at present, be completed before 16–17 weeks of gestation, which is presumably too late for intervention to alter genital masculinization. Thus, therapy would have to be instituted before the sex or the genetic status of the fetus could be known.

Theoretically, with two parents heterozygous for classic CAH, only one of four pregnancies would have a fully affected fetus, but only half of these (the females) would be liable to abnormal genital masculinization. Therefore, only one in eight pregnancies would benefit from hormonal suppressive therapy. We have reported the results of an attempt to treat a pregnant woman carrying a fetus at high risk for classic 21-hydroxylase deficiency in detail, previously.[14]

CASE REPORT

Case and Family History

A 32 year-old pregnant woman with late-onset 21-hydroxylase deficiency[15–19] requested prenatal counseling. Her disease became manifest at puberty with acne, hirsutism, and menstrual irregularities. She had never received glucocorticoid therapy for her condition. Her previous five year-old (46, XX) child was born with severely masculinized genitalia due to classic 21-hydroxylase deficiency associated with salt loss. This child has been followed at the National Institutes of Health (NIH) since infancy, and has required genital reconstructive surgery. At the time the patient presented for prenatal evaluation, she was taking no medications. Her pregnancy was approximately nine weeks gestational age as per ultrasound examination.

The family pedigree and the results of the biochemical and final HLA typing studies on the family members (including the treated fetus) are shown in FIGURE 1. The mother was known to be a compound heterozygote, and, thus, the parents were told that the risk of having another affected female child was between one in four and one in eight. They were also told that prenatal diagnosis was possible by amniocentesis, but not before the second trimester. Both parents said they would elect an abortion for a masculinized female fetus. As an alternative to abortion, the parents were informed about the untested concept of fetal therapy with glucocorticoid suppression of the fetal adrenal gland. They consented to the attempt of this experiment therapy after approval of the protocol was obtained through appropriate bioethical and administrative channels.

Protocol and Results

Dexamethasone 0.25 mg qid (15 μg/kg/d) po, was started on the tenth week of gestation and continued until a normal spontaneous vaginal delivery occurred at 39 weeks. Plasma samples and 24 hour urine collections for measurement of various biochemical parameters of fetal adrenal and maternal adrenal status were obtained before and weekly or biweekly after initiation of therapy. An amniocentesis was performed at 17 weeks of gestation (see below). Plasma, urine, and amniotic fluid hormone concentrations were determined by radioimmunoassay.[14]

To assess the degree of maternal adrenal gland suppression, plasma 17-hydroxyprogesterone and plasma and 24 hour urinary free cortisol were measured throughout pregnancy. These hormone levels fell immediately after treatment and remained low throughout pregnancy, indicating a rapid and sustained suppression of the maternal pituitary-adrenal axis (FIGURE 2a).

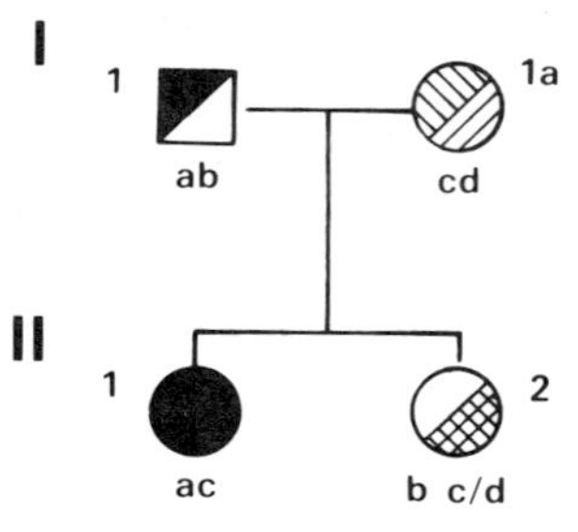

HLA and Complement Phenotypes

I_1 HLA AW30 A28; Bw53; Cw3 Cw4; DR5/8
Complement C4A3,3; C4B 1,2; Bf F,S

I_{1a} HLA A1 A3; B8 B14; Cw-; DR3 DR4
Complement C4A2, Q0; C4B 1,2; Bf S,S

II_1 HLA A1 AW30; B8; Cw3; DR3 DR5/8
Complement C4A3, Q0; C4B 1,2; Bf S,S

II_2 HLA A1 A28; B8 Bw53; Cw4; DR4 DR5/8
Complement C4A3, Q0; C4B1, 1; Bf F,S

a	AW30	Cw3	B-	C4A3	C4B2	Bf S	DR 5/8
b	A28	Cw4	Bw53	C4A3	C4B1	Bf F	DR 5/8
c	A1	Cw-	B8	C4AQ0	C4B1	Bf S	DR3
d	A3	Cw-	B14	C4A2	C4B2	Bf S	DR4

FIGURE 1. Family pedigree. The propositus with classic congenital adrenal hyperplasia due to 21-hydroxylase deficiency who presented with ambiguous genitalia and salt losing is indicated by II_1. Her mother with late-onset mild 21-hydroxylase deficiency (I_{1a}) received dexamethasone therapy during weeks 10–39 of her pregnancy with sister II_2. Infant II_2 appears to be a carrier for 21-hydroxylase deficiency. Moreover, the infant's (II_2) affected haplotype represents a maternal HLA B-DR recombination in which the complotype C4AQ0, C4B1, BfS segregates with HLA B 8 (from reference 14).

To assess fetal adrenal response to therapy, maternal plasma estriol concentrations and 24 hour urinary estriol excretion were measured throughout pregnancy. Estriol is the placental aromatization product of 16-hydroxydehydroepiandrosterone, a metabolite of a fetal adrenal product. Urinary estriol levels immediately declined with therapy and remained lower than normal throughout pregnancy, suggesting suppression of the fetal adrenal gland (FIGURE 2b).

Human placental lactogen (HPL) was measured in maternal plasma throughout pregnancy as a biochemical parameter of syncytiotrophoblastic function (FIGURE 2b). HPL concentrations remained normal throughout pregnancy.

Plasma concentrations of dexamethasone and daily urinary dexamethasone excretion was followed throughout pregnancy (FIGURE 2a). The majority of dexamethasone excreted in urine is in the free form.[20]

Amniocentesis Results

Amniocentesis was performed at 17 weeks gestation for karyotyping, HLA typing, and measurement of hormonal concentrations. Karyotype of cultured amniotic fluid cells was 46, XX. Analysis of amniotic fluid revealed substantial concentrations of dexamethasone. Suppression of the fetal adrenal glands was suggested since concentrations of several steroids, including 17-hydroxyprogesterone, and adrenal androgens were all below or in the low normal range (TABLE 1). In comparison, biochemical fetal adrenal suppression was not shown in a pregnant patient taking prednisone chronically (10 mg BID) for maternal indications.

Obstetric and Pediatric Management

The patient underwent several ultrasound examinations confirming continued fetal growth, appropriate biparietal diameters, femur lengths, total intrauterine volumes, and lack of any discernible phallus. Beginning early in the third trimester, the patient underwent weekly nonstress tests, all of which were normal (reactive). There was continued appropriate fundal growth. An oral glucose tolerance test performed at 24 weeks was normal.

At 39 weeks gestation, the patient went into spontaneous labor. Oxytocin enhance-

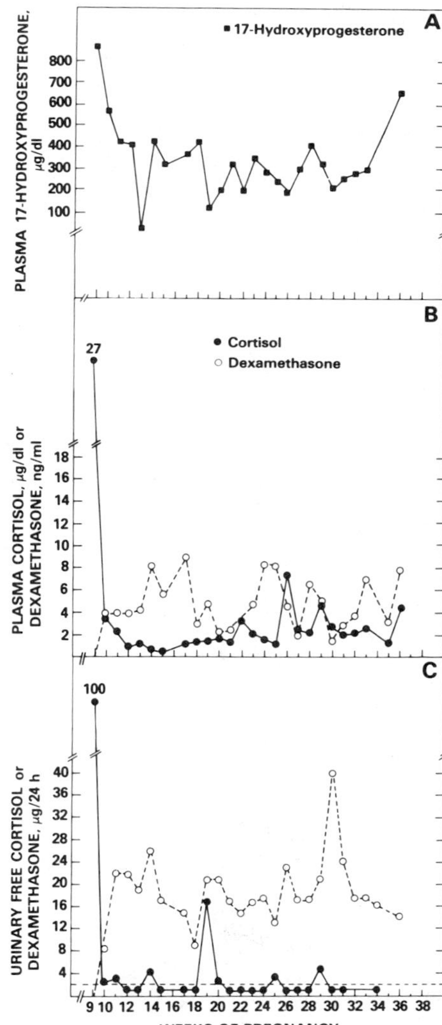

FIGURE 2a. Hormonal profile of I_2 (symbols from FIGURE 1) during pregnancy with II_2. Therapy with dexamethasone was started at the middle of the ninth week of pregnancy. Spontaneous vaginal delivery occurred at 39 weeks (from reference 14). Effects of treatment on maternal hormones: (A) Plasma concentrations of 17-hydroxyprogesterone; (B) Plasma dexamethasone concentrations during therapy; (C) Twenty-four hour urinary free cortisol levels are lowered during therapy. Daily free dexamethasone excretion in the urine is shown.[14]

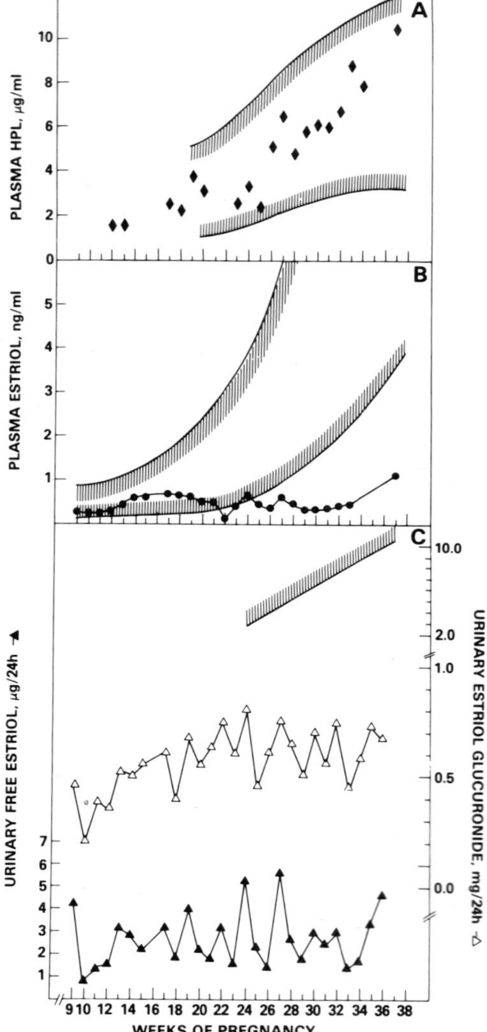

FIGURE 2b. Effects of treatment on fetal-placental parameters with the same hormonal profile and conditions as in FIGURE 2a (Normal ranges included in shaded margins): (A) Plasma concentrations of human placental lactogen (HPL) remained within the normal range indicating sufficient syncytiotrophoblastic function; (B) Plasma estriol concentrations fell below the normal range only after the 21st week of pregnancy; (C) Twenty-four hour urinary estriol excretion (conjugated and unconjugated) fell immediately after initiation of therapy and remained below the normal range throughout pregnancy, indicating immediate and persistent fetal adrenal suppression during therapy. Shaded lines represent the lower limit of normal values of urinary estriol glucuronide.[14]

ment of labor was necessary prior to normal spontaneous vaginal delivery. She was covered with stress doses of glucocorticoids (hydrocortisone 100 mg i.v. q8h) during labor and two days postpartum. Then she was placed on replacement doses of dexamethasone (0.5 mg qHS daily).

The newborn weighed 2430 gm (5th pc) and had a normal physical examination including normal external genitalia. Body length was 47.5 cm (5th pc) and head circumference was 33 cm (25th pc). Placental weight was 480 gm. Apgar scores were 6 and 9 at one and five minutes, respectively, Dubowitz evaluation was compatible with a normal 38–39 weeks gestation, and neurological examination was normal. The infant was started on replacement glucocorticoid (dexamethasone 15 μg/kg/day) for the first four days of life, and hydrocortisone (12.5 mg/m^2/d) and mineralocorticoid therapy (Fluorinef® 50 μg/day) thereafter.

Intravenous ACTH stimulation tests (ACTH 1–24 10 μg/kg/day) at days 2–4, during the seventh week of life, and after complete discontinuation of glucocorticoid therapy at five months, was performed. The results were compatible with the single heterozygote state for either classic or mild 21-hydroxylase deficiency.[4,17]

The early tests also showed subnormal cortisol responses compatible with secondary adrenal insufficiency. Beginning at three months of age, Fluorinef® was discontinued and the patient was slowly weaned from 12.5 mg/m^2/d of hydrocortisone to 3mg/m^2/d by five months of age. At that age, the child's body weight was 6.4 kg (45th pc) and length was 62.4 cm (40th pc). Bone age was appropriate for chronological age. An ACTH stimulation test was performed with a normal cortisol and a slightly elevated 17-hydroxyprogesterone response. At 16 months of age, she has continued good health and normal adrenal function, and there have been no serious medical problems.

DISCUSSION

Maternal therapy with "replacement" doses of dexamethasone in our patient led to suppression of the fetal pituitary-adrenal axis without causing Cushing's syndrome in the mother nor substantial growth retardation in the fetus. Since the first child weighed 2900 gm (35th pc) at birth, it is not clear whether or not the 2430 gm weight of the newborn in the treated pregnancy represents a mild growth retardation secondary to dexamethasone.

Previous experience with glucocorticoid therapy during human pregnancy for maternal disease (systemic lupus erythematosus, rheumatoid arthritis, Crohn's enteritis, etc.) is based mainly on the use of nonfluorinated steroids.[21–25] These compounds, although given at pharmacologic amounts, have very little effect on the fetus; the birth weight of infants born to mothers treated with these steroids is either not affected or

TABLE 1. Amniotic Fluid Steroid Hormone Concentrations in the Second Trimester (from reference 14)

Hormone	This Patient (Dexamethasone Rx)	Different Patient[a] (Prednisone Rx)	Untreated Control Range	(*N*)
Cortisol, μg/dl	0.7	2.5	1.5–6.4	(111)
17-Hydroxyprogesterone, ng/dl	21	129	48–299	(111)
Progesterone, ng/ml	23	31	14–106	(111)
17-Hydroxypregnenolone, ng/dl	125	136	85–164	(9)
Pregnenolone, ng/dl	150	168	53–251	(9)
E_3, pg/ml	396	2580	982–3640	(111)
E_2, pg/ml	15.9	—	12–122	(109)
E_1, pg/ml	92.0	258	111–895	(111)
Dehydroepiandrosterone, ng/dl	28.1	66	39–104	(9)
Δ^4-Androstenedione, ng/dl	60.0	37	41–152	(7)
Δ^5-Androstenediol, ng/dl	64.2	50.0	38–66	(9)
Dexamethasone, ng/ml	1.12[b]	<0.5[c]	<0.5[c]	(7)

[a]Pregnant woman with Crohn's disease receiving prednisone 10 mg bid, 2 hr after last dose.
[b]Maternal plasma concentration of dexamethasone = 5.8 ng/ml.
[c]Assay detection limit.

mildly decreased,[21,23] and this may be a feature of the primary maternal disease rather than the treatment. The cortisol production rate in most of these newborns is normal.[24] The lack of fetal effects with these steroids may be due mainly to their limited transplacental passage, which may be due to their metabolism by placental enzymes.[26] This is also suggested by our amniotic fluid findings in the patient who was receiving pharmacologic doses of prednisone during pregnancy (TABLE 1).

In contrast to these glucocorticoids, however, dexamethasone, at doses equivalent to maternal replacement (approximately 15μg/kg/d),[27,28] had marked effects on the fetal adrenal gland, which was suggested in the present case by the suppression of 24 hour urinary estriol excretion and amniotic fluid estriol, adrenal androgen, and cortisol concentrations. Furthermore, cortisol responses to ACTH stimulation were consistent with secondary adrenal insufficiency in the newborn. These results indicate that the dexamethasone, even when given at rather low doses, enters the fetal circulation, which is an interpretation also consistent with detection of this steroid in the amniotic fluid. Recent studies in rhesus monkeys have shown that dexamethasone suppresses fetal adrenal function in pregnant animals at pharmacologic doses (40–160 μg/kg/day), and that doses above 160 μg/kg/day cause fetal death.[29,30] Transplacental passage of dexamethasone has also been documented previously in rats[31] and in women receiving high doses of dexamethasone in the last trimester of pregnancy for prevention of the respiratory distress syndrome.[25]

Glucocorticoids have been reported to cause malformations, such as cleft palate, and have other effects such as growth retardation, placental degeneration, and fetal demise in some experimental animals.[29,30,32–35] The doses used, however, were very high, and there are marked genetic variations in the susceptibility of various animal strains to cleft palate formation.[36] Congenital malformations associated with glucocorticoid therapy in pregnancy are rare in man, even when pharmacologic doses of steroids are given,[21–25] and may be more frequent than in untreated pregnancies.[21–25] A very few cases of cleft palate were described in early reports, but have not been documented in later studies. In addition, a few cases of pregnancy associated with endogenous Cushing's syndrome have been reported. Spontaneous abortions, stillbirths, and adrenal insufficiency of the newborns were associated with this maternal syndrome, but specific malformations were not described.[21–25]

We consider it likely that fetal adrenal suppression could be achieved in an actual classic CAH fetus. While it appears probable on physiological and embryological grounds, whether such adrenal suppression will prevent external genital masculinization in classic CAH awaits definitive testing. After this investigation was completed, preliminary results from Lyon, France were presented on prevention of external genital masculinization in a female fetus with CAH by maternal therapy with dexamethasone.[37]

Our findings suggest that urinary estriol excretion and amniotic fluid estriol, cortisol, and adrenal androgen concentrations in the second trimester of pregnancy are useful biochemical indicators of fetal adrenal suppression. In addition, plasma concentrations of dexamethasone or daily urinary dexamethasone excretion can be followed for monitoring drug levels during therapy.

Some parents with previously affected children have considered and requested abortion for fetuses with CAH diagnosed at midgestation. However, the only one in eight probability of having an affected female fetus that would benefit from therapy poses ethical questions at the present time. Earlier prenatal diagnosis may be possible in the future, and this will allow treating only fetuses that would benefit without imposing any possible risks on the others. Certainly, while fetal therapy of CAH with dexamethasone does not correct the enzymatic defect and does not eliminate the necessity for lifelong therapy with cortisol, prevention of the masculinization would significantly reduce the difficulties of the social and sexual adjustment of affected children.

REFERENCES

1. Childs, B., M. M. Grumbach & J. J. van Wyk. 1956. Virilizing adrenal hyperplasia: A genetic and hormonal study. J. Clin. Invest. **35:** 213–222.
2. Prader, A. 1958. Die Haufigkeit des kongenitalen adrenogenitalen syndroms. Helv. Paediatr. Acta **13:** 426–431.
3. Pang, S., W. Murphy, L. S. Levine, D. A. Spence, A. Leon, S. LaFranchi, A. S. Surve & M. I. New. 1982. A pilot newborn screening for congenital adrenal hyperplasia in Alaska. J. Clin. Endocrinol. Metab. **55:** 413–420.
4. New, M. I., B. Dupont, S. Pang, M. Pollack & L. Levine. 1981. An update of congenital adrenal hyperplasia. Recent Prog. Horm. Res. **37:** 105–181.
5. Eberlein, R. & A. M. Bongiovanni. 1955. Congenital adrenal hyperplasia with hypertension: Unusual steroid pattern in blood and urine. J. Clin. Endocrinol. **15:** 1531–1534.
6. Hamilton, W. 1972. Congenital adrenal hyperplasia. Inborn errors of cortisol and aldosterone synthesis. J. Clin. Endocrinol. Metab. **1:** 503–547.
7. Bongiovanni, A. M. 1962. The adrenogenital syndrome with deficiency of 3 β-hydroxysteroid dehydrogenase. J. Clin. Invest. **41:** 2086–2092.
8. Villee, D. B. 1969. Development of endocrine function in the human placenta and fetus. I. N. Engl. J. Med. **281:** 473–484.
9. Villee, D. B. 1969. Development of endocrine function in the human placenta and fetus. II. N. Engl. J. Med. **281:** 533–544.
10. Pang, S., L. S. Levine, L. L. Cederqvist, M. Fuentes, V. M. Riccardi, J. H. Holcombe, H. M. Nitowsky, G. Sachs, C. E. Anderson, M. A. Duchon, R. Owens, I. Merkatz & M. I. New. 1980. Amniotic fluid concentrations of Delta 5 and Delta 4 steroids of fetuses with congenital adrenal hyperplasia due to 21-hydroxylase deficiency and anencephalic fetuses. J. Clin. Endocrinol. Metab. **51:** 223–229.
11. Warsof, S. L., J. W. Larsen, S. G. Kent, K. N. Rosenbaum, G. P. August, C. J. Migeon & J. D. Schulman. 1980. Prenatal diagnosis of congenital adrenal hyperplasia. Obstet. Gynecol. Surv. **55:** 751–754.
12. Forest, M. G., H. Betuel, P. Gouillin, A. Boue, M. David, D. Floret, R. Francois, P. Guibaud, H. Plauchu & R. Rappaport. 1981. Prenatal diagnosis of congenital adrenal hyperplasia (CAH) due to 21-hydroxylase deficiency by steroid analysis in the amniotic fluid of mildpregnancy: Comparison with HLA typing in 17 pregnancies at risk for CAH. Prenatal Diagnosis **1:** 197–207.
13. Pollack, M. S., L. S. Levine, S. Pang, R. P. Owens, H. M. Nitowsky, D. Maurer, M. I. New, M. Duchon, I. R. Merkatz, G. Sachs & B. Dupont. 1979. Prenatal diagnosis of congenital adrenal hyperplasia (21-hydroxylase deficiency) by HLA typing. Lancet **1:** 1107–1108.
14. Evans, M. I., G. P. Chrousos, D. Mann, J. W. Larsen, I. Green, J. McCluskey, D. L. Loriaux, J. L. Fletcher, G. Koons, J. Overpeck & J. D. Schulman. 1985. Pharmacologic suppression of the fetal adrenal gland *in utero:* Attempted prevention of abnormal external genital masculinization in suspected congenital adrenal hyperplasia. J. Am. Med. Assoc. **253:** 1015–1020.
15. Blankstein, J., C. Faiman, F. I. Reyes, M. L. Schroeder & J. S. D. Winter. 1980. Adult-onset familial adrenal 21-hydroxylase deficiency. Am. J. Med. **68:** 441–448.
16. Migeon, C. J., Z. Rozenwaks, P. A. Lee. M. D. Urban & W. B. Bias. 1980. The attentuated form of congenital adrenal hyperplasia as an allelic form of 21-hydroxylase deficiency. J. Clin. Endocrinol. Metab. **51:** 647–649.
17. Chrousos, G. P., D. L. Loriaux, D. L. Mann & G. B. Cutler, Jr. 1982. Late-onset 21-hydroxylase deficiency mimicking idiopathic hirsutism or polycystic ovarian disease: An allelic variant of congenital virilizing adrenal hyperplasia with a milder enzymatic defect. Ann. Intern. Med. **96:** 143–148.
18. Chrousos, G. P., D. L. Loriaux, D. L. Mann & G. B. Cutler, Jr. 1982. Late-onset

21-hydroxylase deficiency in an allelic variant of congenital adrenal hyperplasia characterized by attenuated clinical expression and different HLA haplotype associations. Horm. Res. **16:** 193–200.

19. O'Neill, G. J., B. Dupont, M. S. Pollack, L. S. Levine & M. I. New. 1982. Complement C4 allotypes in congenital adrenal hyperplasia due to 21-hydroxylase deficiency: Further evidence for different allelic variants at the 21-hydroxylase locus. Clin. Immunol. Immunopath. **23:** 312–322.
20. Hague, N., K. Thrasher, E. E. Werk, Jr., H. C. Knowles & L. J. Sholiton. 1972. Studies in dexamethasone metabolism in man: Effect of diphenylhydantoin. J. Clin. Endocrinol. Metab. **34:** 44–48.
21. Bongiovanni, A. M. & A. J. McFadden. 1960. Steroids during pregnancy and possible fetal consequences. Fertil. Steril. **11:** 181–186.
22. Green, O. C. 1965. Steroid metabolism in the fetus and the newborn infant. Pediatr. Clin. North Am. **12:** 615–634.
23. Warrell, D. W. & R. Taylor. 1968. Outcome for the fetus of mothers receiving prednisolone during pregnancy. Lancet **1:** 117–118.
24. Kenny, F. M., C. Preeyasombat, J. S. Spaulding & C. J. Migeon. 1966. Cortisol production rate. IV. Infants born of steroid-treated mothers and of diabetic mothers. Infants with Trisomy syndrome and with anencephaly. Pediatrics **37:** 960–966.
25. Shields, J. R. & R. Resnik. 1979. Fetal lung maturation and the antenatal use of glucocorticoids to prevent the respiratory distress syndrome. Obstet. Gynecol. Surv. **34:** 343–363.
26. Murphy, B. E. P., S. J. Clark, I. R. Donald *et al.* 1974. Conversion of maternal cortisol to cortisone during placental transfer to the human fetus. Am. J. Obstet. Gynecol. **118:** 538–541.
27. Nolten, W. E., M. D. Lindheimer, P. A. Rueckert *et al.* 1980. Diurnal patterns and regulation of cortisol secretion in pregnancy. J. Clin. Endocrinol. Metab. **51:** 466–472.
28. Fauci, A. S., D. C. Dale & J. E. Balow. 1976. Glucocorticosteroid therapy: Mechanisms of action and clinical considerations. Ann. Int. Med. **84:** 304–315.
29. Walsh, S. W., R. L. Norman & M. J. Novy. 1979. *In utero* regulation of rhesus monkey fetal adrenals: Effects of dexamethasone, adrenocorticotropin, thyrotrophin releasing hormone, prolactin, human chorionic gonadotropin, and alpha-melanocyte-stimulating hormone on fetal and maternal plasma steroids. Endocrinology **104:** 1805–1813.
30. Novy, M. J. & S. W. Walsh. 1983. Dexamethasone and estradiol treatment of pregnant rhesus macaques: Effects on gestational length, maternal plasma hormones, and fetal growth. Am. J. Obstet. Gynecol. **145:** 920–930.
31. Funkhouser, J. D., K. J. Peevy, P. D. Mockridge & E. R. Hughes. 1978. Distribution of dexamethasone between mother and fetus after maternal administration. Pediatr. Res. **54:** 1053–1056.
32. Motoyama, E. K., M. M. Orzalesi, Y. Kikkawa, M. Kaibara, B. Wu, C. J. Zigas & C. D. Cooke. 1971. Effect of cortisol on the maturation of fetal rabbit lungs. Pediatrics **48:** 547–555.
33. Wellman, K. F. & B. W. Volk. 1972. Fine structure changes in the rabbit placenta induced by cortisone. Arch. Pathol. **94:** 147–157.
34. Kotas, R. V., L. C. Mims & L. K. Hart. 1974. Reversible inhibition of lung cell number after glucocorticoid injection into fetal rabbits to enhance surfactant appearance. Pediatrics **53:** 358–361.
35. Sanfacon, R., F. Possmayer & P. G. R. Harding. 1977. Dexamethasone treatment of the guinea pig fetus: Its effects on the incorporation of ^{3}H-thymidine into deoxyribonucleic acid. Am. J. Obstet. Gynecol. **127:** 745–752.
36. Katsumata, M., M. K. Baker & A. S. Goldman. 1981. An H–2 linked difference in the binding of dexamethasone to murine hepatic cytosol receptor. A possible role of endogenous modifier(s). Biochem. Biophys. Acta **676:** 245–248.
37. Forest, M. & M. David. Prenatal treatment of congenital adrenal hyperplasia (CAH) due to 21-hydroxylase deficiency. 7th International Congress of Endocrinology. Quebec, Canada (abstract 911).

Modern Medical Therapy of Congenital Adrenal Hyperplasia

A Decade of Experience[a]

JEREMY S. D. WINTER AND ROBERT M. COUCH

Department of Paediatrics
University of Manitoba
and
Section of Endocrinology and Metabolism
Children's Hospital of Winnipeg
Winnipeg, Manitoba
R3E OW5 Canada

INTRODUCTION

The fact that glucocorticoid replacement is the cornerstone of medical treatment for 21-hydroxylase deficiency (CAH) has been recognized since 1950.[1,2] For the first two decades, most regimens emphasized the administration of relatively large doses of glucocorticoid sufficient to suppress urinary excretion of 17-ketosteroids and pregnanetriol. In only a minority of patients, those with clinically persistent salt-wasting and dehydration, was mineralocorticoid replacement used, and even then the modalities available, such as intramuscular injections or subcutaneous pellets of desoxycorticosterone acetate, were cumbersome and relatively ineffectual. While this type of therapy certainly saved many lives, it became apparent that treated CAH patients still showed subnormal rates of growth, abnormalities of the timing and progression of puberty, and reduced adult stature.

Analysis of the pathogenesis of virilization and salt-wasting in CAH (FIGURE 1) demonstrates how these two features are inexorably intertwined and why it is difficult or impossible to control one aspect without adequate therapy of the other. Thus, failure to suppress ACTH-mediated 17-hydroxyprogesterone and progesterone secretion will exacerbate salt loss because of the ability of these steroids to compete with aldosterone for the renal tubular mineralocorticoid receptor. Similarly, any degree of sodium depletion could stimulate, via the renin-angiotensin system, adrenocortical synthesis of 17-hydroxyprogesterone, which in turn would be converted to the active androgen, testosterone. Poor growth and abnormal reproductive function can therefore result from any of three situations: inadequate glucocorticoid suppression of ACTH-mediated androgen secretion, leading to virilization and premature epiphyseal closure; inadequate mineralocorticoid replacement, leading to chronic sodium depletion, failure to thrive, and angiotensin-mediated secretion of C_{21} and C_{19} steroids; and, finally, chronic hypercortisolemia, usually a result of vain attempts to suppress adrenal steroidogenesis with large doses of glucocorticoid in sodium-depleted patients.

To correct these problems, it was first necessary to develop better methods for monitoring therapy. By 1974, it appeared that more useful information might be obtained by the measurement of serum 17-hydroxyprogesterone concentrations and

[a]This work was supported by the Medical Research Council of Canada, the Children's Hospital of Winnipeg Research Foundation, and the Manitoba Health Research Foundation.

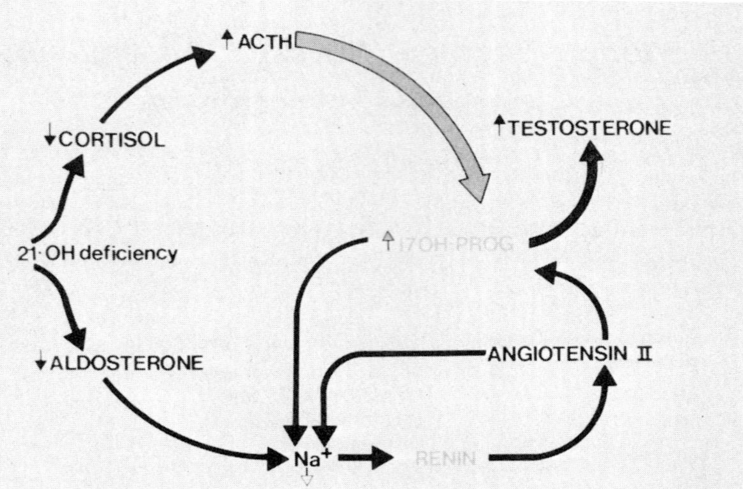

FIGURE 1. The pathogenesis of virilization and salt-wasting in 21-hydroxylase deficiency. The central role for 17-hydroxyprogesterone and renin is emphasized. Note that in the untreated state, attempts to compensate by increased secretion of ACTH and renin in fact exacerbate salt loss.

plasma renin activity,[3] but it was necessary to validate these new assays, to design sampling protocols that were not influenced by circadian rhythms, stress, exercise, and timing of medication, and to evaluate their performance in a busy clinical setting against accepted techniques. It quickly became apparent that, provided blood samples were obtained under carefully controlled circumstances (ambulatory, at 0830 and 1200 hours, with morning medications taken at 0900), assays of 17-hydroxyprogesterone and renin activity could provide a more precise indication of therapeutic control than was possible using 24-hour urine assays.[4] Furthermore, there was an excellent correlation between serum concentrations of 17-hydroxyprogesterone and levels of other relevant steroids such as progesterone, androstenedione, and testosterone.[5] Recently, it has been shown that this type of monitoring may even be carried out at home throughout a 24-hour period using analyses of salivary 17-hydroxyprogesterone,[6,7] although venipuncture is still necessary to assess plasma renin activity.

Since 1974, our treatment of CAH has been based upon the following principles: (1) All patients should receive sufficient mineralocorticoid to maintain plasma renin activity within the age-specific normal range, irrespective of the presence or absence of obvious salt-wasting; (2) The smallest possible dose of oral cortisol should be used which is sufficient to suppress serum 17-hydroxyprogesterone to below 200 ng/dl (6 nmol/L) and to maintain normal age- and sex-specific serum levels of testosterone; (3) Therapy should be monitored by assessments every three months of plasma renin activity and serum 17-hydroxyprogesterone and testosterone, blood pressure, growth velocity, and sexual development, with biennial assessments of skeletal maturation; (4) Treatment must be accompanied and supported by an ongoing program of counseling and education of the parents and child to ensure informed compliance; (5) In affected girls with genital ambiguity, cosmetically and functionally successful vaginoplasty and clitoroplasty should be accomplished by 12–18 months of age. From more than a decade of experience with this approach, we have learned that adequate suppression of adrenal 17-hydroxyprogesterone and testosterone can be maintained at all ages in the compliant patient with oral cortisol 10–20 mg/m^2/day given in three divided doses. (Clinical signs of hypercortisolism do not appear with this dose level.) Surprisingly, we have observed that relatively large doses (as much as 0.15–0.2 mg/day) of the oral mineralocorticoid, 9α-fluorohydrocortisone, are often necessary during infancy to suppress renin activity to normal; during later childhood, however, and at any time that plasma renin activity is suppressed below normal, the dose of mineralocorticoid should be lowered to 0.05–0.1 mg/day or hypertension may ensue. Provided mineralocorticoid replacement is adequate, there is no need to provide supplementary salt, except possibly during the first year of life.

The present report describes the results of this form of medical therapy during the past decade in a standard clinic setting, with particular emphasis upon physical growth.

SUBJECTS

From the 40 patients with 21-hydroxylase deficiency in our clinic, we selected for analysis 29 patients (19 female and 10 male) who had been followed for more than three years. Their present age ranged from 3 to 22 years, with a median of 10 years. Twelve patients, seven female and five male, had entered or completed puberty. Fourteen patients had been treated elsewhere (for 0.5–12 years) before referral to our clinic.

The first group of patients (TABLE 1) consisted of 12 females and 5 males in whom the diagnosis of 21-hydroxylase deficiency was established before a month of age, and in whom therapy was judged in retrospect to have been adequate since that time. All of these patients received oral cortisol 10–20 mg/m^2/day in three doses, and all had received oral 9α-fluorohydrocortisone since diagnosis. Adequate therapy was defined as consistent suppression of plasma renin activity and serum testosterone concentrations to the age-specific normal range. Virtually all of their simultaneous serum 17-hydroxyprogesterone determinations were below 200 ng/dl, but occasional early morning values as high as 1000 ng/dl (30 nM) were observed without any concomitant rise in androgen levels.

TABLE 1. Growth Parameters in CAH Patients Diagnosed before One Month of Age and Treated Adequately

Sex	Age (yr)	Bone Age (yr)	Target Height (percentile)	Present Height (percentile)	Height for Bone Age (percentile)
M	3.0	2.5	75	25	50
F	3.5	3.0	75	25	50
F	3.8	2.0	25	25	>95
F	4.0	3.5	75	25	50
F	4.0	3.0	50	25	90
F	4.8	2.0	5	5	95
F	6.3	4.3	75	50	>95
F	6.5	6.0	50	25	50
M	7.7	6.0	50	25	75
M	8.5	7.0	50	25	50
F	9.0	8.3	50	50	75
F	9.5	9.0	25	50	75
F	10.8	10.0	90	10	25
F	11.3	7.5	50	5	75
M	12.1	10.5	25	10	50
F	13.0	13.0	50	50	50
M	18.0	adult	50	25	25

The second group (TABLE 2) consisted of one female and three males in whom diagnosis and the initiation of therapy was delayed after one month of age, but in whom therapy thereafter was considered to be adequate by the same criteria as the first group. The third group (TABLE 3) consisted of six females and two males in whom therapy was considered to have been inadequate. Two of these had been overtreated initially and were frankly Cushingoid at the time of referral, with significant growth retardation. Three were undertreated for several years prior to referral and presented with advanced skeletal maturation and continued virilization. Three patients (including two siblings) were noncompliant for social reasons and showed continued elevation of serum 17-hydroxyprogesterone and testosterone.

For each patient, skeletal maturation was assessed using the standards of Greulich and Pyle.[8] The target height percentile at any age was judged to be that of the mean parental stature minus 6.5 cm in females, and the mean parental stature plus 6.5 cm in males.[9,10]

RESULTS

The present growth parameters of the CAH patients are summarized in TABLES 1–3. When therapy was initiated soon after birth and maintained adequately, stable growth velocities were maintained (FIGURE 2); the present median height is the 25th percentile for age (range 5 to 50). Thus, as a group, these patients are shorter than their peers. Only two have a height at or below the fifth percentile, but, in one of these, the target height based on parental size is also at the fifth percentile. None of the patients showed advanced skeletal maturation, but in five, the bone age was more than 1½ years behind the chronological age. Although none of them showed any clinical signs of hypercortisolism, it was observed that the two patients with the greatest delay in bone maturation had consistently the greatest degree of suppression of serum 17-hydroxyprogesterone and testosterone levels. The only patient in this group who has finished growing achieved a final height within 5 cm of his estimated target height.

The four patients in the second group, with late diagnosis but adequate therapy subsequently, are all showing normal growth velocity at the present time. However, the three who presented after infancy with simple virilization, including accelerated growth, all demonstrated a "catch-down" initial reduction in growth velocity after therapy was begun. At present, only one shows significant (more than 1½ years) advancement in bone age, but in two, the bone age is now significantly delayed. None have shown any clinical signs of hypercortisolism.

Eight patients have had poor control, either because of over- or undertreatment. Two were frankly Cushingoid when first seen by us because persistent sodium depletion and hyperreninemia caused elevated serum 17-hydroxyprogesterone and urinary 17-ketosteroid levels, which physicians had attempted to suppress with high doses of glucocorticoid. Introduction of mineralocorticoid therapy permitted reduction in oral cortisol dosage to 10–20 mg/m^2/day. One patient is now adult and achieved target height; the other shows significant delay in bone maturation, but is showing some catch-up and is on the 75th height percentile for her bone age. Six patients have been chronically undertreated either because of inadequate therapy prior to referral or because of disruptive family circumstances that have made compliance unsatisfactory. All showed significant acceleration of skeletal maturation; five have now ceased growing, and all of them failed to approach their predicted target heights. The sixth is presently of normal size, but will clearly not achieve her genetic height potential.

None of these patients showed true precocious puberty, in spite of significant bone

TABLE 2. Growth Parameters in Late-Diagnosed CAH Patients Who Were Treated Adequately

Sex	Age at Diagnosis (yr)	Present Age (yr)	Bone Age (yr)	Target Height (percentile)	Present Height (percentile)	Height for Bone Age (percentile)
F	2.0	9.4	5.5	25	10	>95
M	0.3	9.9	7.0	50	25	>95
M	5.3	11.0	13.0	75	50	10
M	3.0	12.0	12.5	25	50	25

TABLE 3. Growth Parameters in CAH Patients in Whom Therapy Was Inadequate

Sex	Age at Diagnosis (yr)	Present Age (yr)	Bone Age (yr)	Target Height (percentile)	Present Height (percentile)	Height for Bone Age (percentile)	Remarks
F	birth	4.5	2.8	50	5	75	early overtreatment
F	birth	19.3	adult	15	10	10	early overtreatment
F	2.0	16.5	adult	50	<5	<5	undertreated to age 8
M	5.5	16.9	adult	25	<5	<5	undertreated to age 11
F	2.0	22.0	adult	50	<5	<5	undertreated to age 13
F	birth	11.5	14	95	50	<5	poor compliance
F	birth	17.5	adult	50	5	<5	poor compliance
M	3.0	15.8	adult	50	10	<5	poor compliance

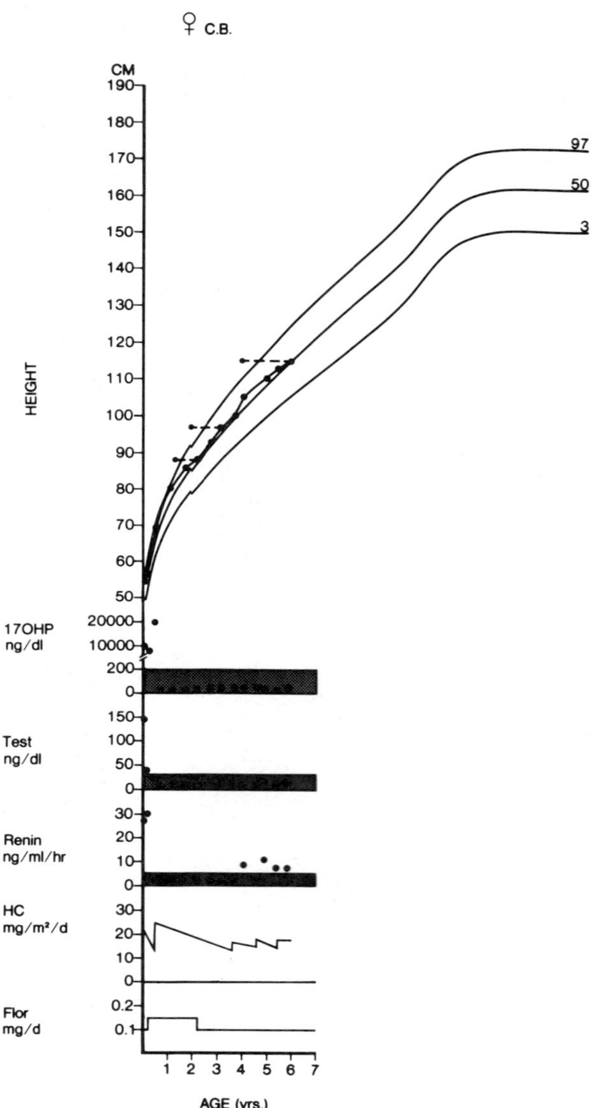

FIGURE 2. Height growth of a girl with 21-hydroxylase deficiency diagnosed and treated at one week of age. Bone age is plotted by dots connected by horizontal lines to the growth chart. Serum concentrations of 17-hydroxyprogesterone (17OHP), testosterone, and plasma renin activity are plotted, with the normal range as a shaded area. The daily doses of oral hydrocortisone (HC) and 9α-fluorohydrocortisone (Flor) are also plotted.

age advance in several. Of the seven pubertal girls, two are still premenarchal. In the rest, the age of menarche ranged from 10.5 to 14 years. Subsequent oligomenorrhea was a problem only in the two noncompliant adolescent girls; one of these improved her control after marriage and has had two successful pregnancies.

DISCUSSION

In order to achieve normal growth and sexual maturation, the goal of therapy in CAH should be to suppress adrenal production of androgens and their precursors, while only using the smallest possible doses of glucocorticoid and mineralocorticoid. Several excellent studies have demonstrated the pivotal role that mineralocorticoid therapy plays in the hyperreninemic CAH patient, regardless of clinically apparent salt-wasting, which permits a reduction in glucocorticoid dosage with a simultaneous improvement in control and in linear growth.[11,12]

In an unpublished analysis of our own CAH patients over a decade ago, at a time when oral cortisol doses of 15–25 mg/m^2/day were commonly used, we showed a significant negative correlation between cortisol dose and short-term growth velocity ($r = -0.93$) with consistent growth suppression when doses exceeded 30 mg/m^2/day. We, therefore, have used and recommended a dose range of 10–20 mg/m^2/day at all ages, a dose which was less than that used in most reported series,[13–15] but which we had demonstrated to be uniformly effective in suppressing serum concentrations of 17-hydroxyprogesterone and testosterone, as well as urinary excretion of 17-ketosteroids and pregnanetriol.[3–5]

The present study presents the results of a decade of experience with this regimen. It provides some reassuring observations, but also raises some disturbing questions. It is clear that very satisfactory growth, normal puberty, and appropriate adult height can be achieved in CAH, even when the diagnosis has been delayed for a year or two, provided there is consistent and rigorous control of plasma renin activity and adrenal androgen secretion. Indeed, it is even possible that periods of significant overtreatment, sufficient to induce clinical hypercortisolism, may be compatible with eventual normal adult stature. Conversely, it is obvious that chronic undertreatment leads to precocious epiphyseal maturation and significant limitation of adult height. The present data are inadequate to permit firm conclusions regarding a possible relationship between prepubertal quality of care and eventual fertility. Obviously, poor control of adrenal androgen secretion in adults causes anovular infertility in females and interferes with spermatogenesis in at least some males.[16] One gains the impression, however, that normal fertility is certainly possible in affected females provided contemporary control is adequate, regardless of whether there was a prolonged period of noncompliance and virilization during childhood.

The disturbing feature of the present data is the evidence that even these modest doses of oral cortisol appear to cause significant delays in bone maturation, along with a tendency to short stature during childhood, even in the absence of clinical signs of overtreatment. The most likely explanation for this is the significant hypercortisolemia, with values as high as 60 μg/dl, which occurs during the absorption phase following each dose of oral cortisol. It does not appear that this phenomenon can be avoided by the use of other glucocorticoids such as cortisone acetate, the absorption and conversion of which is often erratic.[17] In addition to this direct effect of transient hypercortisolemia, it is possible that excessive suppression of adrenal androgen secretion may also affect bone maturation and growth. Certainly, the adrenal atrophy which accompanies satisfactory therapy in CAH routinely leads to a marked suppression of circulating dehydroepiandrosterone sulfate values, but as yet, there is no convincing evidence that this C_{19} steroid has any important direct or indirect influence upon bone growth.

CONCLUSION

The results have been presented of a decade of experience with a modern treatment regimen for CAH based upon both the use of sufficient oral mineralocorticoid to suppress plasma renin activity to normal and the administration of the smallest possible dose of glucocorticoid sufficient to suppress serum testosterone and 17-hydroxyprogesterone values to normal. Clearly, such a regimen is effective, provided that the laboratory services for frequent monitoring of these parameters are available and provided that the clinical services make available the continued counseling and education necessary to ensure informed patient compliance. The only drawback of this regimen appears to be some slowing of growth and skeletal maturation during childhood, which should, however, have no permanent effect upon adult height or reproductive function. Whether a further reduction in oral cortisol dosage, possibly to about 10–15 $mg/m^2/day$ particularly during infancy, might obviate this minor complication remains to be demonstrated over the next decade.

REFERENCES

1. Wilkins, L., R. A. Lewis, R. Klein & E. Rosemberg. 1950. The suppression of androgen secretion by cortisone in a case of congenital adrenal hyperplasia. Bull. Johns Hopkins Hosp. **86:** 249–255.
2. Bartter, F. C., A. P. Forbes & A. Leaf. 1950. Congenital adrenal hyperplasia associated with the adrenogenital syndrome: An attempt to correct its disordered hormonal pattern. J. Clin. Invest. **29:** 797 (abstract).
3. Hughes, I. A. & J. S. D. Winter. 1977. 17-Hydroxyprogesterone and plasma renin activity in congenital adrenal hyperplasia. *In* Congenital Adrenal Hyperplasia. P. A. Lee, L. P. Plotnick, A. A. Kowarski & C. J. Migeon, Eds.: 141–156. University Park Press. Baltimore, Md.
4. Hughes, I. A. & J. S. D. Winter. 1976. The application of a serum 17OH-progesterone radioimmunoassay to the diagnosis and management of congenital adrenal hyperplasia. J. Pediatr. **88:** 766–773.
5. Hughes, I. A. & J. S. D. Winter. 1978. The relationship between serum concentrations of 17OH-progesterone and other serum and urinary steroids in patients with congenital adrenal hyperplasia. J. Clin. Endocrinol. Metab. **46:** 98–104.
6. Otten, B. J., J. J. Wellen, J. C. W. Rijken, G. B. A. Stoelinga & Th. J. Benraad. 1983. Salivary and plasma androstenedione and 17-hydroxyprogesterone levels in congenital adrenal hyperplasia. J. Clin. Endocrinol. Metab. **57:** 1150–1154.
7. Hughes, I. A. & G. F. Read. 1982. Simultaneous plasma and saliva steroid measurements as an index of control in congenital adrenal hyperplasia (CAH). Horm. Res. **16:** 142–150.
8. Greulich, W. W. & S. I. Pyle. 1964. Radiographic Atlas of Skeletal Development of the Hand and Wrist. Stanford Univ. Press. Palo Alto, CA.
9. Tanner, J. M., H. Goldstein & R. H. Whitehouse. 1970. Standards for children's height at ages 2–9 years allowing for height of parents. Arch. Dis. Child. **45:** 755–762.
10. Joss, E., K. Zuppinger, H. P. Schwarz & H. Roten. 1983. Final height of patients with pituitary growth failure and changes in growth variables after long-term hormonal therapy. Pediatr. Res. **17:** 676–679.
11. Kuhnle, U., A. Rosler, J. A. Pereira, P. Gunzcler, L. S. Levine & M. I. New. 1983.

The effect of long-term normalization of sodium balance on linear growth in disorders with aldosterone deficiency. Acta Endocrinol. **102:** 577–582.

12. Jansen, M., J. M. Witt & J. L. van den Brande. 1981. Reinstitution of mineralocorticoid therapy in congenital adrenal hyperplasia. Effects on growth and control. Acta Paediatr. Scand. **70:** 229–233.
13. Brook, C. G. D., M. Zachmann, A. Prader & G. Murset. 1974. Experience with long-term therapy in congenital adrenal hyperplasia. J. Pediatr. **85:** 12–19.
14. Klingensmith, G. J., S. C. Garcia, H. W. Jones, C. J. Migeon & R. M. Blizzard. 1977. Glucocorticoid therapy of girls with congenital adrenal hyperplasia: Effects on height, sexual maturation, and fertility. J. Pediatr. **90:** 996–1004.
15. Duck, S. C. 1980. Acceptable linear growth in congenital adrenal hyperplasia. J. Pediatr. **97:** 93–96.
16. Wischusen, J., H. W. G. Baker & B. Hudson. 1981. Reversible male infertility due to congenital adrenal hyperplasia. Clin. Endocrinol. **14:** 571–577.
17. Hansen, J. W. & D. L. Loriaux. 1976. Variable efficacy of glucocorticoids in congenital adrenal hyperplasia. Pediatrics **57:** 942–947.

LHRH Analog Treatment of Central Precocious Puberty Complicating Congenital Adrenal Hyperplasia

ORA HIRSCH PESCOVITZ, FERNANDO CASSORLA, FLORENCE COMITE, D. LYNN LORIAUX, AND GORDON B. CUTLER, JR.

Developmental Endocrinology Branch
National Institute of Child Health and Human Development
National Institutes of Health
Bethesda, Maryland 20205

INTRODUCTION

Congenital adrenal hyperplasia (CAH) has long been recognized as a major cause of peripheral precocious puberty (premature sexual development without activation of pubertal LH and FSH secretion).[1–4] Elevated adrenal androgen production in this disorder causes pubic and axillary hair development, penile or clitoral enlargement, and rapid growth and advanced bone age. Treatment with hydrocortisone and fludrocortisone usually halts the progression of these findings in children with CAH. Gonadotropin levels are within the normal prepubertal range. Breasts do not develop in the girls and the testes do not enlarge generally in the boys. This is as would be predicted for peripheral precocious puberty due to increased serum concentrations of adrenal androgens.

Some young children with CAH continue to grow rapidly and advance in bone age. However, despite adequate suppression of adrenal androgen overproduction, such children have been found to have elevated gonadotropin levels both basally and in response to LHRH, suggesting that the CAH led to an early activation of pubertal gonadotropin secretion (central precocious puberty).[5–8,12–14]

Recent studies have shown that a long-acting analog of LHRH (D-Trp^6-Pro^9-NEt-LHRH) can suppress pituitary gonadotropins and gonadal sex steroids in children with central precocious puberty.[15,16] This study evaluated the addition of the LHRH analog D-Trp^6-Pro^9-NEt-LHRH to a treatment regimen of hydrocortisone and fludrocortisone in four children with central precocious puberty secondary to congenital adrenal hyperplasia.[17]

PATIENTS

Subjects

Four patients (A, B, C, D), including one female and three males, were 3, 4.5, 5, and 6 years, respectively, at the time of diagnosis of 21-hydroxylase deficiency. They were treated with hydrocortisone and fludrocortisone. They presented with symptoms of true precocious puberty between the ages of 6 and 8 years when their bone ages were between 10.5 and 13.5 years. These children were growing at an excessive rate and all had continuing pubic hair development. The boys had progressive testicular enlargement and the girl had progressive breast development.

Protocol

Patients were admitted to the Clinical Center of the National Institutes of Health. The protocol was approved by the Clinical Research Committee of the National Institute of Child Health and Human Development. We obtained informed consent from either parent and informed assent from older children. Pretreatment evaluation consisted of basal serum gonadotropin determinations every 20 minutes for four hours during the day (1000 to 1400) and night (2200 to 0200). Plasma testosterone (males) or estradiol (females) was measured four times (1000, 1400, 2200, and 0200 hrs].

LHRH 100 μg was administered i.v. at time zero, and serum gonadotropins were measured at −30, −15, 0, 15, 30, 60, 90, 120, and 180 minutes. We measured height ten times using a stadiometer. An X ray of the left hand and wrist was obtained for bone age, which was assessed by a radiologist who had no clinical information about the patients.[18] Computed tomography (CT) of the head and ultrasound study of the adrenal glands were performed in each patient. The female patient had a pelvic ultrasound and the males had testicular ultrasound studies. The patients were photographed to document the stage of pubertal development.[19–21] Patients A and B were treated with $LHRH_a$ for 6 months, patient C for 12 months, and patient D for 18 months at the time this paper was written. The children were reevaluated according to the above protocol, excluding the CT scan, every three to six months. The data presented are from the pretreatment evaluation and the latest treatment evaluation (6–18 months of therapy).

RESULTS

Plasma Gonadotropin and Sex Steroid Levels

All four patients had measurable serum gonadotropin levels prior to therapy and a pubertal response to the LHRH stimulation test. After six months on $LHRH_a$ therapy, both basal and peak gonadotropin levels fell significantly (FIGURE 1).

Sex steroid levels also decreased during treatment (FIGURE 2). Plasma testosterone fell in the three males from a mean of 274 ± 60 to 39 ± 11 ng/dl during treatment ($p < 0.02$), while plasma estradiol fell from a mean of 40 ± 15 to below 12 pg/ml, the detection limit of the assay.

Testicular Volume and Secondary Sexual Characteristics

The clinical data of the four patients is shown in TABLE 1. Testis size decreased in the males from 11 ± 0.4 ml to 7 ± 1 ml ($p < 0.03$). Breast size decreased from Tanner stage III to Tanner stage II in the girl. Pubic hair did not change significantly during this 6–18 month period.

Control of Congenital Adrenal Hyperplasia

Hydrocortisone and fludrocortisone therapy was continued during the 6–18 months of $LHRH_a$ therapy. The degree of adrenal suppression did not change

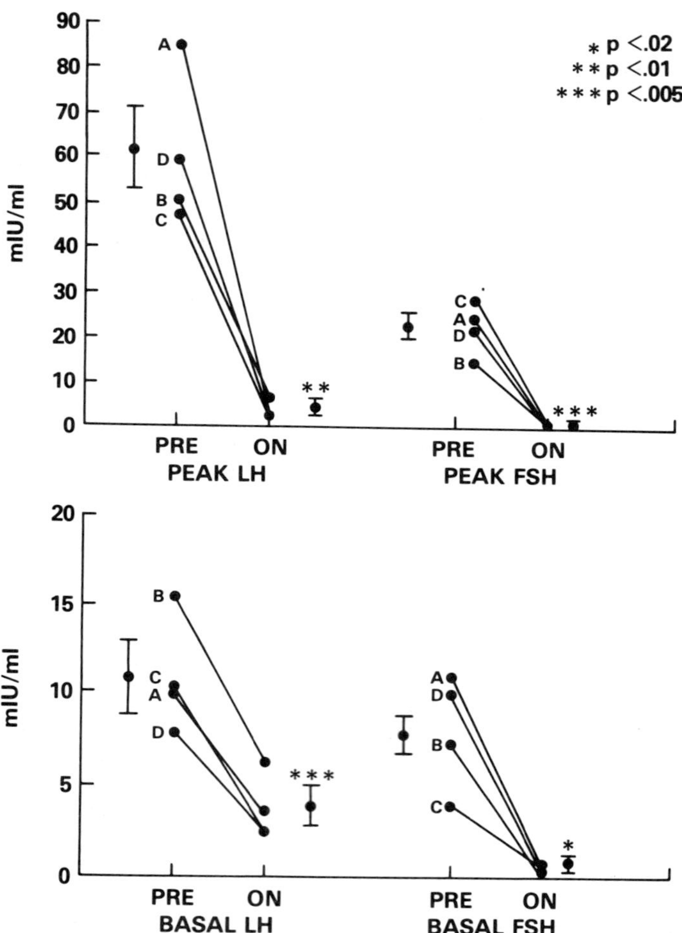

FIGURE 1. Basal and peak (LHRH-stimulated) plasma LH and FSH levels in four children with congenital adrenal hyperplasia and true precocious puberty before ("Pre") and during ("On") 6–18 months of treatment with $LHRH_a$. The basal LH and FSH values for each patient (lower figure) are the mean of 26 measurements obtained at 20 minute intervals from 1000 to 1400 and 2200 to 0200. The peak values (upper figure) are the highest LH and FSH levels achieved during an LHRH stimulation test.[17]

TABLE 1. Clinical Data[17]

Patient No.	Sex	Age at CAH Diagnosis (yr)	Age at Diagnosis of True Precocious Puberty (yr)	Bone Age at Diagnosis of True Precocious Puberty	Height	Testicular Volume (boys) or Breast Stage (girl)	Pubic Hair
A	M	6	8	13.5	153.2 (>99)[c]	11.5 ml[b]	IV[a]
B	F	3	7	10.5	131.7 (>99)[c]	III[a]	III[a]
C	M	4.5	6	11.0	118.6 (>99)[c]	11 ml[b]	III[a]
D	M	5	6	11.5	135.3 (>99)[c]	10 ml[b]	III[a]

[a]According to the classification of Tanner.[20]
[b]Mean of right and left testis volumes estimated with a Prader orchidometer.[21]
[c]Height percentile.

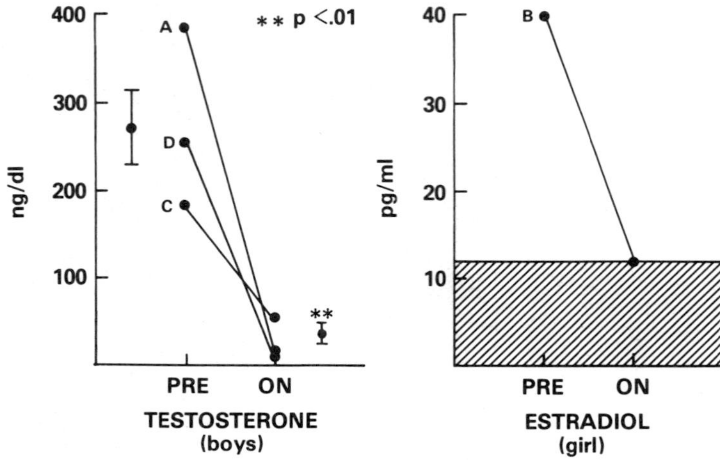

FIGURE 2. Plasma sex steroid concentrations in four children with congenital adrenal hyperplasia and true precocious puberty before and during 6–18 months of treatment with $LHRH_a$. The testosterone levels for each of the three male patients (left panel) are the mean of four measurements at 1000, 1400, 2200, and 0200. The estradiol levels (right panel) are the mean of four measurements at the same times in the female patient. The hatched area represents the detection limit of the assay.[17]

significantly during the $LHRH_a$ treatment period, as shown by the similar levels of plasma 17-hydroxyprogesterone (17OHP), plasma renin activity (PRA), and 24-hour urinary 17-ketosteroid excretion (TABLE 2).

Growth

Linear growth rate decreased from 16 ± 4 cm/yr to 5 ± 1 cm/yr ($p < 0.005$). The rate of bone age advancement (ΔBA/ΔCA) decreased from 1.7 ± 0.1 to 0.7 ± 0.4 during $LHRH_a$ treatment ($p < 0.05$, FIGURE 3).

TABLE 2. Plasma 17-Hydroxyprogesterone, 24-h Urinary 17-Ketosteroid Excretion, and Plasma Renin Activity before and during $LHRH_a$ Therapy[17]

	Treatment	170HP[a] (ng/dl) (n = 4)	24-H Urine 17KS[b] (mg/24 h) (n = 3)	Plasma Renin Activity (ng/ml/hr) (n = 2)
Pre:	Hydrocortisone 14 ± 11 mg/m²/d Fludrocortisone 90 ± 10 μg/d	1540 ± 760[c]	7 ± 1.2	5.9 ± 2.3
During:	Hydrocortisone 18 ± 3 mg/m²/d Fludrocortisone 90 ± 10 μg/d $LHRH_a$ 4 μg/kg/d	2947 ± 906	5 ± 2.0	4.4 ± 2.3

[a]Plasma 17-hydroxyprogesterone.
[b]Urinary 17-ketosteroids.
[c]Mean ± SEM.

DISCUSSION

Congenital adrenal hyperplasia causes precocious puberty because of elevated adrenal androgen production. Several patients with CAH have also had early hypothalamic-pituitary-gonadal maturation (central precocious puberty).[5–8,12–14] One of the earliest such cases was a six year-old girl who presented with rapid growth, pubic hair, and clitoromegaly.[12] She had elevated urinary 17-ketosteroid pregnanediol and urinary gonadotropin levels. Laparoscopy revealed bilaterally enlarged ovaries and developing follicles. It was concluded that the patient had both CAH and central precocious puberty.

Four additional patients are described here. These children initially had rapid growth and pubic hair development. After the diagnosis of 21-hydroxylase deficiency was established, they were treated with hydrocortisone and fludrocortisone. Within two years, they developed signs of hypothalamic-pituitary-gonadal maturation that included testicular or breast development, persistent rapid growth, and bone age advancement, despite treatment, elevated basal and LHRH-stimulated gonadotropin levels, and elevated sex steroid levels.

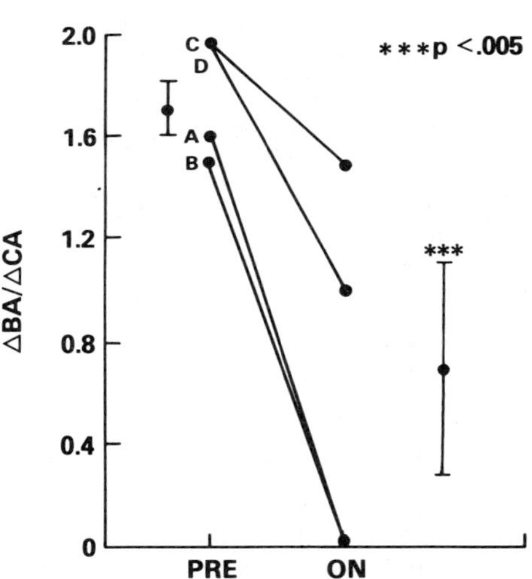

FIGURE 3. Effect of $LHRH_a$ therapy on the rate of bone age advancement in four children with congenital adrenal hyperplasia and true precocious puberty. The rate of bone age advancement ($\Delta BA/\Delta CA$) prior to treatment represents the patient's pretreatment bone age divided by his/her chronologic age. (Serial bone ages were not available during the pretreatment period.) The rate of bone age advancement during treatment represents each child's increase in bone age during treatment divided by the duration of treatment (6–18 months).[17]

Clinical observations suggest several hypotheses about the mechanism of central precocious puberty in CAH. First, the elevated level of sex steroids rather than the CAH itself seems to be the critical factor in causing central precocious puberty. This is so since virilizing adrenal tumors or ovarian granulosa cell tumors have also been associated with premature gonadotropin elevation.[18,19] Second, the likelihood of developing central precocious puberty in CAH may correlate with bone age. Central precocious puberty developed in a 3$^4/_{12}$ year-old boy with a bone age of 11$^6/_{12}$, but did not develop in a 6$^2/_{12}$ year-old boy whose bone age was 10.[7] Our four patients had bone ages of at least 10$^6/_{12}$ years at the time central precocious puberty was recognized. Third, although the decline in androgen levels after hydrocortisone treatment has been postulated to cause the increase in gonadotropin secretion through a decline in negative feedback, the observation of central precocious puberty in untreated CAH suggests that such a mechanism is not obligatory.[8,12,13]

Whatever the mechanism causing central precocious puberty in CAH, though, $LHRH_a$ was highly effective in reversing the hormonal and clinical signs of central precocious puberty. The effects of LHRH analog treatment in CAH were identical to those in idiopathic central precocious puberty or precocious puberty associated with intracranial lesions.[15,16] The improvement was clearly related to suppression of gonadal function since $LHRH_a$ had no effect on the degree of adrenal suppression. As in children with other causes of central precocious puberty, no adverse effects have occurred so far. Although longer term evaluation is necessary, $LHRH_a$ appears to be a safe and effective adjunct to hydrocortisone and fludrocortisone therapy for central precocious puberty complicating congenital adrenal hyperplasia.

REFERENCES

1. SIZONENKO, P. C. 1978. Endocrinology in preadolescents and adolescents I. Hormonal changes during normal puberty. Am. J. Dis. Child. **132:** 704.
2. SIZONENKO, P. C. & L. PAUNIER. 1975. Hormonal changes in puberty. III. Correlation of plasma dehydroepiandrosterone, testosterone, FSH, and LH with stages of puberty and bone age in normal boys and girls and in patients with Addison's disease or hypogonadism or with premature or late adrenarche. J. Clin. Endocrinol. Metab. **41:** 894.
3. OJEDA, S. R., W. W. ANDREWS, J. P. ADVIS & S. SMITH-WHITE. 1980. Recent advances in the endocrinology of puberty. Endocrinol. Rev. **1:** 228.
4. JENNER, M. R., R. P. KELCH, S. L. KAPLAN & M. M. GRUMBACH. 1972. Hormonal changes in puberty. IV. Plasma estradiol, LH, and FSH in prepubertal children, pubertal females, and in precocious puberty, premature thelarche, hypogonadism, and in a child with a feminizing ovarian tumor. J. Clin. Endocrinol. Metab. **34:** 521.
5. GRUMBACH, M. M., G. E. RICHARDS, F. A. CONTE & S. L. KAPLAN. 1978. Clinical disorders of adrenal function and puberty: An assessment of the role of the adrenal cortex in normal and abnormal puberty in man and evidence for an ACTH-like pituitary adrenal androgen stimulating hormone. *In* The Endocrine Function of the Human Adrenal Cortex. V. H. James, M. Serio, G. Giusti & L. Martini, Eds.: 583. Academic Press. London.
6. REITER, E. O., S. L. KAPLAN, F. A. CONTE & M. M. GRUMBACH. 1975. Responsivity of pituitary gonadotropes to luteinizing hormone-releasing factor in idiopathic precocious puberty, precocious thelarche, precocious adrenarche, and in patients treated with medroxyprogesterone acetate. Pediatr. Res. **9:** 111.
7. REITER, E. O., M. M. GRUMBACH, S. L. KAPLAN & F. A. CONTE. 1975. The response of pituitary gonadotropes to synthetic LRF in children with glucocorticoid-treated congenital adrenal hyperplasia: Lack of effect of intrauterine and neonatal androgen excess. J. Clin. Endocrinol. Metab. **40:** 318.
8. BOYAR, R. M., J. W. FINKELSTEIN, R. DAVID, H. ROFFWARG, S. KAPEN, E. D. WEITZMAN & L. HELLMAN. 1973. Twenty-four patterns of plasma luteinizing hormone and follicle-stimulating hormone in sexual precocity. N. Engl. J. Med. **289:** 282.
9. WILKINS, L., J. F. CRIGLER, S. H. SILVERMAN, L. I. GARDNER & C. J. MIGEON. 1952. Further studies on the treatment of congenital adrenal hyperplasia with cortisone. II. The effects of cortisone on sexual and somatic development, with an hypothesis concerning the mechanism of feminization. J. Clin. Endocrinol. Metab. **12:** 277.
10. WILKINS, L. & J. CARA. 1954. Further studies on the treatment of congenital adrenal hyperplasia with cortisone. V. Effects of cortisone therapy on testicular development. J. Clin. Endocrinol. Metab. **14:** 287.
11. MININBERG, D. T., L. S. LEVINE & M. I. NEW. 1979. Current concepts in congenital adrenal hyperplasia. Invest. Urol. **17:** 169.
12. KOVACIC, N. 1959. Congenital adrenal hyperplasia and precocious gonadotropin secretion in a six year-old girl. J. Clin. Endocrinol. Metab. **19:** 844.
13. STEVENS, V. C. & J. W. GOLDZIEHER. 1968. Urinary excretion of gonadotropins in congenital adrenal hyperplasia. Pediatrics **41:** 421.
14. PENNY, R., N. O. OLAMBIWONNU & S. D. FRASIER. 1973. Precocious puberty following

treatment in a six year-old male with congenital adrenal hyperplasia: Studies of serum luteinizing hormone (LH), serum follicle-stimulating hormone (FSH), and plasma testosterone. J. Clin. Endocrinol. Metab. **36:** 920.

15. Comite, F., G. B. Cutler, Jr., J. Rivier, W. N. Vale, D. L. Loriaux & W. F. Crowley. 1981. Short-term treatment of idiopathic precocious puberty with a long-acting analogue of luteinizing hormone-releasing hormone. N. Engl. J. Med. **305:** 1546.
16. Pescovitz, O., K. Reith, K. Hench, A. McNemar, L. Loriaux, G. Cutler & F. Comite. 1983. Luteinizing hormone-releasing hormone analog therapy in boys with true precocious puberty due to hypothalamic hamartoma. Pediatr. Res. **17:** 1698.
17. Pescovitz, O. H., F. Comite, F. Cassorla, A. J. Dwyer, M. A. Poth, M. A. Sperling, K. Hench, A. McNemar, M. Skerda, D. L. Loriaux & G. B. Cutler, Jr. 1984. True precocious puberty complicating congenital adrenal hyperplasia: Treatment with a luteinizing hormone-releasing hormone analogue. J. Clin. Endocrinol. Metab. **58:** 857.
18. Mortimer, J. G., B. T. Rudd & W. R. Butt. 1964. A virilizing adrenal tumor in a prepubertal boy. J. Clin. Endocrinol. Metab. **24:** 842.
19. van Wyk, J. J. 1965. Sexual precocity in the female. *In* The Ciba Collection of Medical Illustrations, vol. 4. F. H. Netter, Ed.: 121. Endocrine System and Selected Metabolic Diseases Ciba Pharmaceutical Company. New York.
20. Tanner, J. M. 1978. *In* Growth at Adolescence. Blackwell. Oxford, England.
21. Zachmann, M., A. Prader, H. P. Kind, H. Haflinger & H. Budlider. 1974. Testicular volume during adolescence. Helv. Pediatr. Acta **29:** 61.

Effect of Hydrocortisone Dose Schedule on Adrenal Steroid Secretion in Congenital Adrenal Hyperplasia

JORG WINTERER, GEORGE P. CHROUSOS,
D. LYNN LORIAUX, AND GORDON B. CUTLER, JR.

Developmental Endocrinology Branch
National Institute of Child Health and Human Development
National Institutes of Health
Bethesda, Maryland 20205

INTRODUCTION

Glucocorticoid therapy has greatly improved the clinical management of children with congenital adrenal hyperplasia (CAH). However, maintaining an optimal therapeutic regimen throughout childhood is difficult because of the narrow therapeutic index of glucocorticoids—overtreatment causes Cushing's syndrome and poor growth, while undertreatment results in accelerated skeletal maturation, premature epiphyseal fusion, and short adult stature. The mean adult height of 30 men with CAH who had received conventional glucocorticoid therapy was only at the fourth percentile of the normal population.[1] The mean adult height of 26 women with CAH in whom treatment was begun during the first year of life approximated the 25th percentile for normal women.[2] This suggests that currently recommended regimens for this disorder might be further improved.

One approach to improve the effectiveness of glucocorticoid therapy has been to seek an optimal dose schedule for glucocorticoid administration. Acute (two day) studies in normal subjects indicated greater effectiveness of nocturnal compared to morning or afternoon dexamethasone doses in suppressing adrenal function.[3] Nocturnal prednisone, however, was less effective in congenital adrenal hyperplasia than the same dose given twice daily.[4]

Most children with congenital adrenal hyperplasia have been treated with hydrocortisone or cortisone acetate, and to our knowledge there has been only one study that examined the comparative effectiveness of different hydrocortisone dose schedules. This study found no benefit from replacing the afternoon dose of a 2-dose hydrocortisone schedule with a nocturnal dose of dexamethasone that contained an equivalent amount of glucocorticoid activity.[5] The current study was undertaken to compare the adrenal-suppressive effect of a constant dose of hydrocortisone while the time of administration was varied systematically among five treatment schedules. The data have been described in detail elsewhere.[6]

MATERIALS AND METHODS

Subjects

Six patients (age 9–20 yr) with 21-hydroxylase deficiency and two patients with 11-hydroxylase deficiency (age 6–12 yr) were studied. The clinical data for these patients are given in TABLE 1.

TABLE 1. Clinical Data (from reference 6)

Case No.	1	2	3	4	5	6	7	8
Sex	M	M	M	M	M	F	M	M
21-Hydroxylase Deficiency[a]	SL	SL	SV	SV	SV	SV		
11-Hydroxylase Deficiency							X	X
Chronologic Age (yrs)	10 4/12	8 11/12	21 7/12	22 11/12	20 4/12	8 5/12	6 4/12	11 11/12
Weight (kg)	18.4	35.1	68.5	68.1	57.6	31.7	22.0	66.1
Height (cm)	102.1	137.0	173.5	158.9	160.0	136.9	110.7	165.6
Surface Area (m^2)	0.71	1.15	1.82	1.70	1.59	1.10	0.81	1.73
Hydrocortisone Dose (mg/day)	8.5	13.8	21.6	20.4	19.1	13.2	9.7	20.8
Fludrocortisone Dose (mg/day)	0.1 mg	0.1 mg	0	0	0	0	0	0
Prestudy Treatment (mg/day)[b]	25 HC	20 HC	10 Pred	0.75 Dex	0.75 Dex	15 HC	10 HC	25 HC

[a]SL, salt-losing variety; SV, simple virilizing form of CAH.
[b]HC, hydrocortisone; Pred, prednisone; Dex, dexamethasone.

Experimental Design

The purpose of this study was to determine whether different hydrocortisone dose schedules have different adrenal suppressive potencies in CAH. A practical obstacle to this goal was that optimal clinical management of CAH requires greater than 90 percent suppression of adrenal function. Differences in potency cannot be assessed as sensitively at this portion of the dose-response curve as at the central linear portion corresponding to 50 percent suppression.[7,8] Thus, for the purpose of this study, patients were temporarily given a total daily dose of hydrocortisone that was less than that usually needed to achieve clinically adequate adrenal suppression.

Protocol

Each patient was removed from his previous treatment regimen and was given 12.5 $mg/m^2/day$ of hydrocortisone acetate dissolved in mineral oil and administered in capsules. This daily dose was then given successively according to five different schedules (I–V). Each schedule was given for a period of 4–6 weeks, and the sequence of schedules was assigned randomly in double-blind fashion. In schedule I, the whole dose was given in the morning and placebos were given at noon and at night; in schedule II, 2/3 of the dose was given in the morning with a placebo at noon and 1/3 at night; in schedule III, the dose was divided equally between morning, noon, and night; in schedule IV, 1/3 of the dose was given in the morning with a placebo at noon and 2/3 at night; and in schedule V, placebos were given in the morning and at noon with the entire dose at night. In general, morning doses were given at 0800, noon doses at 1200, and night doses at 2000. Placebo doses not containing cortisol were administered in capsules indistinguishable from active capsules. Thus, the patients took three capsules daily, from three different bottles labeled morning, noon, and night, throughout each of the five schedules. If a patient was receiving fludrocortisone prior to enrollment in the study, this was continued at the same dose and time during the study.

On the final day of each schedule, the patients were admitted to the endocrine ward at the NIH where samples of venous blood were drawn every two hours to measure 17-hydroxyprogesterone and cortisol, or, in the case of the two patients with 11-hydroxylase deficiency, 11-deoxycortisol and cortisol. Urine was collected over 24h for determination of free cortisol, 17-ketosteroid, 17-hydroxysteroid, and pregnanetriol excretion (subjects with 21-hydroxylase deficiency). At the conclusion of these studies, the patients were discharged on the next randomly assigned schedule until all five schedules were completed.

Assays

Plasma cortisol, 11-deoxycortisol, and 17-hydroxyprogesterone were measured by radioimmunoassay as previously described.[9–11] Urine 17-ketosteroids, 17-hydroxysteroids, and pregnanetriol were measured by Bioscience Laboratories (Columbia, MD).

Statistical Analysis and Calculations

All data are expressed as the mean ± SEM. Differences among dose schedules were assessed by Duncan's Multiple Range Test. Correlations were calculated by Pearson's *r*.

RESULTS

The 24-hour plasma cortisol profile was modified in a predictable fashion by the five treatment schedules (FIGURE 1). Administration of cortisol at night prevented the nocturnal fall in plasma cortisol that was otherwise present. The 24-hour mean plasma cortisol levels, however, did not differ among the schedules and were all within the normal adult range (5.9 to 9.2 μg/dl in our laboratory).

The different schedules also altered the temporal pattern of the plasma 17-hydroxyprogesterone (FIGURE 2) and plasma 11-deoxycortisol (FIGURE 3) profiles. The greatest apparent suppression was observed 2–4 hours after each dose. There was no significant difference in the mean 24-hour levels of either steroid among the five dose schedules, however, and we, therefore, combined the results from both types of congenital adrenal hyperplasia in comparing the relative plasma precursor steroid levels among the different schedules (FIGURE 4).

The 24-hour urine steroid metabolites also did not differ significantly among the five dose schedules (FIGURE 5). The pregnanetriol excretion for the six patients with 21-hydroxylase deficiency did not differ significantly across the five schedules (FIGURE 5). Similarly, no differences between schedules was observed for urine 17-ketosteroids in either the 21-hydroxylase deficiency or the 11-hydroxylase deficiency patients, and the combined data for urine 17-ketosteroids is, therefore, presented in FIGURE 5.

Positive correlations were observed between the 24-hour mean plasma 17-hydroxyprogesterone levels and the corresponding urine pregnanetriol (in the patients with 21-hydroxylase deficiency) ($r = 0.78$, $p < 0.00001$) and urine 17-ketosteroid levels ($r = 0.71$, $p < 0.00001$). Urine pregnanetriol and 17-ketosteroids were also strongly correlated ($r = 0.88, p < 0.00001$).

As expected, the mean levels of plasma 17-hydroxyprogesterone and of urine pregnanetriol and 17-ketosteroids exceeded the values seen in clinically well-controlled patients. Two of the patients with 21-hydroxylase deficiency, however, were well controlled at the hydrocortisone dose of 12.5 $mg/m^2/day$. The data of these two patients also showed no consistent difference among the five dose schedules (TABLE 2).

DISCUSSION

Varying the schedule of hydrocortisone administration offered no advantage in our patients with congenital adrenal hyperplasia since all five schedules caused equivalent adrenal suppression. The different hydrocortisone dose schedules did affect the temporal pattern of adrenal steroid levels, but the mean 24-hour levels of plasma 17-hydroxyprogesterone and 11-deoxycortisol, and of their corresponding urinary metabolites, were not affected by the timing or the number of hydrocortisone doses. Thus, the hypothalamic-pituitary-adrenal axis appeared to integrate the total daily dose of hydrocortisone regardless of the dose schedule.

The hydrocortisone dose in this study (12.5 $mg/m^2/day$) was selected to be within the normal range of cortisol production rate (12–15 $mg/m^2/day$),[11] but was less than the amount that is usually needed to suppress adrenal function in CAH (up to 25 $mg/m^2/day$). The dose of 12.5 $mg/m^2/day$ was chosen to avoid complete suppression of adrenal function since potential differences in potency among the five schedules could not be assessed as accurately at such an extreme position on the dose-response curve.[7,8] The elevated levels of urine pregnanetriol (normal, <2.0 mg/d) indicate that we succeeded in avoiding full adrenal suppression in most of the patients. However, the two patients who had suppression of pregnanetriol into the normal range at the

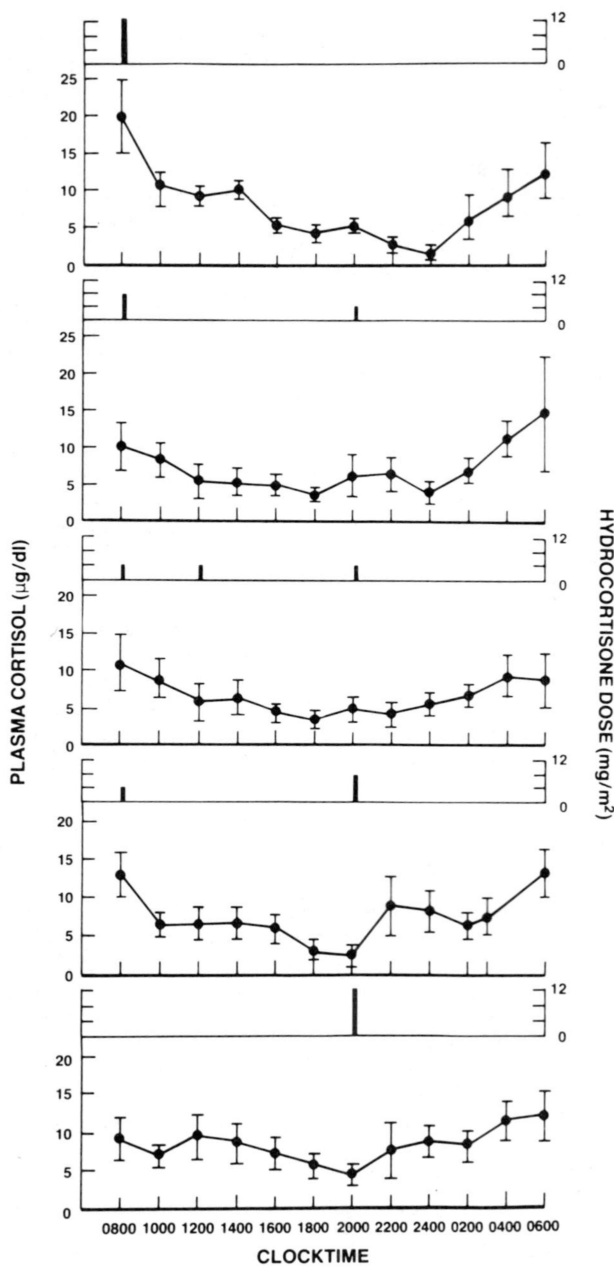

FIGURE 1. The effect of hydrocortisone dose schedule on the 24-hour patterns of plasma cortisol in eight patients with CAH. The timing and dose of hydrocortisone are indicated above each panel (from reference 6).

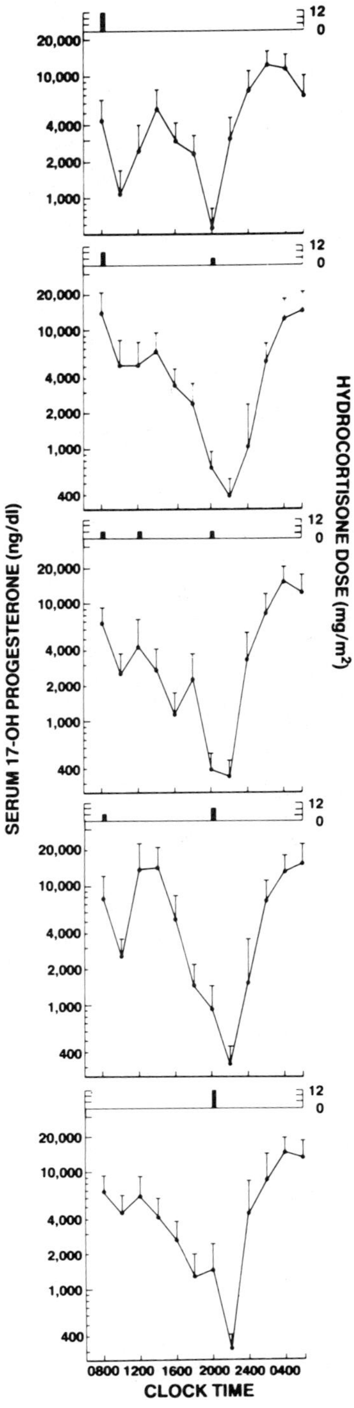

SERUM 17-OH PROGESTERONE (ng/dl)

HYDROCORTISONE DOSE (mg/m²)

CLOCK TIME

0800 1200 1600 2000 2400 0400

20,000 10,000 4,000 2,000 1,000 400

FIGURE 2. The effect of hydrocortisone dose schedule on the 24-hour plasma 17-hydroxyprogesterone profile of six patients with 21-hydroxylase deficiency. The timing and dose of hydrocortisone are indicated above each panel (from reference 6).

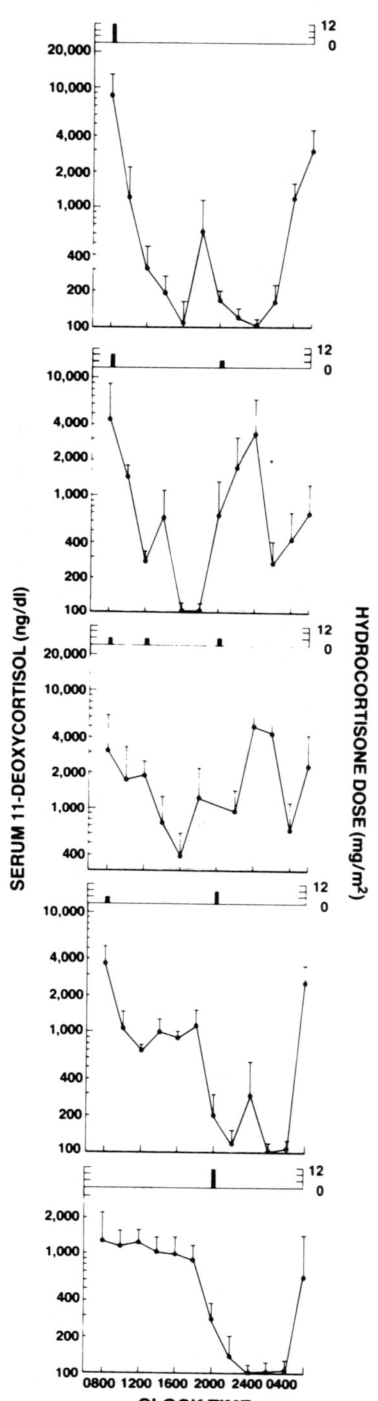

FIGURE 3. The effect of hydrocortisone dose schedule on the 24-hour mean plasma 11-deoxycortisol profile of two patients with 11-hydroxylase deficiency. The timing and dose of hydrocortisone are indicated above each panel (from reference 6).

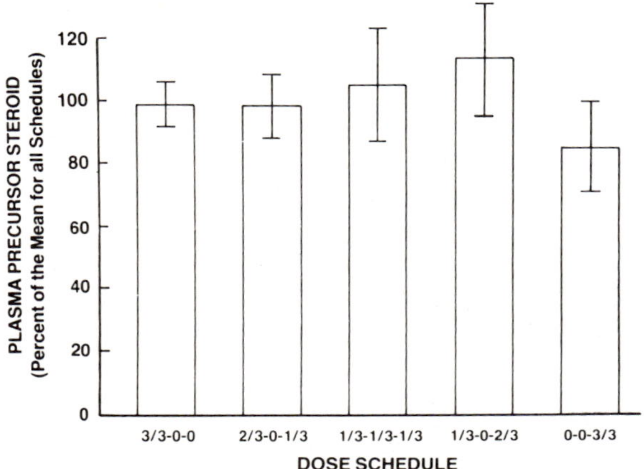

FIGURE 4. The effect of hydrocortisone dose schedule on the level of plasma precursor steroid in congenital adrenal hyperplasia. The precursor steroid was 17-hydroxyprogesterone for the six patients with 21-hydroxylase deficiency and 11-deoxycortisol for the two patients with 11-hydroxylase deficiency. For each patient, the 24-hour mean of the precursor steroid level during each dose schedule was expressed as a percent of the overall mean level from all five schedules. The percentages for each schedule were then averaged over the eight patients in the study. The numbers beneath each bar represent the fractional hydrocortisone dose (in thirds) given at morning, noon, and night, respectively (from reference 6).

hydrocortisone dose of 12.5 mg/m^2/d also showed no consistent difference among the five schedules. Bioassay theory and experience predict this result since potency differences that cannot be detected on the central portion of the dose-response curve should not be detectable at the extreme portions of the curve.[7,8]

The current results are consistent with those of Smith *et al.*[5] They observed no difference in plasma 17-hydroxyprogesterone or ACTH, sampled hourly, in patients

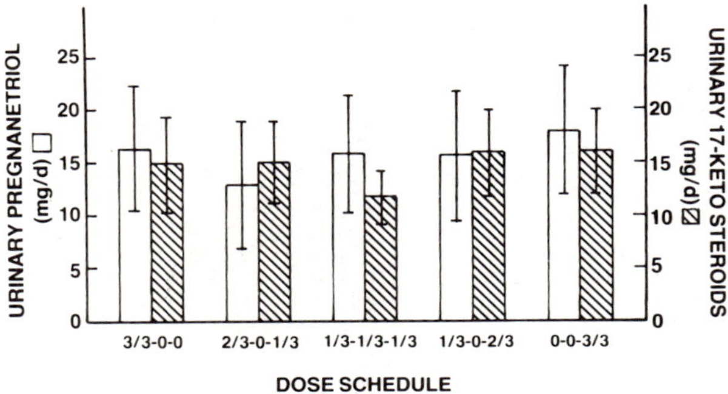

FIGURE 5. The effect of hydrocortisone dose schedule on the average daily urinary pregnanetriol excretion for the six patients with 21-hydroxylase deficiency (light bars) and on the average daily urinary 17-ketosteroid excretion for these six patients and for the two patients with 11-hydroxylase deficiency (shaded bars). The numbers beneath each bar represent the fractional hydrocortisone dose (in thirds) given at morning, noon, and night, respectively (from reference 6).

TABLE 2. The Effect of Hydrocortisone Dose Schedule on Adrenal Suppression in Two Patients with 21-Hydroxylase Deficiency Who Had Adequate Adrenal Suppression on the Hydrocortisone Dose of 12.5 mg/m^2/day[a] (from reference 6)

	Dose Schedule[b]				
Steroid	3/3–0–0	2/3–0–1/3	1/3–1/3–1/3	1/3–0–2/3	0–0–3/3
Plasma 17-Hydroxyprogesterone (ng/dl)[c]	120 ± 39	106 ± 36	74 ± 15	172 ± 104	81 ± 13
Urine Pregnanetriol (mg/d)	0.75 ± 0.3	0.75 ± 0.15	1.20 ± 0.7	0.85 ± 0.15	2.8 ± 2.5
Urine 17-Ketosteroids (mg/d)	6.0 ± 0	7.5 ± 2.5	7.0 ± 3.0	6.0 ± 1.0	9.5 ± 4.6

[a]Mean ± SE of data from patients 4 and 6 (TABLE 1).
[b]Fractional hydrocortisone dose (in thirds) given at morning, noon, and night, respectively.
[c]Twenty-four hour mean of measurements every two hours.

with CAH who were treated with either hydrocortisone (25 mg/m^2/d, given 2/3 at 0800 and 1/3 at 1600) or a similar morning dose coupled with a midnight dose of dexamethasone calculated to have "equivalent" glucocorticoid activity. The duration of each treatment schedule ranged from 3–6 months. They concluded that substitution of nocturnal dexamethasone for the afternoon hydrocortisone dose does not significantly improve plasma ACTH or 17-hydroxyprogesterone secretion in patients with CAH.

One previous study observed that twice daily prednisone treatment was more effective than a single nocturnal dose in normalizing 24-hour mean plasma 17-hydroxyprogesterone.[4] Possible explanations for the differences between this observation and those of the current study include the different glucocorticoid used and the different study design. Specifically, the randomized, double-blind design and the use of placebos in the current study were intended to eliminate the potential effects of investigator or patient bias on adrenal function or on compliance with the treatment schedules.

Nichols *et al.* observed that nocturnal administration of 0.5 mg dexamethasone suppressed adrenal function of normal volunteers more effectively than morning or afternoon administration.[3] We did not observe preferential suppression by nocturnal hydrocortisone compared to the same dose in the morning. Our studies were chronic (4–6 weeks), however, while the above studies were acute (two days). The current study also differed in the subject population, the glucocorticoid administered, and the study design.

This study was designed to detect effects of different dose schedules solely on adrenal suppression. It included adults as well as children, and employed a treatment duration that was too short to examine the effects of the treatment on growth rate and bone age advancement. Thus, although the different dose schedules did not differ in their degree of adrenal suppression, it would be premature to conclude that they might not differ in their growth-inhibiting side effects. As an example, morning administration of hydrocortisone, which mimics the normal circadian pattern of cortisol secretion, would avoid the theoretical possibility that nighttime administration might blunt the nocturnal surge of growth hormone secretion or might impair the growth response to it.

The current study did not indicate any therapeutic advantage of varying the dose schedule of glucocorticoid replacement in CAH. Nocturnal administration of all or a part of the daily dose did not improve adrenal suppression. These observations suggest that treatment of CAH with a once-a-day regimen may be as effective as conventional multiple-dose schedules. Until this hypothesis has been tested by more extended clinical studies, however, we do not recommend a once-a-day approach. We also emphasize that, regardless of which schedule is used, the amount of the daily dose should be carefully adjusted in each patient to achieve a normal rate of growth and bone age advancement.[13,14]

ACKNOWLEDGMENTS

We thank Ms. Penny Colbert, Ms. Mary Hall, and Ms. Mary Beth McCole for typing the manuscript.

REFERENCES

1. URBAN, M. D., P. A. LEE & C. J. MIGEON. 1978. Adult height and fertility in men with congenital virilizing adrenal hyperplasia. N. Engl. J. Med. **229:** 1392–1396.

2. Klingensmith, G. J., C. G. Garcia, H. W. Jones, C. J. Migeon & R. M. Blizzard. Glucocorticoid treatment of girls with congenital adrenal hyperplasia: Effects on height, sexual maturation, and fertility. J. Pediatr. **90:** 996–1004.
3. Nichols, T., C. A. Nugent & F. H. Tyler. 1965. Diurnal variation in suppression of adrenal function by glucocorticoids. J. Clin. Endocrinol. Metab. **25:** 343–348.
4. Huseman, C. A., M. M. Varma, R. M. Blizzard & A. Johanson. 1977. Treatment of congenital virilizing adrenal hyperplasia patients with single and multiple daily doses of prednisone. J. Pediatr. **90:** 538–542.
5. Smith, R., R. A. Donald, E. A. Espiner, C. Glathaar, G. Abbott & M. Scandrett. 1980. The effect of different treatment regimens on hormonal profiles in congenital adrenal hyperplasia. J. Clin. Endocrinol. Metab. **51:** 230–236.
6. Winterer, J., G. P. Chrousos, D. L. Loriaux & G. B. Cutler, Jr. 1985. The adrenal suppressive effect of hydrocortisone in congenital adrenal hyperplasia is not affected by varying the dose schedule. J. Pediatr. **106:** 137–142.
7. Finney, D. J. 1964. Statistical Method in Biological Assay, 2nd edition. Griffin. London.
8. Colquhoun, D. 1971. Lectures on Biostatistics. Clarendon Press. Oxford.
9. Kao, M., S. Voina, A. Nichols & R. Horton. 1975. Parallel radioimmunoassay for plasma cortisol and 11-deoxycortisol. Clin. Chem. **21:** 1644–1647.
10. Glass, A. R., C. E. Smith, G. S. Kidd & R. A. Vigersky. 1978. Response of the hypothalamic-pituitary-testicular axis to surgery. Fertil Steril. **30**(5): 560–563.
11. Schiebinger, R. J., B. D. Albertson, K. M. Barnes, G. B. Cutler, Jr. & D. L. Loriaux. 1981. Developmental changes in rabbit and dog adrenal function: A possible homologue of adrenarche in the dog. Am. J. Physiol. **240:** E694–E699.
12. Kenny, F. M., C. Preegasombat & C. J. Migeon. 1966. Cortisol production rate. II. Normal infants, children, and adults. Pediatrics **37:** 34–42.
13. Bongiovanni, A. M. 1982. The adrenal cortex. *In* Clinical Pediatric and Adolescent Endocrinology. S. A. Kaplan, Ed. WB Saunders. Philadelphia.
14. Brook, C. G. D. 1981. Congenital adrenal hyperplasia. *In* Clinical Pediatric Endocrinology. C. G. D. Brook, Ed. Blackwell Scientific Publications. Oxford.

Monitoring Treatment in Congenital Adrenal Hyperplasia

Use of Serial Measurements of 17-OH-Progesterone in Plasma, Capillary Blood, and Saliva[a]

I. A. HUGHES, J. DYAS, J. ROBINSON,
R. F. WALKER, AND D. R. FAHMY

Department of Child Health
and
Tenovus Institute
University of Wales College of Medicine
Heath Park
Cardiff CF4 4XN United Kingdom

INTRODUCTION

Congenital adrenal hyperplasia (CAH) is caused by a deficiency of 21-hydroxylase enzyme in more than 90 percent of the cases.[1] Measurement of 17-OH-progesterone (17P) in plasma or serum is established clearly as a reliable test for the rapid and early diagnosis of this condition in a newborn infant.[2–4]

Once treatment is started, control needs to be monitored carefully to ensure normal growth and reproductive potential, while only using the minimum dose of steroid replacement for therapy. To supplement the clinical criteria used to assess control, some biochemical marker is required, which would be a sensitive index of short-term changes in therapy. Single, random measurements of plasma 17P are unreliable as an index of therapeutic control. Interpretation of the data is complicated by a pronounced diurnal rhythm in 17P concentrations, which is established by three months of age,[5] episodic fluctuations in secretion of the steroid, and the timing of sample collection relative to glucocorticoid dosage.[3,6] Consequently, it is necessary to perform daily profiles as well as serial, single measurements of 17P concentrations to assess control adequately.

PATIENTS AND METHODS

All studies were performed on patients with CAH attending the Pediatric Endocrine Clinic at the University Hospital of Wales, Cardiff. The first part of this paper reports results on serial plasma 17P and testosterone (T) concentrations in infants at diagnosis and in response to glucocorticoid replacement therapy. All infants had 21-hydroxylase deficiency, except one female infant with 11β-hydroxylase deficiency. Preliminary data on growth velocity during the first year of life also are

[a]Part of this work was supported by a grant from the Welsh Scheme for the Development of Health and Social Research.

presented. The second part describes the value of detailed profiles of 17P concentrations using capillary blood spots and saliva as the matrices for 17P analysis. These studies were performed in older children ranging in age from 5–18 years (n = 25).

The concentration of 17P in plasma, capillary blood spots, and saliva was determined by radioimmunoassay using an antiserum coupled to a magnetizable, solid-phase support and an ^{125}I-radioligand.[7] Plasma testosterone was determined by radioimmunoassay as previously described.[8]

INFANTS: DIAGNOSIS AND EARLY TREATMENT

FIGURE 1 shows the plasma 17P concentrations in five infants at diagnosis and serially thereafter in response to glucocorticoid replacement. An additional infant with

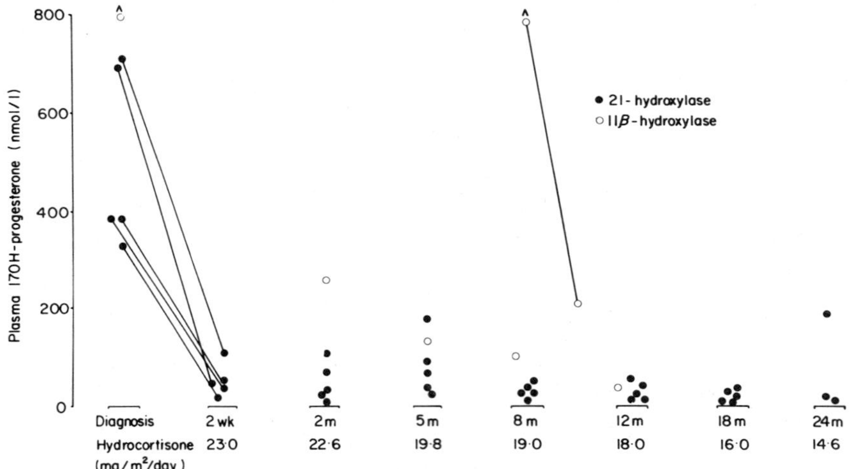

FIGURE 1. Plasma concentrations of 17P (five infants) and 11-deoxycortisol (one infant) before and after treatment for CAH during the first two years of life. Also shown is the change in hydrocortisone dose during this period, related to surface area. The symbol ô denotes a plasma 11-deoxycortisol concentration of 2000 nmol/L.

11β-hydroxylase deficiency, indicated by the open circle symbol, had plasma 11-deoxycortisol concentrations determined. Concentrations of plasma 17P are markedly elevated in untreated infants. Using a mean hydrocortisone replacement dose of 23.0 mg/m^2/day (range 20–25 equivalent to 4–5 mg hydrocortisone daily given in three divided doses), there was a marked decline in 17P concentrations after only two weeks of treatment. Normal values were maintained subsequently during the first one to two years of life using a hydrocortisone replacement dose as low as 15 mg/m^2/day. Similar results were obtained in the infant with 11β-hydroxylase deficiency. There was noncompliance with treatment at eight months of age, leading to an elevated plasma 11-deoxycortisol concentration (2000 nmol/L). However, when hydrocortisone was restarted, plasma 11-deoxycortisol concentrations returned to normal within a few months.

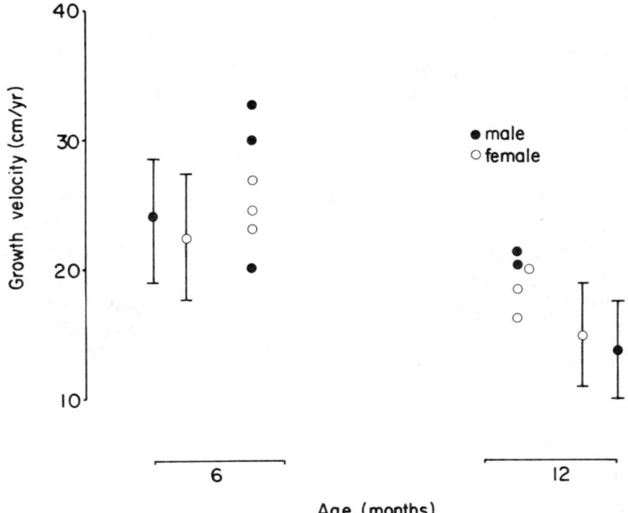

FIGURE 2. Growth velocity during early infancy in treated CAH patients. The bars represent the mean ± 2SD growth velocity in normal infants, after Smith.[9]

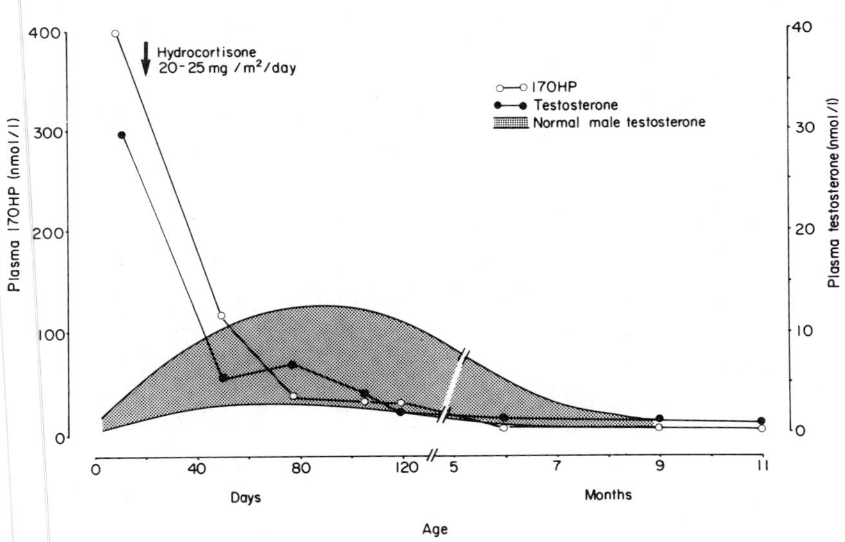

FIGURE 3. Serial plasma 17P and T concentrations in male infants with CAH before and after treatment in early infancy. The shaded area represents the rise in plasma T concentrations in normal male infants, after Winter *et al.*[10]

The growth velocity during the first six and twelve months of life for these infants is illustrated in FIGURE 2. For comparison, the mean ± 2SD for growth velocity in normal male and female infants adapted from Smith[9] is also shown. There was no evidence of growth suppression during the first year of life, which is a critical period of rapid growth in the human.

More frequent serial measurements of plasma 17P and T concentrations were performed in three male infants with CAH (FIGURE 3). At diagnosis, concentrations of 17P and T were greatly increased; the latter within the lower range of T concentrations in the adult male. The concentration of both steroids decreased rapidly with hydrocortisone replacement. However, at about 2–3 months of age, there was a transient rise in plasma T, but not 17P concentrations. A similar increase in T concentrations is also observed in normal male infants at this age, which is indicated in FIGURE 3 by the shaded area. The testes are the source of testosterone production that is associated with an increase in luteinizing hormone secretion.[10] This normal phenomenon was observed

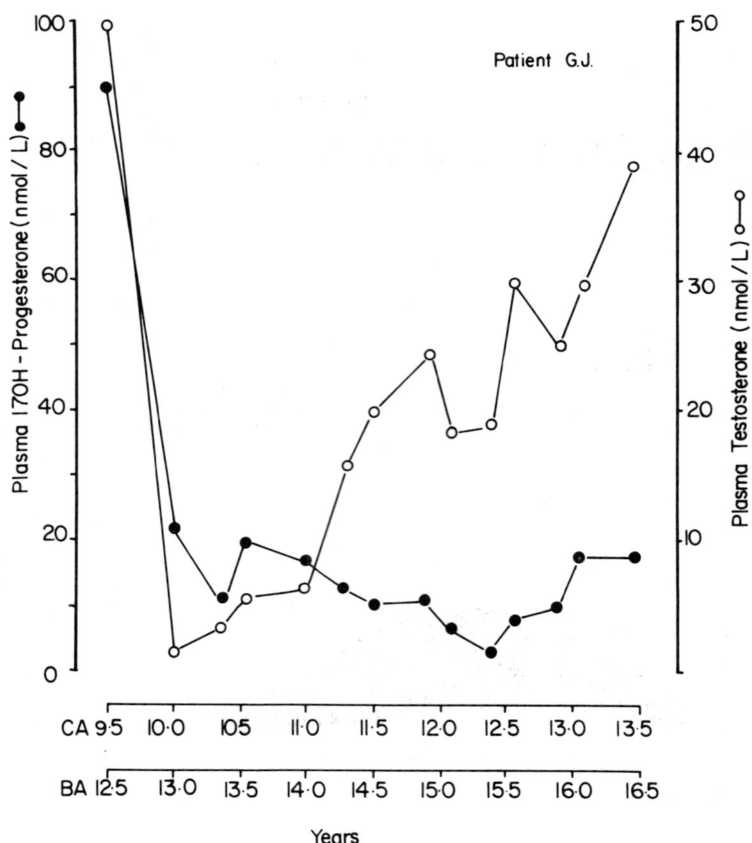

FIGURE 4. Plasma 17P and T concentrations determined longitudinally during puberty in a treated male patient with CAH. Age is depicted both in chronological years (CA) and in bone age years (BA).

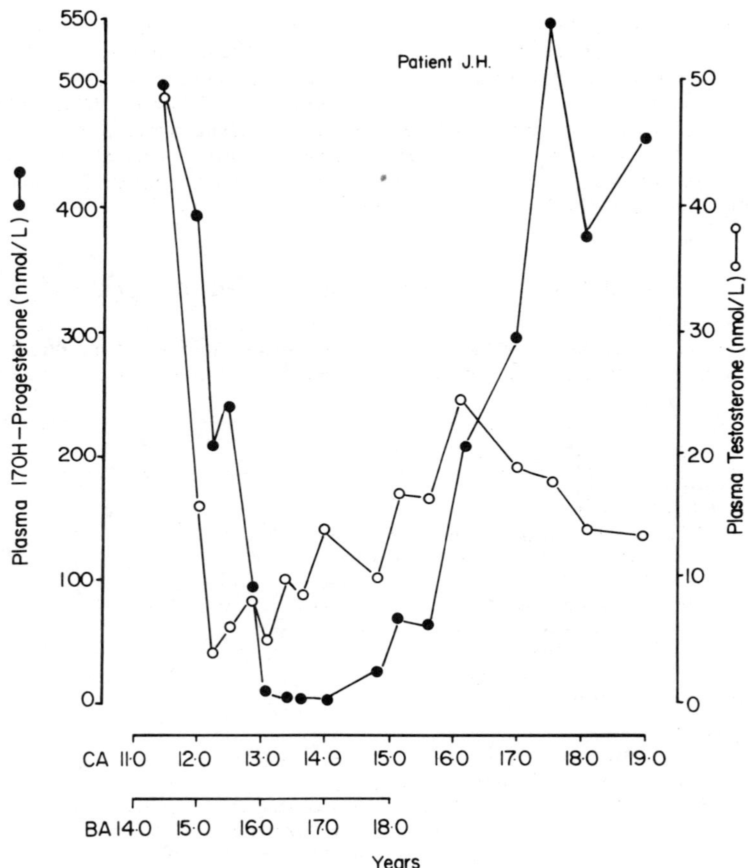

FIGURE 5. Plasma 17P and T concentrations determined longitudinally during puberty in a male patient with CAH, initially on treatment and subsequently stopped. The BA years axis is shortened as the value 18.0 represents epiphyseal fusion.

in male CAH infants given glucocorticoid in replacement doses from the time of diagnosis.

A similar divergence in plasma 17P and T concentrations is observed in males with CAH at the time of puberty. FIGURES 4 and 5 show data on serial measurements of plasma 17P and T in two boys during puberty. Age is shown both as chronological years and as bone age years, which is determined by the TW2 method of Tanner *et al.*[11] In patient, G.J. (FIGURE 4), both plasma 17P and T were elevated just before puberty (CA 9.5, BA 12.5 year). With improved treatment, the concentration of both steroids decreased initially. However, while plasma 17P levels continued to fall, there was a rise in plasma T levels coinciding with the development of clinical signs of puberty (CA 11.0, BA 14.0). By age 13.0 yr (BA 16.0 yr), plasma T concentrations were in the adult male range, whereas 17P concentrations were suppressed, indicating adequate control. In contrast, the results shown in FIGURE 5 on patient J.H. illustrate the effect of noncompliance with treatment. Initially, there was a rise in plasma T levels at age 12.0 (BA 15.0) when treatment was satisfactory. At age 16.0 years, plasma T levels were in the adult range, but increasing noncompliance with therapy resulted in grossly elevated plasma 17P levels and a decrease in plasma T levels. Presumably, the excessive testosterone production from adrenal steroid substrates was sufficient to suppress testicular testosterone production both by a direct gonadal effect and by inhibition of luteinizing hormone secretion. Continued noncompliance with treatment is likely to result in infertility in this patient.[12]

MONITORING TREATMENT: USE OF STEROID PROFILES

Previous studies have reported the value of salivary steroid measurements as an index of changes in plasma steroid concentrations.[13] The application of this technique

TABLE 1. Criteria Used to Assess Control in CAH

I. Clinical
(a) Growth velocity—
birth–2 yr, determined every three months
2 yr–adult, determined every six months
(b) Skeletal maturity—
determined every six months by TW2 method[11]
(c) Clinical signs of glucocorticoid overdosage—
weight gain, striae, hypertension
(d) Signs of puberty at appropriate CA/BA
(e) Regularity of menses (postmenarchal girls)
II. Biochemical
(a) Plasma 17P, T, saliva 17P at 0800–1000 hr (omit morning steroid dose)
(b) Plasma renin activity at 0800–1000 hr
(c) Saliva 17P profiles on two consecutive weekend days (8–10 samples)
(d) Capillary blood spot 17P profile on one day (four samples) in infants
(e) 24-hour 17P profiles in saliva/blood spot (indicated occasionally)
(f) Monthly saliva progesterone profiles as index of ovulation[15] (postmenarchal girls)

for the measurement of saliva 17P concentrations to monitor control in patients with CAH has been reviewed in a recent Workshop on Immunoassays of Steroids in Saliva.[14]

All children with CAH older than two years of age routinely collect saliva samples 4–5 times daily for two consecutive weekend days prior to a clinic visit. At this time, control is assessed according to the criteria in TABLE 1. Serial, single measurements of simultaneous plasma 17P and T, and saliva 17P are still useful when analyzed longitudinally, just as long as the samples are collected at a regular time of day and in relation to previous glucocorticoid dosage. However, more detailed information is afforded by the use of frequent daily profiles of saliva 17P. FIGURE 6 illustrates the data on weekend saliva 17P profiles in treated patients in relation to the degree of control based on the criteria used in TABLE 1. There is considerable overlap between the three groups if a single 17P value is used, particularly during the latter part of the day. This is due to the pronounced 17P diurnal rhythm that is still present when the concentrations are elevated. The pattern observed in a daily profile of 17P con-

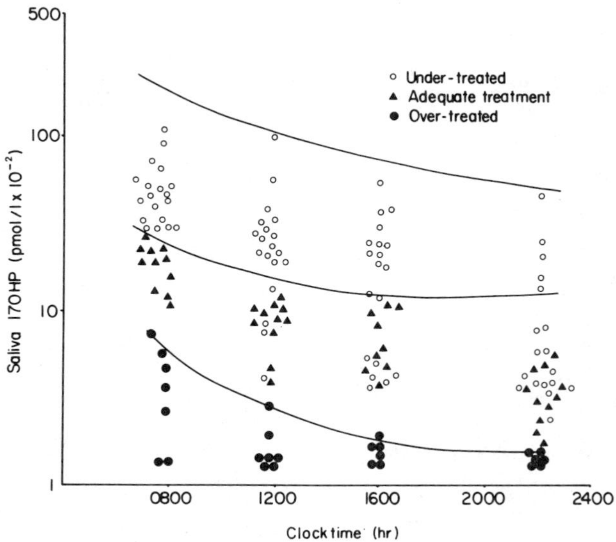

FIGURE 6. Consecutive 2-day saliva 17P profiles in treated CAH patients classified according to adequacy of control. Saliva 17P concentrations are plotted on a logarithmic scale. (From Hughes and Read,[14] with permission from Alpha Omega Publishing Ltd.)

centrations permits discrimination between the three groups of patients, and in particular, provides a reliable index of overtreatment. Typical daily 17P profiles depicting the overtreated, adequately treated, and undertreated patients are shown in TABLE 2.

Since frequent saliva sample collection is technically difficult in children below two years of age, the method for 17P radioimmunoassay was applied to capillary blood collected onto filter paper. Parents were instructed to collect sufficient capillary blood with an Autolet device as used by diabetic patients for home blood glucose monitoring. To evaluate the accuracy of this method, plasma 17P and T, blood spot 17P, and saliva 17P were measured every two hours over a 12-hour period in two patients with CAH (see FIGURES 7a and 7b). Identical results were obtained for 17P concentrations determined in plasma and capillary blood spots. The daily profile of 17P was similar in all three matrices in which the steroid was measured. Both patients showed a

TABLE 2. Typical Daily Saliva 17P Profiles in Treated CAH Patients[a]

	Saliva 17P (pmol/L)		
Time(h)	Undertreated	Adequately Treated	Overtreated
0800	6000	1500	500
1200	4000	1000	<150
1600	2500	600	<150
2200	2000	400	<150

[a]From Hughes and Read,[14] with permission.

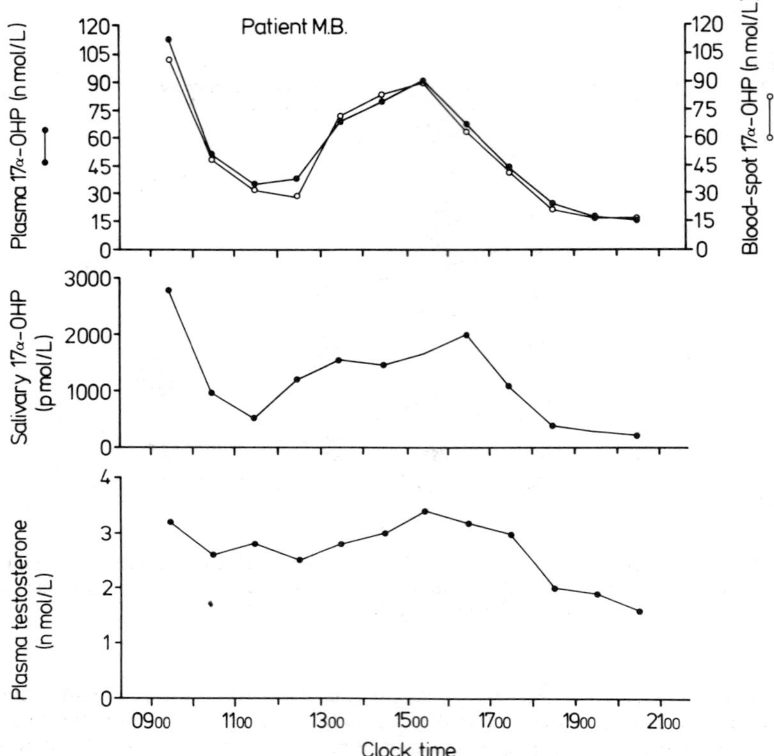

FIGURE 7a. Simultaneous measurements of plasma, capillary blood spot and saliva 17P, and plasma T over 12 hours in patient M.B. with CAH.

pronounced 17P diurnal rhythm; there was also a suggestion of a similar rhythm in plasma testosterone.

CONCLUSIONS

Following diagnosis, infants with congenital adrenal hyperplasia require only a replacement dose of hydrocortisone, which should be calculated based on the normal cortisol secretion rate (12 mg/m^2/day).[16] When given in three divided doses, this lowers 17P concentrations towards the normal range in a matter of weeks. Continuing therapy during the first year of life with hydrocortisone in dosages ranging between 15–20 mg/m^2/day ensures a normal growth velocity for age. In male infants, this treatment regimen does not affect the normal testis-derived testosterone surge characteristic of early infancy. This phenomenon recurs during puberty, so the plasma T measurements at this time do not reflect adrenal precursor steroid secretion. It is recommended that glucocorticoid replacement should be continued in adult males in order to prevent endogenous suppression of Leydig cell function and possible later infertility.

Since concentrations of 17P can fluctuate widely, daily profiles of this steroid should be used to monitor treatment effectively. Serial saliva sampling is a simple and painless method, particularly applicable to children. For infants and young children, collection of capillary blood onto filter paper is also practical; analysis of 17P levels in both matrices accurately reflects changes observed in plasma. These newer techniques ensure that adjustments in treatment schedules based on detailed 17P profiles are now possible in all CAH patients from early infancy to adult life.

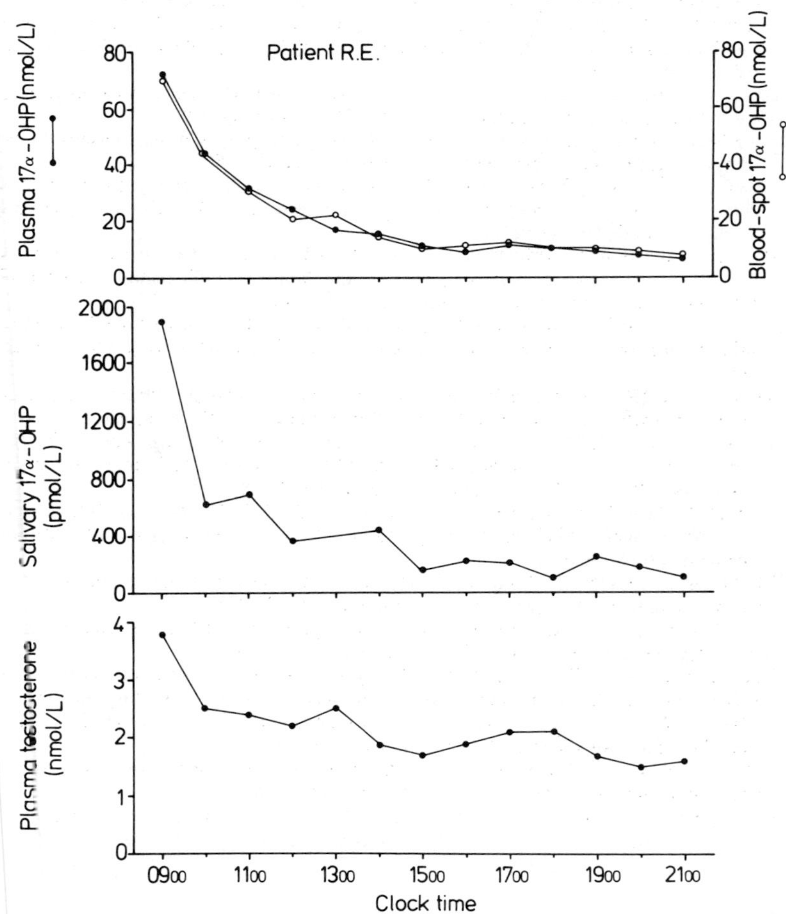

FIGURE 7b. Same measurements and conditions as in FIGURE 7a, except with patient R.E. with CAH.

ACKNOWLEDGMENTS

I. A. Hughes gratefully acknowledges the award of a Travel Grant from the Wellcome Trust to attend the Symposium on Advances in Congenital Adrenal Hyperplasia.

REFERENCES

1. HUGHES, I. A. 1982. Congenital and acquired disorders of the adrenal cortex. Clin. Endocrinol. Metab. **11:** 89–125.
2. YOUSSEFNEJADIAN, E. & R. DAVID. 1975. Early diagnosis of congenital adrenal hyperplasia by measurements of 17-hydroxyprogesterone. Clin. Endocrinol. **4:** 451–454.
3. HUGHES, I. A. & J. S. D. WINTER. 1976. The application of a serum 17-OH-progesterone radioimmunoassay to the diagnosis and management of congenital adrenal hyperplasia. J. Pediatr. **88:** 766–773.
4. HUGHES, I. A., D. RIAD-FAHMY & K. GRIFFITHS. 1979. Plasma 17-OH-progesterone concentrations in newborn infants. Arch. Dis. Child. **54:** 347–349.
5. SOLYOM, J. 1984. Diurnal variation in blood 17-hydroxyprogesterone concentrations in untreated congenital adrenal hyperplasia. Arch. Dis. Child. **59:** 743–747.
6. FRISCH, H., K. PARTH, E. SCHOBER & W. SWOBODA. 1981. Circadian patterns of plasma cortisol, 17-hydroxyprogesterone, and testosterone in congenital adrenal hyperplasia. Arch. Dis. Child. **56:** 208–213.
7. DYAS, J., G. F. READ, T. J. MAULIK, I. A. HUGHES & D. RIAD-FAHMY. 1984. A rapid assay for 17OH-progesterone in plasma, saliva, and amniotic fluid using a magnetisable solid-phase antiserum. Ann. Clin. Biochem. **21:** 417–424.
8. TURKES, A., A. O. TURKES, B. G. JOYCE, G. F. READ & D. RIAD-FAHMY. 1979. A sensitive solid-phase enzyme immunoassay for testosterone in plasma and saliva. Steroids **33:** 347–359.
9. SMITH, D. W. 1977. Growth and its disorders. *In* Major Problems in Clinical Pediatrics, vol. 15. A. J. Schaffer & M. Markowitz, Eds. W.B. Saunders Co. Philadelphia.
10. WINTER, J. S. D., C. FAIMAN & F. REYES. 1981. Sexual endocrinology of fetal and perinatal life. *In* Mechanisms of Sex Differentiation in Animals and Man. C. R. Austin & R. G. Edwards, Eds: 205–253. Academic Press. New York.
11. TANNER, J. M., R. H. WHITEHOUSE, W. A. MARSHALL, M. J. R. HEALY & H. GOLDSTEIN. 1975. Assessment of Skeletal Maturity and Prediction of Adult Height (TW2 Method). Academic Press. London.
12. WISCHUSEN, J., H. W. G. BAKER & B. HUDSON. 1981. Reversible male infertility due to congenital adrenal hyperplasia. Clin. Endocrinol. **14:** 571–577.
13. RIAD-FAHMY, D., G. F. READ, R. F. WALKER & K. GRIFFITHS. 1982. Steroids in saliva for assessing endocrine function. Endocrinology Rev. **3:** 367–395.
14. HUGHES, I. A. & G. F. READ. 1984. Salivary 17α-hydroxyprogesterone and congenital adrenal hyperplasia. *In* Immunoassays of Steroids in Saliva. Proceedings of the Ninth Tenovus Workshop. G. F. Read, D. Riad-Fahmy, R. F. Walker & K. Griffiths, Eds.: 274–295. Alpha Omega Publishing Ltd. Cardiff, U.K.
15. HUGHES, I. A. & G. F. READ. 1982. Menarche and subsequent ovarian function in girls with congenital adrenal hyperplasia. Horm. Res. **16:** 100–106.
16. KENNY, F. M., C. PREEYASOMBAT & C. J. MIGEON. 1966. Cortisol production rate. II. Normal infants, children, and adults. Pediatrics **37:** 34–42.

Cytochrome P-450: Physiology of Steroidogenesis[a]

PETER F. HALL

Worcester Foundation for Experimental Biology
222 Maple Avenue
Shrewsbury, Massachusetts 01545

INTRODUCTION

The conversion of cholesterol to steroid hormones requires a number of cytochromes P-450. Two of these are found in the inner mitochondrial membrane (C_{27} side-chain cleavage and 11β-hydroxylase) and four are microsomal (C_{21} side-chain cleavage, 17α-hydroxylase, 21-hydroxylase, and aromatase). All cytochromes P-450 (including those not involved in steroid synthesis) possess many properties in common—these properties arise from the heme, from the specific mode of attachment of heme to the protein via a thiolate bond, and from parts of the protein structure conserved during evolution. In addition, the protein moiety provides specificity for specific substrates. The steroidogenic cytochromes P-450 catalyze two types of reactions clearly indicated by their common names—hydroxylation and cleavage of C–C bonds. What follows is designed to relate the properties of these enzymes to the regulation of the steroidogenic pathway in order to facilitate interpretation of abnormalities in steroid metabolism that result from congenital defects in these enzymes.

CYTOCHROMES P-450

The cytochromes P-450 constitute a family of heme proteins that catalyze monooxygenase reactions. These enzymes are reduced by electrons from reduced pyridine nucleotides. The reduced enzyme activates molecular oxygen, and one atom of the oxygen is inserted into the product as a hydroxyl group, in contrast to dioxygenases, which insert both atoms of oxygen into the product. The fundamental statement of monooxygenation is as follows:

$$\mathrm{NADPH} + \mathrm{H}^+ + {}^{18}\mathrm{O}_2 + \mathrm{RH} \xrightarrow{[\text{P-450}]} \mathrm{R}\text{-}{}^{18}\mathrm{OH} + \mathrm{H}_2{}^{18}\mathrm{O} + \mathrm{NAD}^+$$

The equation shows reduction of atmospheric or molecular oxygen (labeled as ${}^{18}O_2$) by electrons from NADPH; the oxygen of water is not used in this reaction. The substrate is a lipophilic substance that becomes more hydrophilic as the result of the new hydroxyl group. The more hydrophilic product is readily removed from the body by filtration. In this way, monooxygenation removes lipophilic drugs from the body and promotes metabolism of a number of substances made in the body, e.g., steroids and

[a]This work was supported by grants from the National Institutes of Health, Bethesda, Maryland (HD–16525, CA–29497, AM–32236, and AM–28113).

prostaglandins, after these substances have been formed in the normal course of cell and organ function. The equation also reveals the characteristic stoichiometry of monooxygenation, namely, one mole of NADPH, one mole of H^+, and one mole of oxygen per mole of product formed. The equation does not show one important feature of the reaction, namely, that one or two electron carriers are necessary for the transport of electrons from NADPH to P-450 depending upon whether the P-450 is microsomal (one flavoprotein containg FAD and FMN) or mitochondrial (one flavoprotein containing FAD and an iron-sulfur protein).

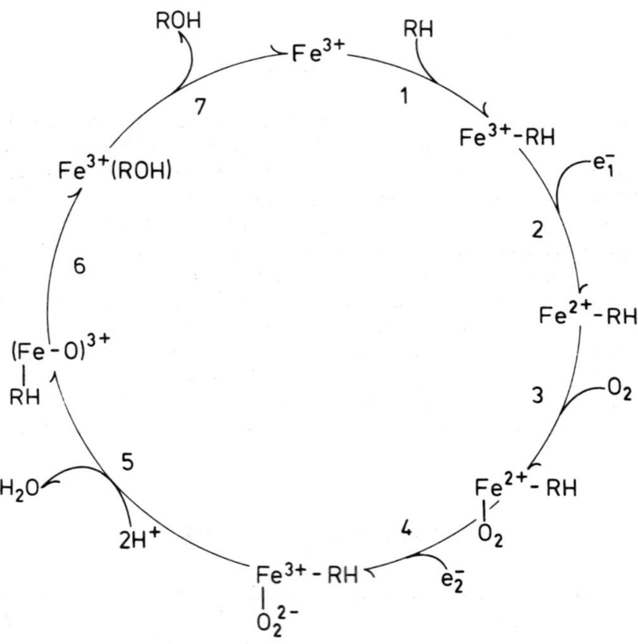

FIGURE 1. The P-450 cycle.[22]

The P-450 Cycle

Since P-450 is reduced to activate oxygen and is reoxidized during the formation of the products of the reaction, there is clearly a cycle of changes in which the heme protein is alternately oxidized and reduced. The cycle (FIGURE 1) takes place in seven steps as follows:

(I) Substrate binds to P-450 to form an enzyme-substrate complex.
(II) The first electron passes from NADPH to reduce the iron of the enzyme-substrate complex. Step (I) facilitates this step, i.e., the enzyme-substrate complex is reduced more rapidly than enzymes without substrate.
(III) The reduced heme of the enzyme-substrate complex binds oxygen.
(IV) The second electron from NADPH activates the oxygen and rearrangement of electrons causes the iron to revert to the ferric form.

(V) The oxygen molecule undergoes cleavage and one atom is reduced to water.

(VI) Hydroxylation of the substrate occurs. It is worth mentioning one of the most plausible mechanisms for this step, although agreement has not been reached about the detailed mechanism of this reaction. The mechanism in question is called oxygen rebound. The enzyme abstracts hydrogen from the substrate to form a carbon radical plus a hydroxyl radical. The hydroxyl radical attacks the carbon radical (it is said to "rebound"), resulting in the formation of a hydroxylated product.

(VII) The hydroxylated product leaves the enzyme and free oxidized P-450 is again ready to start the cycle. The hydroxyl group presumably facilitates this step by providing repulsive forces in the hydrophobic crevice of the enzyme that houses the heme group and the substrate.

Several variations in the types of reactions catalyzed by P-450 are known. For example, metabolism of certain drugs includes N-dealkylation reactions catalyzed by P-450, using some variation of the scheme described above. More important, in the present context, are two reactions in the steroidogenic pathway in which C–C bonds are split, i.e., side-chain cleavage of cholesterol to give pregnenolone and side-chain cleavage of C_{21} steroids to give C_{19} steroids. These general properties of P-450 have been reviewed elsewhere.[1–3]

Spectral Properties of Cytochrome P-450

Cytochromes P-450 all show three spectral properties that are important in studying these enzymes:

(I) Absorbance Spectra

The absolute absorbance spectra of all heme proteins are characterized by a pronounced peak at approximately 420nm known as the Soret peak (FIGURE 2). Since heme is characterized by conjugated double bonds and since the π electrons of conjugated double bonds are susceptible to influences exerted by charged molecules in the immediate environment, the heme group of heme proteins can be regarded as a sensitive reporter group for the occurrence of subtle changes in the structure of the surrounding protein. In view of its high extinction coefficient, the position and intensity of the Soret peak provide important information concerning the protein moieties of heme proteins including cytochrome P-450.

In addition to the Soret peak, cytochrome P-450 shows two peaks in the region of 550nm which are replaced by a single peak when the heme iron is reduced.

(II) CO-Reduced P-450 Complex

When heme proteins are reduced, the heme group combines with carbon monoxide. As one would expect, this treatment results in a shift in the Soret peak, which in most heme proteins is confined to a few nm. In cytochromes P-450, the shift in the Soret peak produced by CO is extensive and the peak now appears at 450nm (FIGURE 2), which gives these enzymes the name "pigment 450" (P-450). This spectral shift provides the most generally useful method for detecting and measuring P-450 in biological samples. For such purposes, it is convenient to measure the spectral shift by

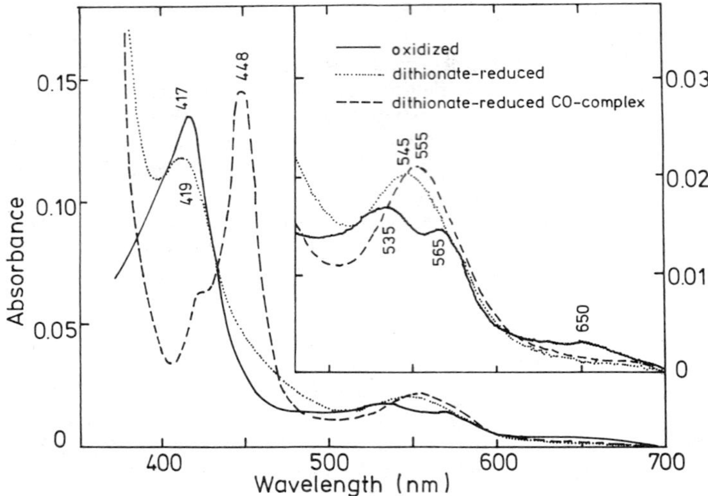

FIGURE 2. Spectral properties of cytochrome P-450.

difference spectroscopy ($A_{390-450}$) because background absorbance seen in turbid samples is thereby subtracted.

(III) Substrate-Induced Difference Spectra

When a substrate binds an enzyme to form an enzyme-substrate complex (ES), the conformation of the enzyme changes, and, in the case of P-450, this causes a shift in the Soret peak to approximately 390nm. This change is associated with a change in the valence state of iron from hexacoordinate to pentacoordinate and concomitant displacement of iron from the plane of the heme ring. It is convenient to measure this change in the Soret peak by difference spectroscopy ($A_{390-420}$) (FIGURE 3). Difference spectroscopy involves subtraction of absorbance in the reference cuvette (no substrate), from that in the sample cuvette (enzyme plus substrate). Subtraction converts the peak at 420 to a trough and the $A_{390-420}$ expands the absorbance signal form a peak to a peak and trough, i.e., for a given number of bound molecules (ES), the absorbance signal is greater when measured by difference (FIGURE 3).

All three of these spectral properties are seen with all cytochromes P-450 because they result from features that are common to all members of this family. The main features of the absolute spectrum, including the Soret peak, are common to all heme compounds (i.e., whether the protein is present or not). The shift in the Soret peak by CO arises from the nature of the attachment of the heme to protein, which is the same in all members of the P-450 family; this attachment takes the form of a thiolate bond. The spectral shift induced by binding of substrate results from the change in the iron (hexacoordinate to pentacoordinate) and the concomitant alteration of the d orbital electrons—once again these changes are common to all cytochromes P-450. Further details of these features of P-450 are to be found in references 1 and 2.

The Steroidogenic Pathway

When one compares the structure of cholesterol to those of the major steroid hormones (FIGURE 4), it is clear that the following changes must occur in the structure of cholesterol in order to convert this substance to active steroid hormones:

Activity	C Atoms or Steroid Ring
Hydroxylation*	22, 20, 17α, 21, 11β, 18
Dehydrogenation	3β
Oxidase Activity*	18
C–C Cleavage*	20, 22 and 17, 20
Aromatization*	A ring
Isomerization	$\Delta^{4,5}$

The reactions marked "*" require cytochrome P-450 enzymes.

The conversion of cholesterol to the steroid hormones (steroidogenesis), requires the coordination of mitochondrial and microsomal enzyme systems (FIGURE 5). Steroidogenesis begins in the inner mitochondrial membrane when cholesterol binds to the first enzyme in the pathway, i.e., the C_{27} side-chain cleavage enzyme which converts cholesterol to pregnenolone, from which the various steroid hormones are synthesized. The pregnenolone so formed moves to the microsomal membrane where the 3β-hydroxyl group is converted to a ketone and isomerization of the double bond completes the Δ^4-3-ketosteroid structure. The two enzymes involved in the conversion of pregnenolone to progesterone are not cytochromes P-450. The right half of FIGURE 4

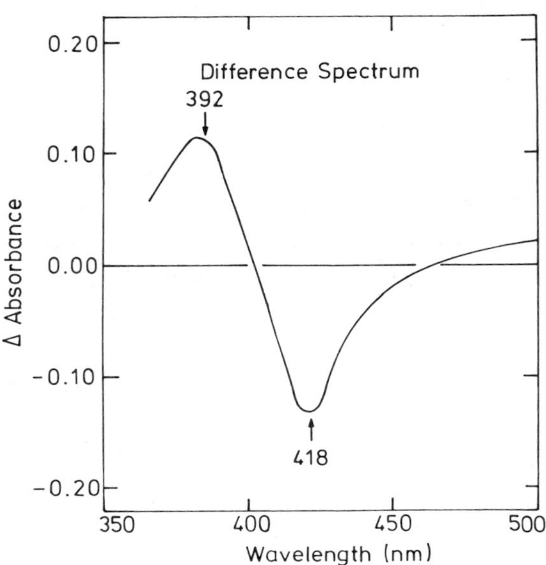

FIGURE 3. Substrate-induced difference spectrum of P-450.[23]

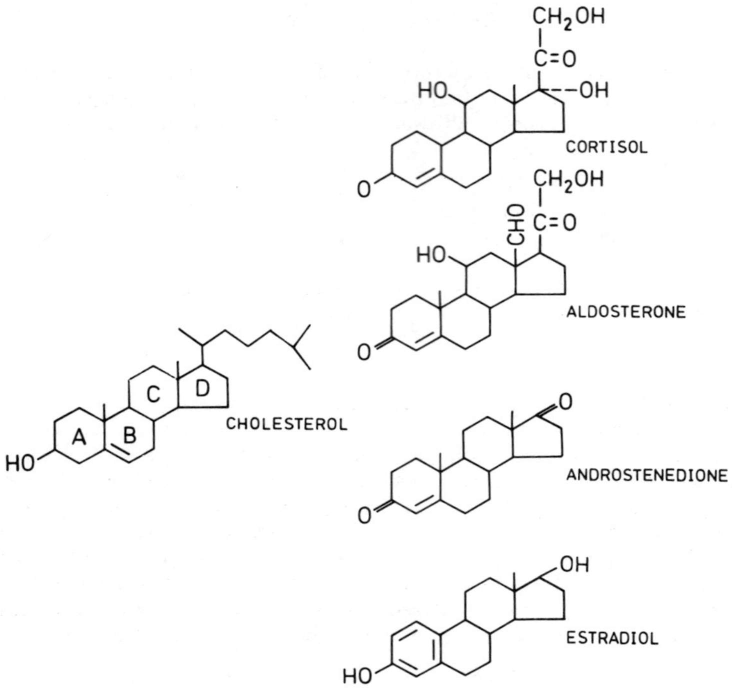

FIGURE 4. Structures of cholesterol and steroid hormones.[22]

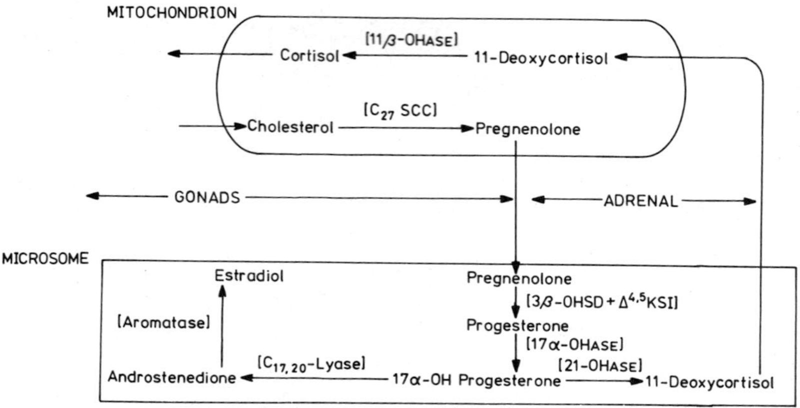

FIGURE 5. The organization of steroidogenesis.[23]

shows the synthesis of corticosteroids. This involves two microsomal cytochromes P-450, namely, 17α-hydroxylase and 21-hydroxylase in that order. These are two distinct enzymes that are typical microsomal cytochromes P-450. The 17α-, 21-dihydroxy steroid is conveniently called 11-deoxycortisol. This intermediate must return to the inner mitochondrial membrane for the last step in the pathway, namely 11β-hydroxylation catalyzed by a typical mitochondrial P-450.

Two modifications of this pathway should be considered at this point. In the first place, the rat and several other species do not show 17α-hydroxylation of adrenocortical steroids. Therefore, the 21-hydroxyl derivative of progesterone, which is known as 11-deoxycorticosterone (or DOC), returns to the inner mitochondrial membrane for 11β-hydroxylation to corticosterone. Secondly, the *zona glomerulosa* makes the 17-deoxysteroid aldosterone, so that the cells of this zone do not catalyze 17α-hydroxylation. Therefore, DOC is once again the intermediate that returns to mitochondria for 11β-hydroxylation to corticosterone, which is then subjected to 18-hydroxylation followed by an oxidase reaction to produce aldosterone. These reactions all take place in the inner mitochondrial membrane (FIGURE 6).

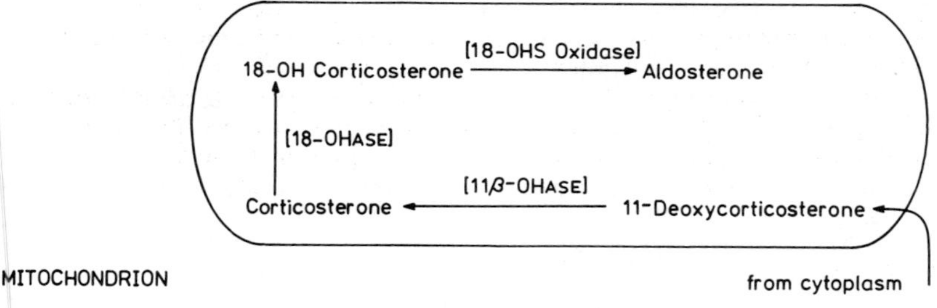

FIGURE 6. The biosynthesis of aldosterone.[23]

The left half of FIGURE 5, shows what happens in the gonads where 17α-hydroxylation occurs in all species since it is a necessary step in the side-chain cleavage of C_{21} steroids to give C_{19} steroids. The commonest situation appears to be that shown in FIGURE 5, where progesterone is the substrate for two reactions catalyzed by microsomal cytochrome P-450 to give androstenedione, which is converted to testosterone in the microsomal membrane. The two androgens are converted to estrogens by a P-450 aromatase that is also microsomal. We can now consider these enzymes in turn. A general review of steroidogenesis is to be found in reference 3.

THE STEROIDOGENIC CYTOCHROMES P-450

C_{27} Side-Chain Cleavage

The conversion of cholesterol to pregnenolone takes place in three steps although the whole process is coordinated into three catalytic cycles of P-450 that occur in rapid succession (FIGURE 7).

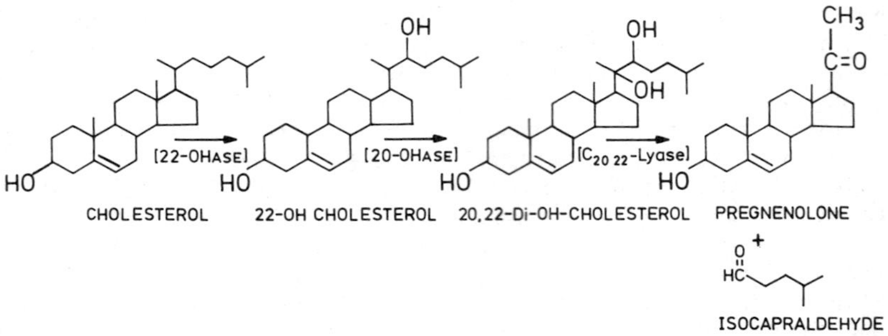

FIGURE 7. The side-chain cleavage of cholesterol.[23]

The enzyme was first purified by Shikita and Hall in 1973 from beef adrenal cortex.[4] The enzyme behaves as a single protein on electrophoresis in SDS gels and by immunochemical criteria. Evidently, all three steps are catalyzed by a single protein. Although the catalysis of two hydroxylation reactions by a single enzyme was not very surprising, the cleavage of a C–C bond by cytochrome P-450 appeared to be without precedent. It was, therefore, important to determine the stoichiometry of this step catalyzed by the pure enzyme. By measuring enzyme activity and the consumption of NADPH, H^+, and O_2 simultaneously, it was found that all three steps in side-chain cleavage show the typical stoichiometry of monooxygenation (FIGURE 8).[5]

It was, therefore, essential to determine whether or not the heme moiety of the

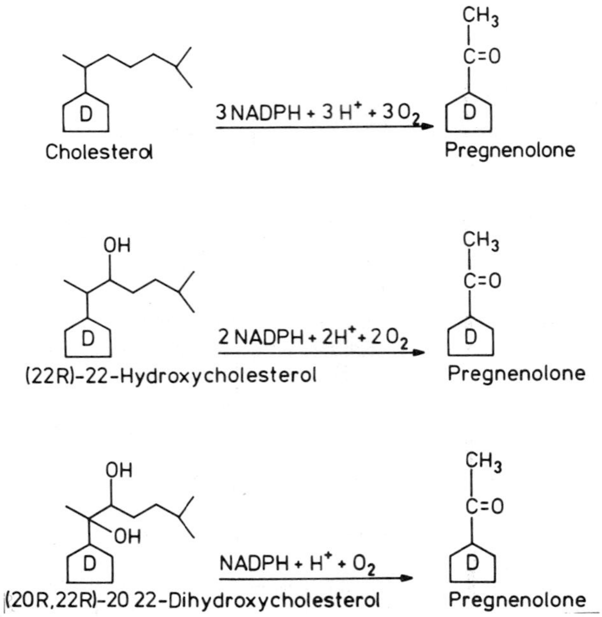

FIGURE 8. Stoichiometry of the three steps of side-chain cleavage of cholesterol.

enzyme is involved in the cleavage of the C_{20}–C_{22} bond. This was done by determining photochemical action spectra for the overall reaction (cholesterol → pregnenolone) and for the cleavage step (20,22-dihydroxycholesterol → pregnenolone). This procedure depends upon three properties of the reduced P-450–CO complex. The complex is enzymatically inactive, it is light sensitive, and it shows (as we have seen) an absorbance maximum at 450nm. The procedure involves measuring the affinity of the enzyme for CO in the dark and in the presence of light of various wavelengths. Since the CO complex is light sensitive, light releases the enzyme from inhibition, which is readily measured by determining the production of pregnenolone. FIGURE 9 shows photochemical action spectra for the whole reaction and the last step.[6]

The term on the ordinate of FIGURE 9 represents the effect of light of wavelength X divided by that of 451nm. When X equals 451nm, this term reduces to 1 and light of shorter and longer wavelengths are less effective in reserving inhibition by CO. Clearly, heme is involved in the last step, i.e., the C–C cleavage. This was the first demonstration that cytochrome P-450 can cleave a C–C bond by monooxygenation.[6]

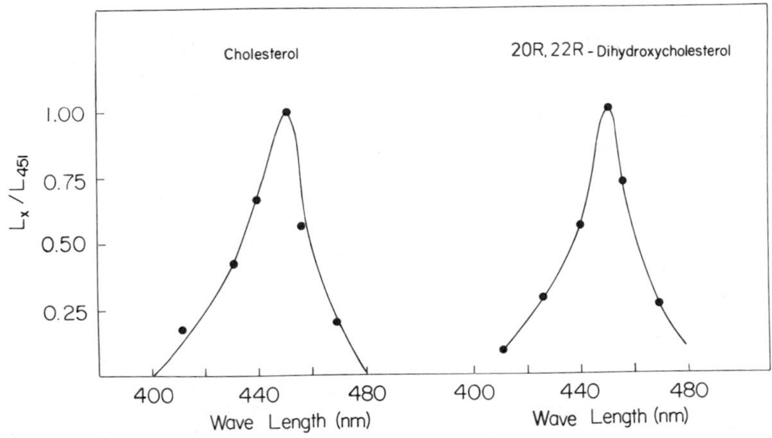

FIGURE 9. Photochemical action spectra of side-chain cleavage (for adrenal P-450).

Since there is only one heme group per molecule of enzyme and since all three steps require heme, we were forced to give serious consideration to the possibility that the enzyme possesses a single active site. Indeed, kinetic and binding studies suggest that the enzyme behaves as though it possesses a single active site.[7]

When the enzyme is purified, it contains bound phospholipid and remains soluble in water in the absence or near-absence of detergent. When the enzyme is incubated with electron carriers and NADPH, it aggregates to a hexadecamer of MW 850,000. It remains to determine whether this is the active form of the enzyme *in vivo*.[8]

C_{21} Side-Chain Cleavage

It has never been seriously considered that the hydroxylase and lyase could be two activities of a single enzyme (FIGURE 10). The reasons for this were at least two in number: (I) The adrenal cortex must possess a 17α-hydroxylase that does not show

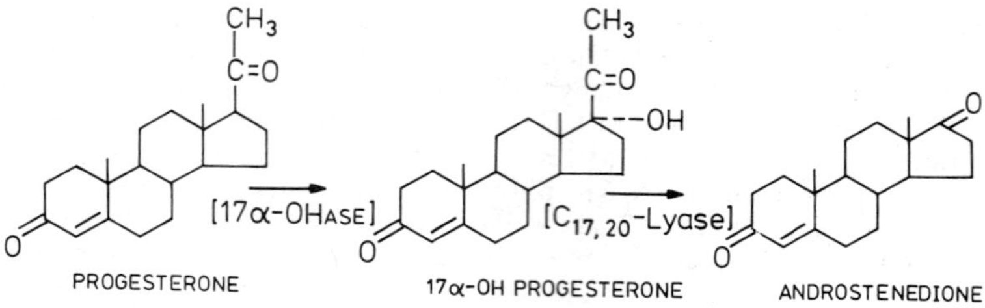

PROGESTERONE [17α-OHASE] 17α-OH PROGESTERONE [$C_{17,20}$-Lyase] ANDROSTENEDIONE

FIGURE 10. Conversion of progesterone to androstenedione.[23]

lyase activity since the cortex produces 17α-hydroxy-C_{21} steroids (e.g., cortisol); (II) In contrast to the C_{27} system in which significant accumulation of hydroxylated intermediates does not occur, 17α-hydroxyprogesterone is readily detected in the testis and ovary suggesting that the side-chain cleavage of C_{21} steroids proceeds in two discrete steps.

When the 17α-hydroxylase was purified from pig testicular microsomes, it was found to possess vigorous lyase activity and no other lyase activity could be detected in the microsomal fractions.[9] The enzyme is homogeneous by acrylamide gels and by immunochemical criteria.[9,10] Moreover, half of the amino acid sequence is now available and each residue appears as a single amino acid.[10] Clearly, the enzyme is homogeneous.[9,10]

Again, the photochemical action spectra shows that the cleavage step requires P-450.[11] Moreover, two competitive inhibitors of the SU series inhibit both hydroxylase and lyase activities with the same value for K_i with each inhibitor, but with an order of magnitude difference in K_i values for the two inhibitors. In addition, an antibody (IgG) inhibits both hydroxylase and lyase activities with the same concentration dependence, and gentle heating at 30°C inhibits both activities with the same time dependence.[10] When the binding of substrate (progesterone or pregnenolone) is studied by spectral or equilibrium dialysis methods, these substrates and their 17α-hydroxy counterparts behave as though they are competing for a single active site.[12]

Obviously, there are many similarities between the two side-chain cleavage enzymes. The differences seem to arise from the occurrence of one in mitochondria and the other in microsomes.

Adrenal 17α-Hydroxylase

When this enzyme was studied, it was expected that it would show hydroxylase but not lyase activity. Surprisingly, the pure enzyme very closely resembles the testicular hydroxylase/lyase. Enzyme activities, values for K_m, and binding affinities measured by spectral and dialysis methods gave almost the same values for the two enzymes.[13] The immune system of the rabbit failed to distinguish between the two enzymes, both of which cross-reacted with antibodies raised by each of the pure enzymes. The amino acid compositions of the two enzymes were very similar (although not identical).[13] Again, the amino acid sequence of the first 16 residues from the NH_2 terminus show only two conservative changes (ser → thr in testis and adrenal, respectively).[13]

If one incubates pig adrenal microsomes with progesterone, no C_{19} steroids are formed, and with 17α-hydroxyprogesterone, very little androstenedione is formed.[14]

Clearly, the lyase activity of the pure enzyme, which is as active as that of the testicular enzyme, must be inhibited in the adrenal microsomes, but not in testicular microsomes which form large amounts of androstenedione from progesterone.

21-Hydroxylase

This enzyme which converts 17α-hydroxyprogesterone to 11-deoxycortisol has been purified from beef[15] and pig adrenal.[16] Both enzymes show lower K_m with 17α-hydroxyprogesterone than with progesterone.[14,15] This is interesting in view of the fact that 17α-hydroxylation precedes 21-hydroxylation in the synthesis of cortisol. Perhaps, the sequence of these two reactions results from the preference of the 21-hydroxylase for the 17α-hydroxy substrate.

11β-Hydroxylase

The last enzyme in the pathway to cortisol is 11β-hydroxylase. When this enzyme was purified from beef adrenal, it was found that the homogeneous enzyme catalyzes both 11β-hydroxylation of DOC and 18-hydroxylation of DOC, but not that of corticosterone.[17,18] It is generally believed that 18-hydroxy DOC is not an intermediate in the synthesis of aldosterone (FIGURE 11).

The 11β-hydroxylase that we have purified may be that of the fasciculata because its properties are consistent with that of cells that do not form aldosterone. Moreover, 18-hydroxy DOC is found in the fasciculata of the rat, and losses of glomerulosa when the cortex is scraped from the adrenal capsule during preparation of the tissue are clearly considerable. It seems likely that the glomerulosa possesses another 11β-hydroxylase that may be capable of 18-hydroxylation of corticosterone. It is of interest that the fasciculata enzyme is also capable of 19-hydroxylase activity.[19]

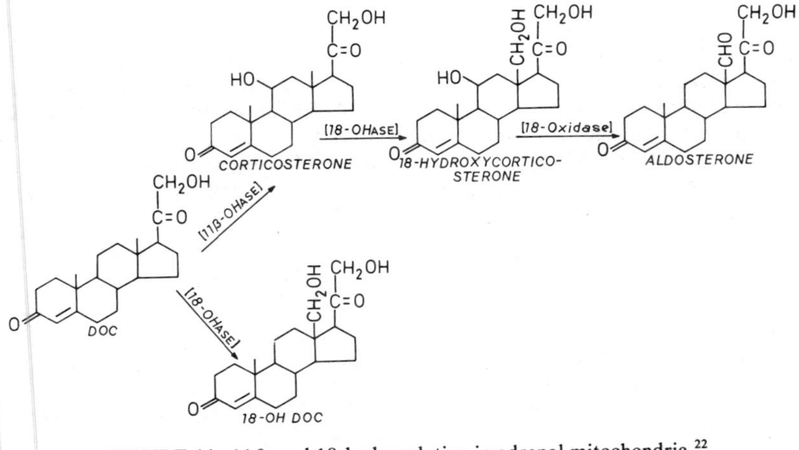

FIGURE 11. 11β- and 18-hydroxylation in adrenal mitochondria.[22]

Aromatase

The enzyme responsible for converting testosterone to estradiol is a microsomal cytochrome P-450. This enzyme has not been purified, but it is believed to act by inserting three hydroxyl groups into the substrate.[20,21]

Multiple Hydroxylation by Cytochromes P-450

The two C atoms of the side-chain of cholesterol that are hydroxylated by the C_{27} side-chain cleavage enzyme (C_{20} and C_{22}) are separated by the usual C–C bond distance (1.54Å). In a molecular model, the 11 carbon is equidistant from C_{18} and C_{19} (3.1Å), i.e., approximately twice that of the usual C–C bond (FIGURE 12).

The work of Skeets and Vickery suggests that the C_{22} comes within 5Å of the heme iron. The question arises as to whether the C_{20} is close enough to undergo the second hydroxylation or whether the substrate must move relative to the enzyme to bring the

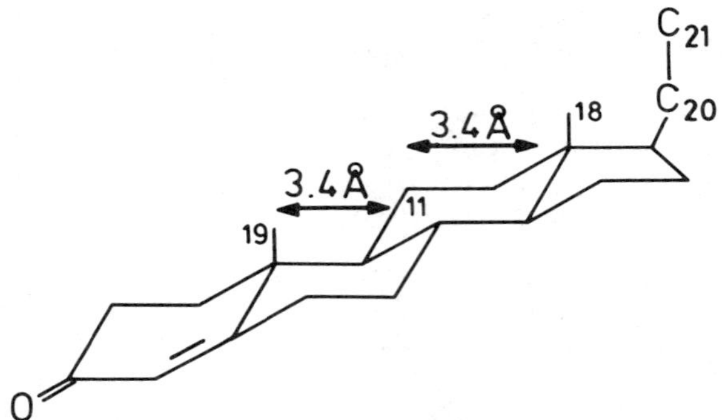

FIGURE 12. The relationship between C_{11}, C_{18}, and C_{19}.

C_{20} closer to the heme iron. It seems likely that 3.4Å would be too great a distance for a second hydroxylation. Thus, that movement of the substrate may be necessary for the second hydroxylation. Perhaps, the fact that the monohydroxylated intermediates accumulate whereas the two hydroxylated intermediates in C_{27} side-chain cleavage (FIGURE 7) do not accumulate results from the fact that the intermediates are tightly bound and do not move relative to the enzyme during the three catalytic cycles necessary for side-chain cleavage.

REFERENCES

1. WHITE, R. E. & M. J. COON. 1980. Ann. Rev. Biochem. **49:** 315–356.
2. GRIFFIN, B. W., J. A. PETERSON & R. W. ESTABROOK. 1979. *In* The Porphyrins, vol. VII. D. Dolphin, Ed.: 333–368. Academic Press. New York.
3. GUNSALUS, I. C., J. R. MEEKS, J. D. LIPSCOMB, P. DEBRUNNER & E. MINICK. *In* Molecular Mechanisms of Oxygen Activation. O. Hayiashi, Ed.: 550–567. Academic Press. New York, London.
4. SHIKITA, M. & P. F. HALL. 1973. J. Biol. Chem. **248:** 5598–5604.

5. SHIKITA, M. & P. F. HALL. 1974. Proc. Natl. Acad. Sci. USA **71:** 1441–1445.
6. HALL, P. F., J. LEE LEWES & E. D. LIPSON. 1975. J. Biol. Chem. **250:** 2283–2286.
7. DUQUE, C., M. MORISAKI, N. IKEKAWA & M. SHIKITA. 1978. Biochem. Biophys. Res. Commun. **82:** 174–178.
8. TAKAGI, Y., M. SHIKITA & P. F. HALL. 1975. J. Biol. Chem. **250:** 8445–8448.
9. NAKAJIN, S. & P. F. HALL. 1981. J. Biol. Chem. **256:** 3871–3876.
10. NAKAJIN, S., J. SHIVELY, P-M. YUAN & P. F. HALL. 1981. Biochemistry **20:** 4037–4042.
11. NAKAJIN, S. & P. F. HALL. 1983. J. Steroid Biochem. **1**(3): 1345–1348.
12. NAKAJIN, S., P. F. HALL & M. ONODA. 1981. J. Biol. Chem. **256:** 6134–6139.
13. NAKAJIN, S., M. SHINODA, M. HANUI, J. E. SHIVELY & P. F. HALL. 1984. J. Biol. Chem. **259:** 3971–3976.
14. NAKAJIN, S., M. SHINODA & P. F. HALL. 1983. Biochem. Biophys. Res. Commun. **111:** 512–517.
15. KAMINAMI, S., H. OCHI, Y. KOBAYASHI & S. TAKAMORI. 1980. J. Biol. Chem. **255:** 3386–3391.
16. YUAN, P-M., S. NAKAJIN, M. HANIU, P. F. HALL & J. E. SHIVELY. 1983. Biochemistry **22:** 143–149.
17. WATANUKI, M., B. E. TILLEY & P. F. HALL. 1977. Biochim. Biophys. Acta **483:** 236–247.
18. WATANUKI, M., B. E. TILLEY & P. F. HALL. 1978. Biochemistry **17:** 127–130.
19. MOMOI, J., M. OKAMOTO, S. FUJII, C. Y. KIM, Y. MIYAKE & T. YAMANO. 1983. J. Biol. Chem. **258:** 8855–8859.
20. FISHMAN, J. & M. S. RAJU. 1981. J. Biol. Chem. **256:** 4472–4476.
21. THOMPSON, E. A. & E. A. SIITERI. 1974. J. Biol. Chem. **249:** 5364–5371.
22. HALL, P. F. 1985. The role of cytochromes P-450 in the biosynthesis of steroid hormones. *In* Vitamins and Hormones, vol. 42. G. D. Aurbach, Ed. Academic Press. New York. In press.
23. HALL, P. F. 1984. Cellular organization of steroidogenesis. Int. Rev. Cytol. **86:** 153–95.

Biochemical Properties of Cytochrome P-450 in Relation to Steroid Oxygenation[a]

MINOR J. COON AND KUNIYO INOUYE

Department of Biological Chemistry
Medical School
The University of Michigan
Ann Arbor, Michigan 48109-0010

The studies from other laboratories presented at this symposium indicate that congenital adrenal hyperplasia is a complex human disorder involving, among other factors, a genetic deficiency in 21-hydroxylase. This hydroxylase is one of many oxygenases that function in lipid transformations and are recognized from their biochemical properties as falling in the class of red, carbon-monoxide-binding pigments known as cytochrome P-450.[1]

As judged by the reversal of CO inhibition by exposure to monochromatic light at about 450 nm, the 21-hydroxylation of 17α-hydroxyprogesterone was the first such steroid conversion shown to be P-450-catalyzed.[2] The 21-hydroxylase is required in the synthesis of adrenal corticosteroids since it catalyzes the oxygenation of Δ^4-3-ketosteroids such as progesterone, as well as of 17α-hydroxyprogesterone. Like the 17α-hydroxylase, the 21-hydroxylase is located in the endoplasmic reticulum of the adrenal cortex and requires NADPH-cytochrome P-450 reductase to effect electron transfer. The 21-hydroxylase has been purified from bovine adrenal cortical microsomes.[3–6] The minimal molecular weight of the purified enzyme has been stated to range from 47,000 to 52,000, and P-450 specific contents as high as 20 nmol of heme per milligram of protein have been reported. The enzyme is apparently immunochemically distinct from the other adrenal P-450's and contains some carbohydrate. It is low-spin as isolated, but can be converted to the high-spin form by the addition of substrate. The purified enzyme is specific for 21-hydroxylation and, unlike the 17α-hydroxylase, shows no change in activity upon the addition of cytochrome b_5.[5] The purified preparation catalyzes the demethylation of benzphetamine at about 30% of the rate of 21-hydroxylation of 17α-hydroxyprogesterone, but is apparently inactive in the demethylation of *p*-nitroanisole or the *p*-hydroxylation of aniline. Of particular interest, the 21-hydroxylase has also been purified to homogeneity from porcine adrenocortical microsomes, and the cysteine-containing peptides from this protein have been sequenced as described below.[7]

In the present paper, some fundamental biochemical properties of cytochrome P-450 are reviewed and some implications for steroid hydroxylation are discussed. In particular, the known influence of environmental factors on the activities of many P-450's raises the question of whether some of the puzzling variables in congenital adrenal hyperplasia and in the function of the 21-hydroxylase are related to exposure to xenobiotics.

[a]This work was supported by grant no. AM–10339 from the National Institutes of Health.

BIOCHEMISTRY OF CYTOCHROME P-450

The active site of cytochrome P-450, as proposed previously,[8] is pictured in FIGURE 1. This site contains iron protoporphyrin IX, presumably in a hydrophobic cleft in the protein, which favors the binding of a variety of nonpolar substrates or portions of substrates, including physiologically occurring compounds such as steroids, prostaglandins, and fatty acids. In some instances, a variety of foreign compounds, including drugs, carcinogens, anesthetics, pesticides, organic solvents, etc., are also accommodated by the substrate-binding site. The heme, which is not bound covalently to the protein, is always penta- or hexacoordinate, four of the ligands being provided by the N atoms of the porphyrin ring. The fifth ligand is believed to be a specific cysteine residue of the polypeptide chain,[9,10] and the sixth coordination position of the iron is occupied by an easily exchangeable oxygen ligand, perhaps from water[11,12] or the hydroxyl group of a serine or threonine residue,[13,14] in the native, ferric state of the enzyme. A tyrosine residue is also a likely candidate.[15,16] Upon reduction of the iron, the sixth position

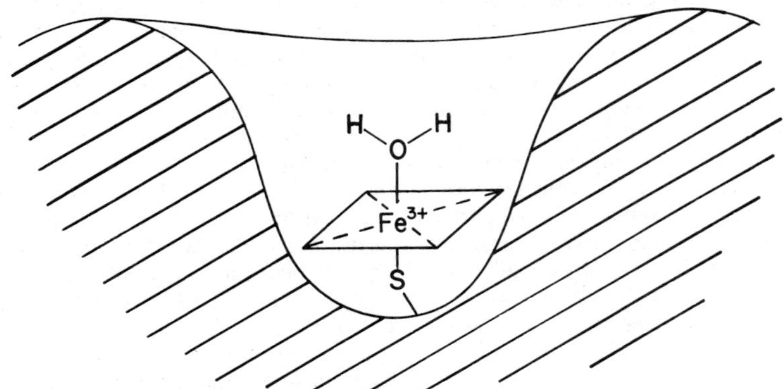

FIGURE 1. Model of the active site of cytochrome P-450.

becomes the site of O_2 binding, but other diatomic ligands such as CO, NO, and CN^- may compete with oxygen. Thus, CO gives the complex an absorption maximum at about 450 nm and is inhibitory to the action of the enzyme because of the competitive displacement of O_2. Although it would be highly desirable in all instances to name these enzymes by their function, as is often done with the steroid hydroxylases, some forms of the cytochrome have such broad substrate specificity that there is no easy way of choosing among the hundreds of potential substrates and the catalyst is therefore designated by its relative electrophoretic mobility or some other arbitrary system.

The existence of tyrosyl and histidyl residues as well as a cysteinyl residue in the substrate-binding site was suggested by chemical modification studies.[17,18] In this connection, we have recently observed that upon the binding of the detergent CHAPS [3-(3-cholamidopropyl)-dimethylammonio-1-propanesulfonate] to rabbit P-450 isozyme 2, a unique absorption difference spectrum is seen that is similar to that induced by the binding of type II substrates (amines). This detergent may disaggregate the P-450 into monomers without denaturation of the protein structure and concomitant conversion of P-450 to P-420. The formation of the CHAPS-induced difference spectrum is slower than that caused by substrate binding, which is probably diffusion-controlled. Since CHAPS is not a type II substrate, the difference spectrum observed may be due to the movement of the amine moiety of some amino acid residue such as histidine, tryptophan, lysine, or arginine to the heme, with resulting ligation. The results obtained by chemical modification[17,18] implicate a histidyl residue. We have observed also that CO binding to reduced P-450 isozyme 2 is inhibited in the presence of CHAPS and that the extent of this inhibition with varying detergent concentrations corresponds to that which would be predicted by CHAPS-induced spectral changes. These results strongly suggest that a conformational change occurs in the substrate-binding site, followed by ligation of the amine moiety (presumably of histidine) to the heme.

NADPH is the primary electron donor to the cytochrome, with NADPH-cytochrome P-450 reductase serving as the carrier in microsomes, which is shown in FIGURE 2. The microsomal system, in contrast to that in mitochondria, does not utilize an intermediate iron-sulfur protein and, thus, makes use of direct flavin-heme electron transfer. The flavoprotein in this case is unusual in containing both FMN and FAD;[19] studies with FMN-depleted enzyme have shown that FMN directly reduces the P-450 heme.[20] The proposed scheme in FIGURE 3 accounts for the evidence presently available from this and other laboratories for electron transfer and oxygen activation in this enzyme system. The steps, beginning with the resting ferric enzyme at the top of the cycle and going clockwise, are as follows: (1) substrate binding; (2) transfer of the first electron from the reductase; (3) dioxygen binding to the ferrous protein, thus giving the ferrous-dioxygen complex, which can also be written as a resonance form, ferric-superoxide; (4) transfer of the second electron, which may come via the reductase or, in some instances, from ferrous cytochrome b_5; (5) splitting of the oxygen-oxygen bond to yield H_2O and a postulated iron-oxene species; (6) hydrogen abstraction from the substrate to give a transient substrate carbon radical and an iron-bound hydroxyl radical or equivalent, as proposed by Groves *et al.*[21] on the basis of the intramolecular isotope effect observed with deuteronorbornane; and (7) radical recombination followed by (8) dissociation from the enzyme of the resulting product, ROH. Thus, the resting ferric state of the enzyme is regenerated prior to another cycle. The reactions proposed are compatible with the known stoichiometry of the hydroxylation reaction:[22]

$$RH + O_2 + NADPH + H^+ \rightarrow ROH + H_2O + NADP^+$$

An interesting variation is now recognized in which a peroxy compound (XOOH),

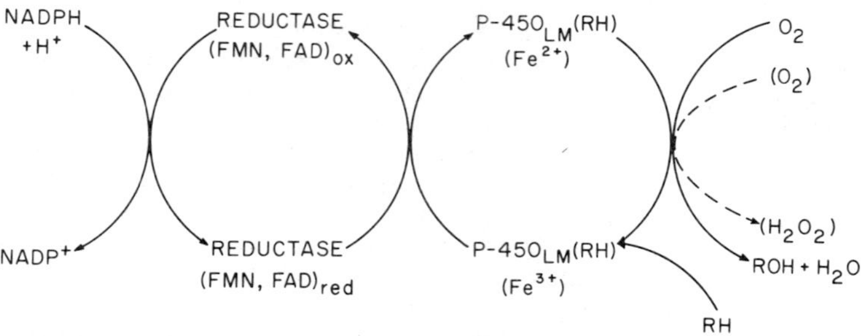

FIGURE 2. Electron transfer reactions in the microsomal P-450 system.

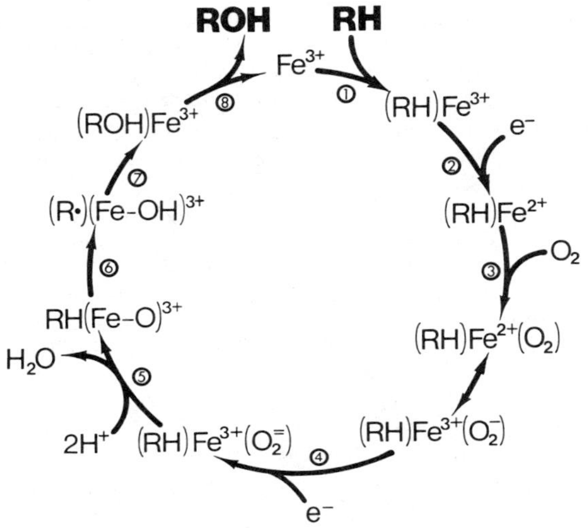

FIGURE 3. Proposed scheme for mechanism of action of cytochrome P-450 in hydroxylation reactions, taken from reference 8. RH represents a substrate and ROH represents the corresponding product.

which may be hydrogen peroxide or an organic peracid or hydroperoxide, provides the oxygen atom in a reaction not requiring molecular oxygen, NADPH, or the reductase:[23]

$$RH + XOOH \rightarrow ROH + XOH$$

Clearly, variations on this scheme are required to account for the mechanism of other reactions involving P-450, in processes as diverse as dehalogenation and desulfuration.

MULTIPLICITY OF P-450 ISOZYMES

As reviewed elsewhere,[1,24,25] the total number of mammalian P-450 isozymes is large, but not yet known exactly, and they appear to fall into two large groups: those concerned primarily with lipid metabolism, and those having very broad substrate specificity and acting both on lipids and foreign compounds. The latter type, which is found in hepatic microsomes, includes those P-450 cytochromes that were the first to be purified to homogeneity and characterized biochemically. For example, the presently known isozymes from rabbit liver microsomes are shown in TABLE 1. Since the discovery and determination of the properties of a number of these has already been reviewed,[1,24,25] only more recent developments will be emphasized here. Isozyme 1 has been obtained from uninduced animals,[26] isozyme 3a from alcohol- and imidazole-induced animals[27,28] (and, apparently, also after benzene treatment[29]), and isozyme 5 from lung[30] (corresponding to a form in liver[31]); in addition, the polymorphism of isozyme 3b has been demonstrated.[32] Furthermore, Wikvall and associates[33] have purified liver microsomal isozymes with the same electrophoretic behavior as isozyme

4,[34] but apparently otherwise distinct, and they have also recently isolated a liver mitochondrial P-450 with cholesterol 26-hydroxylase activity.[35] Some of the forms isolated by Sato and colleagues[36] are believed to be different from those studied in other laboratories; two of these are listed here as P-450_7 and P-450_8, but it should be noted that in this case the nomenclature with substrates is not based on electrophoretic mobility. Many of these isozymes are active in the hydroxylation of testosterone and androstenedione in various positions[25] or of other steroids.

P-450 isozyme 1 has been reported to have the highest activity of six rabbit liver cytochromes examined for the 21-hydroxylation of progesterone.[26,37] The purified enzyme has a minimal molecular weight of about 48,000 and exhibits a K_m and V_{max} for progesterone hydroxylation of 1.6 μM and 7.1 min^{-1}, respectively, which are values very similar to those reported for the adrenal 21-hydroxylase.[5] However, unlike the adrenal enzyme, which hydroxylates only at the 21-position, isozyme 1 also hydroxylates at the 6β and 16α positions of progesterone.[5] In addition, this purified hepatic enzyme catalyzes the 2-hydroxylation of estradiol.[38] Whether the liver enzyme has

TABLE 1. Rabbit Liver Microsomal Cytochrome P-450 Isozymes and Inducers

Isozyme	Inducer
P-450 1	
P-450 2	Phenobarbital
P-450 3a (A1c)	Ethanol, imidazole, pyrazole, acetone, trichloroethylene, isoniazid
P-450 3b	
P-450 3c	Troleandomycin, erythromycin
P-450 4	Isosafrole, β-naphthoflavone, TCDD, 3-methylcholanthrene
P-450 (4) (Chol. 7α)	Cholestyramine
P-450 (4) (Chol. 12α)	
P-450 (4) (Chol. 26)	
P-450 5	Phenobarbital
P-450 6	TCDD, imidazole
P-450_7	
P-450_8	

extensive structural features in common with those of the adrenal microsomal enzyme and, furthermore, whether it provides a physiological pathway for the extra-adrenal production of hormones remains to be determined.

Since a cysteine residue is known to provide the sulfur ligand to the heme iron atom in the P-450 cytochromes, as indicated above, it may be presumed that the cysteine peptide providing this function would be conserved during evolution. A comparison of three conserved cysteine peptides in various P-450 cytochromes is given in FIGURE 4, in which the areas of structural homology with respect to isozyme 4 are enclosed in boxes. The P-450's represented are rabbit isozymes 4[39] and 2,[40,41] rat isozymes b,[42,43] c,[44–46] and d,[47,48] mouse isozymes 1 and 3,[49,50] bacterial P-450_{cam},[51] and pig adrenal microsomal 21-hydroxylase.[7,44] The most extensive homology is seen with the cysteine near the COOH-terminus, which may possibly be the source of the sulfur ligand. Recent experiments in this laboratory (unpublished results, S. D. Black and M. J. Coon), in which rabbit isozyme 2 was treated with DTNB or monobromobimane appear to rule

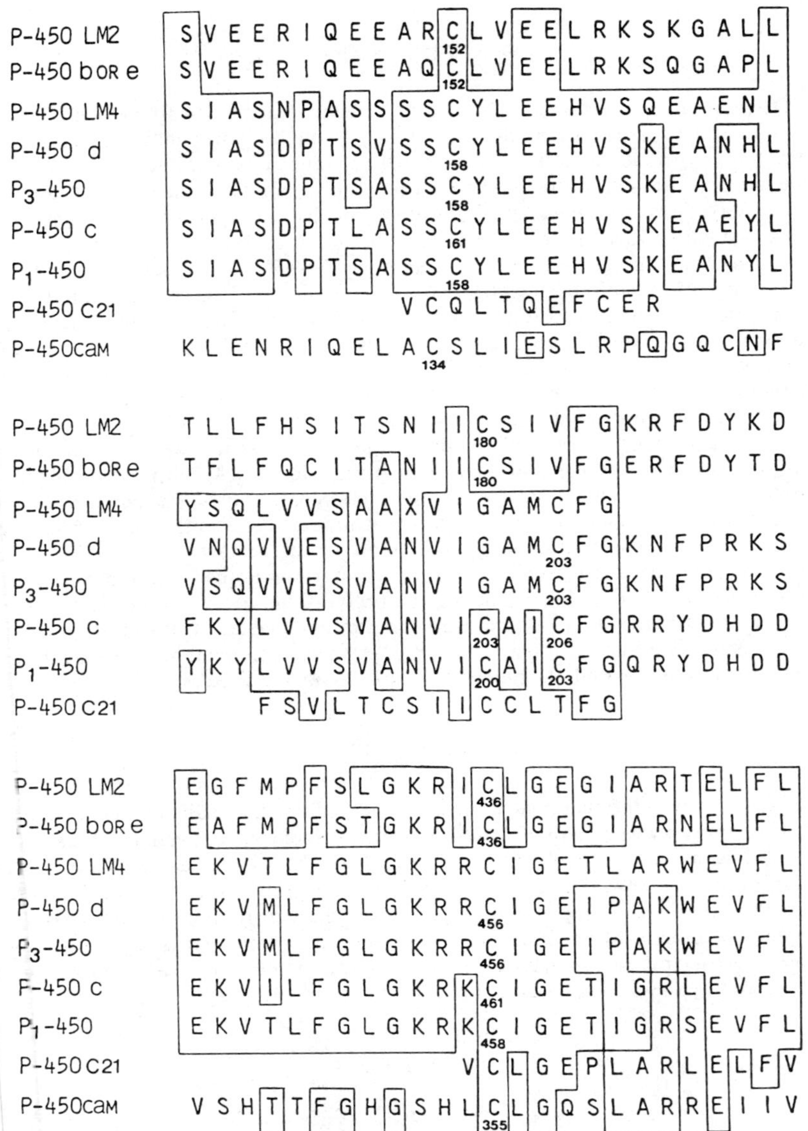

FIGURE 4. Comparison of the sequence of cysteine peptides in various P-450 cytochromes. Homology with respect to rabbit liver microsomal P-450 isozyme 4 is shown by the enclosed areas.

out Cys-152 as providing the sulfur ligand, but it may have some other function such as the binding of NADPH-cytochrome P-450 reductase.

DISCUSSION

Despite some differences in substrate specificity, the P-450 cytochromes are believed to be similar both in their ability to bind hydrophobic substrates and molecular oxygen at the active site and in the manner in which they bring about the reductive activation of molecular oxygen, with one atom going to water and the other being inserted into the substrate. The activity of this family of enzymes in substrate monooxygenation varies in some instances with the age and sex of the animal, as well as with exposure to foreign chemicals. Indeed, with those P-450's having broad substrate specificity, a myriad of physiologically occurring lipids and foreign compounds serves both as substrates and as mutually competitive inhibitors. The more specific, extrahepatic steroid-hydroxylating cytochromes are known in some instances to oxygenate certain xenobiotics as well, but the possibility has not been studied adequately as to whether those xenobiotics that are not oxygenated at a significant rate are nevertheless bound at the active site and, thus, serve as nonmetabolizable, competitive inhibitors. Thus, an enormous variety of foreign organic compounds in our environment (estimated to be at least 100,000 in technologically advanced countries) could play an indirect role in adrenal hyperplasia. Exposure to foreign compounds, including ethanol, food additives, insecticides, dyes, combustion products, solvents, and so forth, may occur both pre- and postnatally. The P-450 system is also sensitive to dietary alterations, not only because of the need for vitamins and other building blocks in the biosynthesis of NADPH, flavins, heme, and proteins, but also because the integrity of biological membranes in which these enzymes are embedded is subject to nutritional influences.[52–57] Thus, genetic control of the 21-hydroxylase and other steroid hormone-metabolizing P-450's could be modified by environmental factors.

SUMMARY

The P-450 cytochromes have been characterized biochemically in recent years as a family of monooxygenases that reductively activate molecular oxygen for insertion into steroids and other physiologically occurring lipids. Many of these enzymes are also known to bind and oxygenate a host of foreign compounds, including alcohol, drugs, pesticides, anesthetics, and mutagens. Some of the poorly understood variations in congenital adrenal hyperplasia may represent nutritional effects on the P-450 oxygenase systems or the ability of xenobiotics to interfere with normal steroid metabolism by these versatile cytochromes.

REFERENCES

1. Coon, M. J. & D. R. Koop. 1983. P-450 oxygenases in lipid transformation. *In* The Enzymes, vol. XVI. P. D. Boyer, Ed.: 645–677. Academic Press. New York.
2. Estabrook, R. W., D. Y. Cooper & O. Rosenthal. 1963. Biochem. Z. **338:** 741–755.
3. Kominami, S., S. Mori & S. Takemori. 1978. FEBS Lett. **89:** 215–218.
4. Kominami, S., H. Ochi, Y. Kobayashi & S. Takemori. 1980. J. Biol. Chem. **255:** 3386–3394.
5. Hiwatashi, A. & Y. Ichikawa. 1981. Biochim. Biophys. Acta **664:** 33–48.

6. Bumpus, J. A. & K. M. Dus. 1982. J. Biol. Chem. **257:** 12696–12704.
7. Yuan, P-M., S. Nakajin, M. Haniu, M. Shinoda, P. F. Hall & J. E. Shively. 1983. Biochemistry **22:** 143–149.
8. White, R. E. & M. J. Coon. 1980. Annu. Rev. Biochem. **49:** 315–356.
9. Champion, P. M., B. R. Stallard, G. C. Wagner & I. C. Gunsalus. 1982. J. Am. Chem. Soc. **104:** 5469–5472.
10. Hahn, J. E., K. O. Hodgson, L. A. Anderson & J. H. Dawson. 1982. J. Biol. Chem. **257:** 10934–10941.
11. Griffin, B. W. & J. A. Peterson. 1975. J. Biol. Chem. **250:** 6445–6451.
12. Philson, S. B., P. G. Debrunner, P. G. Schmidt & I. C. Gunsalus. 1979. J. Biol. Chem. **254:** 10173–10179.
13. Ullrich, V., H. Sakurai & H. H. Ruf. 1979. Acta Biol. Med. Ger. **39:** 287–297.
14. Ruf, H. H., P. Wende & V. Ullrich. 1979. J. Inorg. Biochem. **11:** 189–204.
15. Ruckpaul, K., H. Rein, D. P. Ballou & M. J. Coon. 1980. Biochim. Biophys. Acta **626:** 41–56.
16. Jänig, G-R., A. Makower, H. Rabe, R. Bernhardt & K. Ruckpaul. 1984. Biochim. Biophys. Acta **787:** 8–18.
17. Dus, K. M. 1980. Application of covalent affinity and photoaffinity probes targeted for the active site of P-450 hemeproteins. *In* Biochemistry, Biophysics, and Regulation of Cytochrome P-450. J-Å. Gustafsson, J. Carlstedt-Duke, A. Mode & J. Rafter, Eds.: 129–132. Elsevier. Amsterdam.
18. Gibson, G. G. & P. P. Tambrini. 1980. Spin equilibrium of purified cytochrome P-450. Effect of substrate and histidine modification. *In* Biochemistry, Biophysics, and Regulation of Cytochrome P-450. J-Å. Gustafsson, J. Carlstedt-Duke, A. Mode & J. Rafter, Eds.: 133–136. Elsevier. Amsterdam.
19. Vermilion, J. L. & M. J. Coon. 1978. J. Biol. Chem. **253:** 2694–2704.
20. Vermilion, J. L. & M. J. Coon. 1978. J. Biol. Chem. **253:** 8812–8819.
21. Groves, J. T., G. A. McClusky, R. E. White & M. J. Coon. 1978. Biochem. Biophys. Res. Commun. **81:** 154–160.
22. Nordblom, G. D. & M. J. Coon. 1977. Arch. Biochem. Biophys. **180:** 343–347.
23. Nordblom, G. D., R. E. White & M. J. Coon. 1976. Arch. Biochem. Biophys. **175:** 524–533.
24. Lu, A. Y. H. & S. B. West. 1980. Pharmacol. Rev. **31:** 277–295
25. Coon, M. J., S. D. Black, D. R. Koop, E. T. Morgan & G. E. Tarr. 1982. Structural and catalytic studies with purified microsomal enzymes. *In* Microsomes, Drug Oxidations, and Drug Toxicity. R. Sato and R. Kato, Eds.: 13–23. Japan Scientific Societies Press. Tokyo.
26. Dieter, H. H., U. Muller-Eberhard & E. F. Johnson. 1982. Biochem. Biophys. Res. Commun. **105:** 515–520.
27. Koop, D. R., E. T. Morgan, G. E. Tarr & M. J. Coon. 1982. J. Biol. Chem. **257:** 8472–8480.
28. Koop, D. R. & M. J. Coon. 1984. Mol. Pharmacol. **25:** 494–501.
29. Ingelman-Sundberg, M. & I. Johansson. 1984. J. Biol. Chem. **259:** 6447–6458.
30. Slauter, S. R., C. R. Wolf, J. P. Marciniszyn & R. M. Philpot. 1981. J. Biol. Chem. **256:** 2499–2503.
31. Serabjit-Singh, C. J., P. W. Albro, I. G. C. Robertson & R. M. Philpot. 1983. J. Biol. Chem. **258:** 12827–12834.
32. Dieter, H. H. & E. F. Johnson. 1982. J. Biol. Chem. **257:** 9315–9323.
33. Bostrom, H. & K. Wikvall. 1982. J. Biol. Chem. **257:** 11755–11759.
34. Haugen, D. A. & M. J. Coon. 1976. J. Biol. Chem. **251:** 7929–7939.
35. Wikvall, K. 1984. J. Biol. Chem. **259:** 3800–3804.
36. Aoyama, T., Y. Imai & R. Sato. 1982. Multiple forms of cytochrome P-450 from liver microsomes of drug-untreated rabbits: Purification and characterization. *In* Microsomes, Drug Oxidations, and Drug Toxicity. R. Sato & R. Kato, Eds.: 83–84. Japan Scientific Societies Press. Tokyo.
37. Dieter, H. H., U. Muller-Eberhard & E. F. Johnson. 1982. Science **217:** 741–743.
38. Johnson, E. F., H. H. Dieter, G. E. Schwab, I. Reubi & U. Muller-Eberhard. 1982.

Rabbit microsomal cytochrome P-450 1: A cytochrome linked to variations in rabbit steroid hormone metabolism. *In* Cytochrome P-450: Biochemistry, Biophysics, and Environmental Implications. E. Hietanen, M. Laitenen & O. Hänninen, Eds.: 337–340. Elsevier Biomedical Press. Amsterdam.
39. FUJITA, V. S., S. D. BLACK, G. E. TARR, D. R. KOOP & M. J. COON. 1984. Proc. Natl. Acad. Sci. USA **81:** 4260–4264.
40. TARR, G. E., S. D. BLACK, V. S. FUJITA & M. J. COON. 1983. Proc. Natl. Acad. Sci. USA **80:** 6552–6556.
41. HEINEMANN, F. S. & J. OZOLS. 1983. J. Biol. Chem. **258:** 4195–4201.
42. FUJII-KURIYAMA, Y., Y. MIZUKAMI, K. KAWAJIRI, K. SOGAWA & M. MURAMATSU. 1982. Proc. Natl. Acad. Sci. USA **79:** 2793–2797.
43. YUAN, P-M., D. E. RYAN, W. LEVIN & J. E. SHIVELY. 1983. Proc. Natl. Acad. Sci. USA **80:** 1169–1173.
44. HANIU, M., P-M. YUAN, D. E. RYAN, W. LEVIN & J. E. SHIVELY. 1984. Biochemistry **23:** 2478–2482.
45. YABUSAKI, Y., M. SHIMIZU, H. MURAKAMI, K. NAKAMURA, K. OEDA & H. OHKAWA. 1984. Nucl. Acids Res. **12:** 2929–2938.
46. SOGAWA, K., O. GOTOH, K. KAWAJIRI & Y. FUJII-KURIYAMA. 1984. Proc. Natl. Acad. Sci. USA **81:** 5066–5070.
47. KAWAJIRI, K., O. GOTOH, K. SOGAWA, Y. TAGASHIRA, M. MURAMATSU & Y. FUJII-KURIYAMA. 1984. Proc. Natl. Acad. Sci. USA **81:** 1649–1653.
48. HANIU, M., D. E. RYAN, W. LEVIN & J. E. SHIVELY. 1984. Proc. Natl. Acad. Sci. USA **81:** 4298–4301.
49. KIMURA, S., F. J. GONZALEZ & D. W. NEBERT. 1984. Nucl. Acids Res. **12:** 2917–2928.
50. KIMURA, S., F. J. GONZALEZ & D. W. NEBERT. 1984. J. Biol. Chem. **259:** 10705–10713.
51. HANIU, M., L. G. ARMES, K. T. YASUNOBU, B. A. SHASTRY & I. C. GUNSALUS. 1982. J. Biol. Chem. **257:** 12664–12671.
52. LU, A. Y. H. 1976. Fed. Proc. **35:** 2460–2463.
53. ZANNONI, V. G. & P. H. SATO. 1976. Fed. Proc. **35:** 2464–2469.
54. CAMPBELL, T. C. & J. R. HAYES. 1976. Fed. Proc. **35:** 2470–2474.
55. WADE, A. E. & W. P. NORRED. 1976. Fed. Proc. **35:** 2475–2479.
56. BECKING, G. C. 1976. Fed. Proc. **35:** 2480–2485.
57. ALVARES, A. P., E. J. PANTUCK, K. E. ANDERSON, A. KAPPAS & A. H. CONNEY. 1979. Drug Metab. Rev. **9:** 185–205.

Evidence that Variation among Untreated Rabbits in Hepatic Progesterone 21-Hydroxylase Activity Is Indicative of Enzyme Heterogeneity Rather than a Transient Inductive Effect[a]

U. MULLER-EBERHARD,[b] L. GHIZZONI,[b] H.H. LIEM,[b]
M. NEW,[b] M. FINLAYSON,[c] AND E.F. JOHNSON[c]

[b]*Department of Pediatrics*
Cornell School of Medicine
New York, New York 10021

[c]*Division of Biochemistry*
Department of Basic and Clinical Research
Scripps Clinic and Research Foundation
La Jolla, California 92037

INTRODUCTION

Rabbit liver microsomes catalyze the 21-hydroxylation of progesterone as a major route of metabolism that results in the formation of the mineralocorticoid, deoxycorticosterone (DOC).[1] This is unusual because the synthesis of DOC in most species is restricted primarily to the endoplasmic reticulum of the zona fasciculata of the adrenal cortex. Moreover, liver microsomes prepared from different rabbits exhibit a greater than tenfold variation in progesterone 21-hydroxylase activity.[2] This variation among animals exceeds that generally seen for other activities catalyzed by cytochrome P-450 isozymes and exceeds the magnitude of the changes in activity elicited by compounds that induce P-450's.[3]

In most cases, extra-adrenal production of DOC by the 21-hydroxylation of progesterone is not thought to alter the circulating levels of this mineralocorticoid. However, two situations exist in human subjects where an extra-adrenal enzyme could provide a source of DOC. It has been noted that patients expressing congenital adrenal hyperplasia due to an adrenal 21-hydroxylase deficiency exhibit elevated serum progesterone levels. Antonipillai *et al.*[4] have shown that the extra-adrenal conversion of progesterone to DOC, which provides approximately one-third the levels of this mineralocorticoid in these patients, contributes to the maintenance of normal plasma levels of DOC in some of those individuals not exhibiting the salt-wasting form of this condition. Furthermore, Winkel *et al.*[5] have reported that extra-adrenal 21-hydroxylation of progesterone contributes to the rise in plasma DOC levels observed when plasma progesterone levels increase during pregnancy. In addition, they found that the fractional conversion of progesterone to DOC varied more than tenfold among human

[a]This work was supported by United States Public Health Service grant nos. HD-04445 and GM-31001.

subjects. Repeated testing of some of these subjects demonstrated this conversion to be relatively constant for a given patient.

This raised a question as to whether differences observed in the rabbit constituted a temporary event or were reflective of enzyme heterogeneity in the outbred population. Therefore, we measured the hepatic 21-hydroxylase activity of a group of rabbits in serial biopsy samples taken over a two-month period. In addition, single determinations are reported from a larger group of rabbits in order to establish more precisely the distribution of this activity.

METHODS

One of the two groups of outbred New Zealand White (NZW) rabbits that were studied was composed of 19 male and 34 female animals. The activities for the hepatic metabolism of progesterone were determined for microsomes prepared from these animals as previously described.[2] The other group of rabbits included 10 males and 18 females. These rabbits were purchased from CAMM Research Laboratories (NJ). Hepatic progesterone metabolism was determined on serial biopsy samples taken from the latter group of animals. At least three liver punch biopsies (TRU-CUT disposable needle, Travenol Laboratories Inc., Deerfield, IL) were taken from sodium pentobarbital (22 mg/kg) anesthetized rabbits at intervals of at least three weeks. The biopsy samples (approximately 20 mg) were homogenized (10 μl/mg) in 50 m*M* potassium phosphate buffer, pH 7.4, containing 20% glycerol.

A postmitochondrial fraction was prepared by centrifugation at 8700 g for 30 minutes. The resulting supernatant was diluted 1:10 in the homogenization buffer prior to assay. Microsomes were not isolated from these fractions due to the small amount of homogenate that was obtained.

Progesterone hydroxylase activities were determined on the microsomal and postmitochondrial fractions as described.[3] The final volume of the reaction mixture contained 50 μmol of potassium phosphate, pH 7.4, 30 μg microsomal protein or 50 μl of the postmitochondrial fraction, and 10 nmol [^{14}C]progesterone (Amersham; 56 mCi/mmol) added in 10 μl of methanol. The reaction was initiated by addition of 0.5 μmol NADPH following a three-minute preincubation at 37°C. After an incubation of five minutes, the reaction was terminated by extraction of substrate and products into chloroform. Progesterone and metabolites were visualized by autoradiography following sequential thin-layer chromatography in isopropyl ether/acetone (80:20) and benzene/ethyl acetate/acetone (80:10:10). The procedure resolves and detects three major metabolites; 6β-, 16α-, and 21-hydroxyprogesterone. Henceforth, metabolites will be referred to as 6-, 16-, and 21-hydroxyprogesterone, respectively. Total recovery of progesterone and its metabolites was $85 \pm 10\%$.[2] Results are expressed as mean ± S.D. The statistical significance between two measurements was determined using Student's t test at $p < 0.05$.

RESULTS

In order to establish the incidence of the high 21-hydroxylase activity in outbred NZW rabbits, hepatic progesterone 21-hydroxylase activities were determined for microsomes prepared from whole liver homogenates obtained from each of 53 animals. As shown in FIGURE 1, we observed a bimodal distribution of the 21-hydroxylase activity in both male and female rabbits. Approximately one-third of these animals exhibited a 21-hydroxylase activity of greater than 1.0 nmol/min/mg protein, and these rabbits were classified as 21-H (high 21-hydroxylase). The frequency of the 21-H phenotype was found to be independent of the sex of the animal. Of the 19 males assayed, 6 were phenotyped as 21-H with a mean 21-hydroxylase activity of 4.3 ± 1.3 nmol/min/mg protein, and 11 of 34 female rabbits exhibited the 21-H phenotype with a mean 21-hydroxylase activity of 3.3 ± 1.4 nmol/min/mg protein. The mean 21-hydroxylase activities of those 13 males and 23 females classified as 21-L (low 21-hydroxylase) were 0.35 ± 0.17 and 0.27 ± 0.1 nmol/min/mg protein, respectively. Differences between male and female rabbits in the mean 21-hydroxylase activities were not statistically significant for either the 21-H or 21-L animals. In addition, the rate of formation of 16-hydroxyprogesterone was determined. The 16-hydroxylase

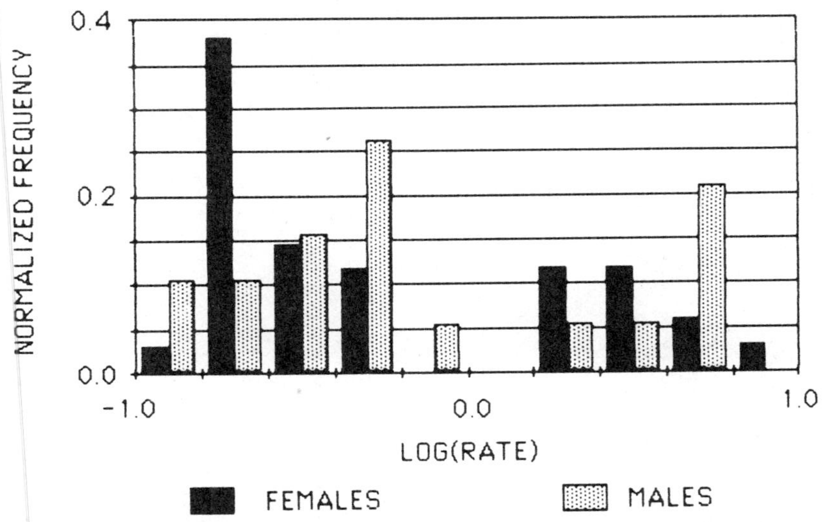

FIGURE 1. Distribution of hepatic progesterone 21-hydroxylase activity for 19 male and 23 female adult New Zealand White rabbits. Activities were determined for liver microsomes prepared as described in MATERIALS AND METHODS. The number of animals exhibiting rates within intervals of 0.2 log units were determined, and normalized frequency distributions were plotted as histograms for males (dotted bars) and females (solid bars).

activity varied over a smaller range than that of the 21-hydroxylase and the two rates were not correlated ($r < 0.5$) (FIGURE 2).

To address whether the elevation of the 21-hydroxylase activity was the result of a transient phenomenon, serial biopsy samples were taken at least three times at about three-week intervals from a second group of rabbits. Due to the small size of the punch biopsy sample, rates of progesterone metabolism were determined on a postmitochondrial fraction supernatant rather than microsomes, and the activities were expressed per mg wet tissue weight. It is important to note that values expressed in this manner always will be lower than those determined for microsomes, which were expressed per mg microsomal protein.

Sequential measurements of the 21- and 16-hydroxylase activities for seven rabbits that showed the highest levels of 21-hydroxylase activity (greater than 0.1 nmol/

min/mg wet wt) are illustrated in FIGURE 3. We observed variation for the 21-hydroxylase activity between biopsy samples for individual rabbits. In each case, differences in the 16-hydroxylase activity paralleled those of the 21-hydroxylase (FIGURE 3). As illustrated in FIGURE 4, the ratio of 21/16-hydroxylase activities was relatively constant for each animal throughout the duration of the study. Since the rate of the 16-hydroxylase was not correlated to the rate of the 21-hydroxylase in the other group of animals, the variations observed in the rates between biopsy samples for individual rabbits probably reflected differing amounts of fatty and connective tissue in the biopsy samples. This was supported by the observation that variable amounts of red blood cell contamination were present in the samples. Contamination of this nature would affect the specific activity expressed in terms of sample weight without altering

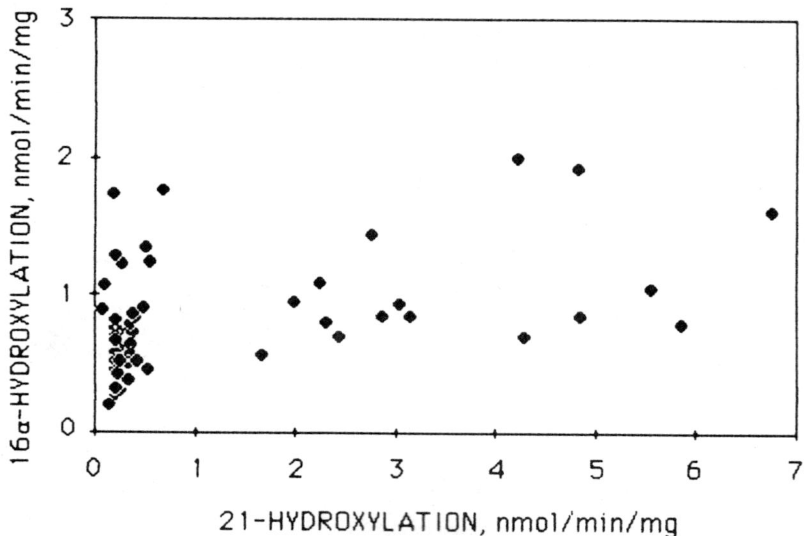

FIGURE 2. Scatter diagram for hepatic progesterone 21- and 16-hydroxylase activities of microsomes prepared for 19 male and 34 female New Zealand White rabbits. Each symbol designates the values obtained for a single animal. The variation of the two activities is not significantly correlated ($r < 0.5$).

the ratio of the two activities. Therefore, we employed the ratio of 21/16-hydroxylation to phenotype these animals.

As shown in FIGURE 4, seven rabbits exhibit the 21-H phenotype throughout the course of the study with an average ratio determined for the biopsy samples of 3.2 ± 1.2. The remaining 21 animals exhibited the 21-L phenotype with a mean ratio of 0.8 ± 0.3. These values are similar to the 21/16-hydroxylase ratios determined for the 21-H (3.8 ± 1.6) and 21-L (0.5 ± 0.2) rabbits of the other group.

The internal variability of the values was also computed and expressed as the coefficient of variation for each of the two populations of rabbits. The 21-H population was found to have a coefficient of variation of 15.2 ± 6.94, whereas that for the 21-L population was 25.11 ± 17.94. This difference was not statistically significant.

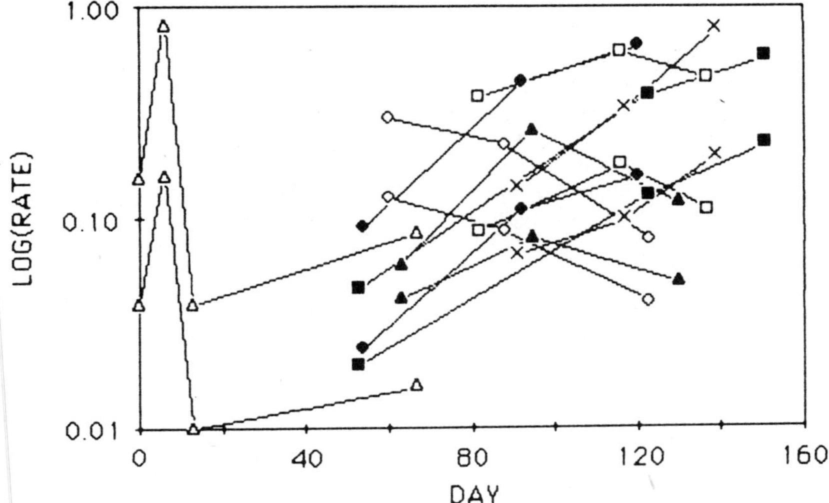

FIGURE 3. The progesterone 21- and 16-hydroxylase activity of serial liver biopsy samples prepared from New Zealand White rabbits exhibiting elevated 21-hydroxylase activity. Each symbol denotes one of seven rabbits, with the upper line depicting the 21-hydroxylase and the lower line depicting the 16-hydroxylase activity. The variation of the 16-hydroxylase activity among biopsy samples for individual rabbits was found to parallel that of the 21-hydroxylase activity in each case.

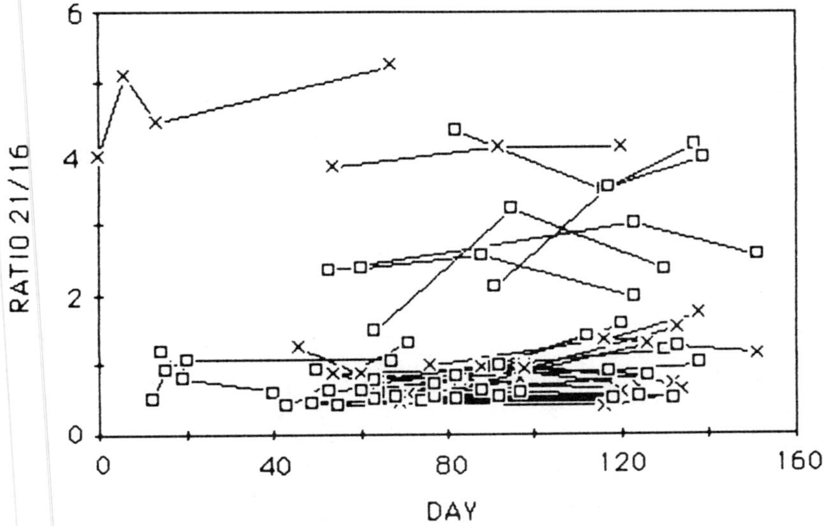

FIGURE 4. The ratio of 21- to 16-hydroxylase activity determined for serial liver biopsy samples from 10 male (X) and 18 female (□) New Zealand White rabbits. Seven rabbits exhibiting high ratios of 21/16-hydroxylase activity can be clearly distinguished from the remaining rabbits, which exhibit lower ratios.

DISCUSSION

Previous work has established that hepatic 21-hydroxylase activity varies considerably among outbred NZW rabbits.[2] It has also been shown that this activity in the rabbit is largely catalyzed by P-450 1.[6] Immunoquantitation studies have demonstrated that the microsomal concentration of P-450 1 is highly correlated with the level of progesterone 21-hydroxylase activity.[7] In contrast, the 16-hydroxylation of progesterone is catalyzed by several forms of P-450. Of these, P-450 3b is responsible for more than 50% of this activity.[8] Compared to P-450 1, the hepatic concentration of P-450 3b is relatively constant in the rabbit. The possibility exists that the variability in the expression of P-450 1, the 21-hydroxylase, may be the result of an environmental or dietary factor(s) producing a transient induction of this activity. Alternatively, these differences would reflect an underlying genetic polymorphism regulating expression of the 21-hydroxylase activity.

The results of the serial biopsy study suggest the latter to be the case. When the ratio of the 21- to 16-hydroxylase activity was determined, a relatively constant value for each animal resulted, which was sustained over the two-month period. This suggested that the variation in each activity between biopsies was largely due to the contamination of the samples with blood, fatty, and connective tissue. When all animals in the study were considered together on the basis of the 21/16-hydroxylase ratio, the 21-H phenotype was seen in approximately 30% of the rabbits. The phenotypic difference is of interest as it may provide an animal model for variations of the extra-adrenal conversion of progesterone to DOC observed in humans.

Differences of up to tenfold have been reported for the extra-adrenal conversion of plasma progesterone to DOC in male and pregnant and nonpregnant female subjects.[5] In late pregnancy, parallel increases in plasma DOC and progesterone are observed in humans,[9,10] which are attributable to the extra-adrenal conversion of progesterone to DOC.[5] In addition, impairment of the adrenal 21-hydroxylase activity in patients with congenital adrenal hyperplasia results in high plasma progesterone levels. It would be expected, therefore, that serum DOC levels would be low in these individuals. This is generally the case although a number of patients express DOC levels that fall in the normal range,[11] which is due in part to extra-adrenal conversion of progesterone producing 25–40% of plasma DOC.[4] Thus, an analogy may exist between human and rabbit populations that express elevated levels of extra-adrenal 21-hydroxylase activity.

SUMMARY

Previous work has shown that the 21-hydroxylation of progesterone in the hepatic microsomal fraction of outbred NZW rabbits varies over a tenfold range. In contrast, the 16-hydroxylase activity is relatively constant and is not correlated to the activity of the 21-hydroxylase. The distribution of the 21-hydroxylase activity is roughly bimodal with about one-third of the animals (21-H) exhibiting 21-hydroxylase activity exceeding 1 nmol/min/mg microsomal protein, whereas the remainder (21-L) generally exhibit an activity that is less than 1 nmol/min/mg protein.

To determine if this was due to a transient phenomenon, liver punch biopsies were collected from 28 rabbits at intervals of approximately three weeks for at least three serial samples. The 21- and 16-hydroxylase activities were determined in the post-mitochondrial fractions of these biopsy samples. A substantial variability in both 21- and 16-hydroxylase activities was observed for serial biopsy samples from individual rabbits. The variation of the 16-hydroxylase activity paralleled, however, that of the

21-hydroxylase, thus suggesting that the variation between biopsy samples for individual rabbits was due to factors such as contamination with blood and connective tissue, which would affect both activities equally. Rabbits, therefore, were phenotyped as 21-H or 21-L on the basis of the 21/16-hydroxylase ratio. The mean ratio was 3.2 ± 1.2 and 0.8 ± 0.3 for rabbits phenotyped as 21-H and 21-L, respectively. Similar values for this ratio were obtained for the other group of rabbits phenotyped as 21-H and 21-L, 3.8 ± 1.6 and 0.5 ± 0.2, respectively, following the isolation of microsomes from whole liver homogenates. The ratio of 21/16-hydroxylase activity was found to be relatively constant for biopsy samples obtained from the same animal over the course of this study, thus indicating that the elevated 21-hydroxylase activity is not a transient phenomenon.

REFERENCES

1. SENCIALL, I. R., G. BULLOCK & S. RAHAL. 1982. Can. J. Cell Biol. **61:** 722–730.
2. DIETER, H. H., U. MULLER-EBERHARD & E. F. JOHNSON. 1982. Science **217:** 741–743.
3. DIETER, H. H., U. MULLER-EBERHARD & E. F. JOHNSON. 1982. Biochem. Biophys. Res. Comm. **105:** 515–520.
4. ANTONIPILLAI, I., E. MOGHISSI, S. D. FRASIER & R. HORTON. 1983. J. Clin. Endocrinol. Metab. **57:** 580–584.
5. WINKEL, C. A., L. MILEWICH, C. R. PARKER, JR., N. F. GANT, E. R. SIMPSON & P. C. MACDONALD. 1980. J. Clin. Invest. **66:** 803–812.
6. REUBI, I., K. J. GRIFFIN, J. R. RAUCY & E. F. JOHNSON. 1984. J. Biol. Chem. **259:** 5887–5892.
7. JOHNSON, E. F. & K. J. GRIFFIN. 1984. Arch. Biochem. Biophys. In press.
8. REUBI, I., K. J. GRIFFIN, J. RAUCY & E. F. JOHNSON. 1984. Biochemistry **23:** 4598–4603.
9. BROWN, R. F., C. A. STROTT & G. W. LIDDLE. 1972. J. Clin. Endocrinol. Metab. **35:** 736–742.
10. NOLTEN, W. E., M. D. LINDHEIMER, S. OPARIL, P. A. REUCHERT & E. N. ERLICH. 1978. Am. J. Obstet. Gynecol. **132:** 414–420.
11. KUHNLE, U., D. CHOW, R. RAPAPORT, S. PANG, L. E. LEVINE & M. I. NEW. 1981. J. Clin. Endocrinol. Metab. **52:** 534–544.

Deoxycorticosterone Biosynthesis in Kidney Tissue of Experimental Animals

Characterization of Steroid 21-Hydroxylase Activity in Guinea Pig Kidney[a]

M. LINETTE CASEY,[b] PAUL C. MACDONALD,[b]
AND CRAIG A. WINKEL[c]

[b]*The Cecil H. and Ida Green Center
for Reproductive Biology Sciences
and the
Departments of Biochemistry and Obstetrics-Gynecology
The University of Texas
Southwestern Medical School
Dallas, Texas 75235*

[c]*Department of Obstetrics-Gynecology
The Uniformed Services University for the Health Sciences,
School of Medicine
Bethesda, Maryland 20814*

INTRODUCTION

In the human, plasma progesterone is converted to the mineralocorticosteroid, deoxycorticosterone (DOC), in extra-adrenal tissues.[1,2] Indeed, the rate of production of DOC from plasma progesterone is proportional to the plasma concentration of progesterone in pregnant women[1] and in women during the luteal phase of the ovarian cycle.[2] Importantly, extra-adrenal steroid 21-hydroxylase activity is stimulated by estrogen[3] and is known to be present in human kidney,[4,5] aorta,[6] and thymus.[7] Interestingly, these are tissues that are believed to be sites of DOC action.[8–10] In addition, steroid 21-hydroxylase activity has been demonstrated in spleen and other tissues of the human fetus,[7] and in spleen of adult guinea pig.[11] Also, it may be of considerable importance that extra-adrenal 21-hydroxylase activity, *in vivo,* is not reduced in persons with apparent congenital adrenal hyperplasia that is believed to be caused by adrenal 21-hydroxylase deficiency.[12,13]

Since there may be profound physiologic or pathophysiologic consequences of the *in situ* formation of DOC in kidney, which is a tissue responsive to mineralocorticosteroid action, we have conducted a search for an animal model in which these possibilities could be investigated. Microsome-enriched preparations of kidney tissue of the rat, mouse, rabbit, hamster, sheep, pig, cow, and guinea pig, as well as fetal kidney tissue of the rabbit, sheep, and rhesus monkey, were evaluated for steroid 21-hydroxylase

[a]This investigation was supported in part by USPHS grant nos. 5–P50–HD11149 and 5–P01–AG00306. The views expressed herein are strictly those of the authors and are not to be construed as representing those of the United States or the Department of Defense.

activity. We found that the enzyme activity was low or undetectable, or else did not follow Michaelis-Menten kinetics except in kidney of adult guinea pig, fetal rhesus monkey, and fetal rabbits. For these reasons, we conducted a systematic study of the reaction catalyzed by steroid 21-hydroxylase in microsome-enriched preparations of kidney tissue of guinea pigs.

METHODS

Steroids

[1,2,6,7-^{3}H(N)]Progesterone (98 Ci/mmol), purchased from New England Nuclear (Boston, MA), was purified by column chromatography on ethylene-glycol celite.[7] [4-^{14}C]DOC (40 mCi/mmol), also purchased from New England Nuclear, was purified by liquid-liquid partition chromatography on celite with the solvent system isooctane:*tert*-butanol:methanol:water (25:10:9:6, by vol). Nonradiolabeled progesterone, DOC, and DOC-acetate were purchased from Steraloids (Wilton, NH).

Animals

Rats (Sprague-Dawley), mice (C57/black), rabbits (New Zealand White), hamsters (Syrian Golden), and sheep (Merino) were obtained from our own colonies or from local suppliers. Kidneys from rhesus monkeys (*Macaca mulatta*) were provided by Dr. Robert Jaffe, University of California, San Francisco, CA. Bovine and swine kidneys were obtained from a local abattoir immediately after sacrifice of the animals. Guinea pigs (Hartley) were obtained from Hazelton-Dutchland (Denver, CO).

Tissue Preparation and Incubation

Kidneys were removed and placed in an ice-cold solution of NaCl (0.15 M). The kidney tissues, dissected free of capsule, adventitious tissue, and major calyces, were homogenized in 5 vol (w/v) of an ice-cold solution of sucrose (0.25 M) with a Teflon-glass homogenizer. Microsome-enriched fractions of homogenates of kidney tissues were prepared and incubated with [^{3}H]progesterone (3×10^{7} dpm), in various concentrations, in the presence of an NADPH-generating system as described previously (Winkel *et al.*, 1980c). In some experiments, namely, those in which we utilized fetal rabbit kidney as the source of steroid 21-hydroxylase, microsome-enriched preparations were prepared from kidneys obtained from several fetal rabbits.

Aliquots of the microsome-enriched suspension were assayed for protein content by the method of Lowry *et al.*,[14] with bovine serum albumin as the standard.

Purification of Radiolabeled DOC Formed in Incubation Mixtures

The [^{3}H]DOC formed in incubations of microsome-enriched preparations of kidney tissue with [^{3}H]progesterone was isolated and purified as described previously.[5] Briefly, [^{14}C]DOC (10,000–14,000 dpm) and nonradiolabeled DOC (10 μg) were added to the incubation mixtures prior to extraction of steroids with ethyl acetate. The [^{3}H]DOC formed during the incubation period was purified by successive thin-layer chromatographic (TLC) steps with the solvent systems, ethyl acetate:isooctane (7:3, v/v), methylene chloride:diethyl ether (4:1, v/v), and ethyl acetate:isooctane:acetic acid (9:9:2, by vol).

The purified DOC was acetylated by treatment with a mixture of pyridine and acetic anhydride (1:1, v/v) for 2 h at 37°C. The DOC-acetate was isolated by TLC with the solvent system, ethyl acetate:isooctane (1:1, v/v). The radiolabeled DOC-acetate, together with 40 mg nonradiolabled DOC-acetate, was crystallized from the solvent pair diethyl ether and petroleum ether (b.p. 20 to 40°C) until the ^3H:^{14}C ratios of the isolated DOC-acetate in successive mother liquors and crystals were similar.

RESULTS

Steroid 21-hydroxylase activities in kidney tissues of the animals studied are presented in TABLE 1. We found that this enzyme activity was present in kidney tissue of the rhesus monkey fetus, adult and fetal rabbit, adult rat, and adult guinea pig. The enzyme activity in kidney tissue of the adult and fetal sheep, adult mouse, pig, and cow was very low or undetectable. The apparent K_m of steroid 21-hydroxylase for progesterone in microsome-enriched preparations of homogenates of adult rabbit kidney, however, was extraordinarily high (FIGURE 1); indeed, saturation of the enzyme was not achieved even at progesterone concentrations in the incubation medium of 100 μM. In addition, saturation was not achieved when incubations were conducted in the presence of a variety of detergents [Triton WR1339, Emulgen 913, cetyl pyridinium chloride, and Brij 96 (data not shown)].

Steroid 21-hydroxylase activity also was demonstrated in microsome-enriched preparations of rat kidney. In this tissue, as in that of the rabbit, however, the enzyme was not saturable even with high concentrations of progesterone, i.e., the fractional conversion of substrate (progesterone) to product (DOC) was greater when the substrate concentration was 5 μM than when the concentration was 1 μM or less.

On the other hand, the kinetics of steroid 21-hydroxylase activity in microsome-enriched preparations of adult guinea pig kidney did conform to those of Michaelis-Menten (FIGURE 2). The apparent K_m of the enzyme for progesterone in this tissue was

TABLE 1. Specific Activity of Steroid 21-Hydroxylase in Microsome-Enriched Preparations of Kidney Tissues of Experimental Animals

Species	Specific Activity[a] (pmol × h^{-1} × mg^{-1} protein)
Guinea Pig	1.64
Rabbit (adult)	6.10
Rabbit (fetus)	7.90
Rhesus Monkey (fetus)	5.80
Rat[b]	5.03
Mouse	No Activity
Hamster	No Activity
Sheep (adult)	No Activity
Sheep (fetus)	No Activity
Cow	No Activity
Pig	No Activity

[a]Specific activity calculated from data obtained after incubation of microsome-enriched preparations of renal tissue with [^{3}H]progesterone (5 μM).

[b]Steroid 21-hydroxylase not saturated at progesterone concentration of 5 μM.

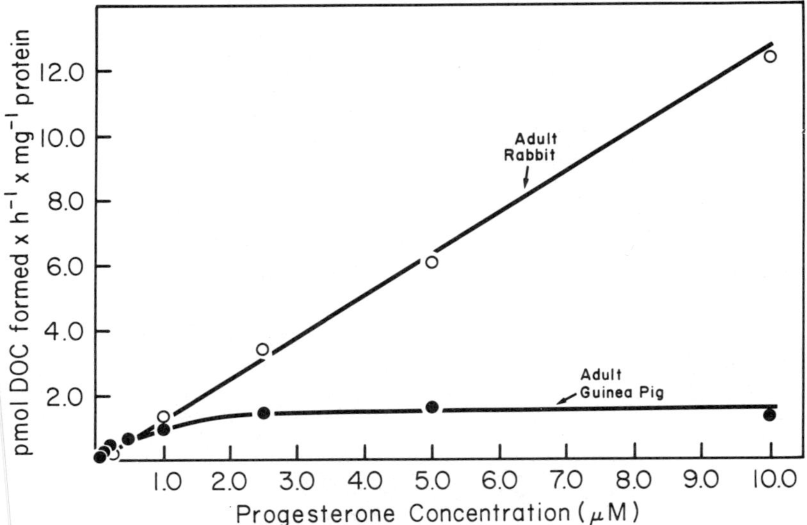

FIGURE 1. Steroid 21-hydroxylase activity in kidney of adult rabbit (○) and adult guinea pig (●). Microsome-enriched preparations were incubated for 30 min in medium that contained progesterone in various concentrations.

0.5 μM (FIGURE 2). In this tissue, the reaction was linear with time for 2 h and with protein concentration up to 2.5 mg. The specific activity of steroid 21-hydroxylase in kidney tissue of adult guinea pig, however, was much lower than that in kidney of adult rabbit (FIGURE 1).

Steroid 21-hydroxylase activity in microsome-enriched preparations of fetal rabbit kidneys also conformed to Michaelis-Menten kinetics (FIGURE 3).

DISCUSSION

In search of an animal model that is appropriate for an investigation of the regulation of and physiologic role of steroid 21-hydroxylase activity in kidney tissue, we assayed this enzyme activity in kidney tissue of a number of mammalian species. The activity of the enzyme was low or undetectable in microsome-enriched preparations of kidney tissue of the adult pig, cow, and mouse, and of the adult and fetal sheep, under the conditions of our assay. Steroid 21-hydroxylase activity was demonstrated in kidney tissue of the adult rat and rabbit; under the conditions we employed, however, the enzyme in these tissues was not saturable with substrate, namely, progesterone, in contrast to that in human adult[4] and fetal[5] kidney tissues.

Steroid 21-hydroxylase activity also was demonstrated in fetal rabbit and in adult guinea pig kidney tissues. In these tissues, the enzyme reaction did follow Michaelis-Menten kinetics and the apparent K_m of the enzyme, 0.5 μM, was similar to that found in microsome-enriched preparations of guinea pig spleen[11] and in human adult[4] and fetal[5] kidney tissues. Thus, the guinea pig may be an appropriate model for use in a study of the role of progesterone as a circulating prehormone for DOC formation in

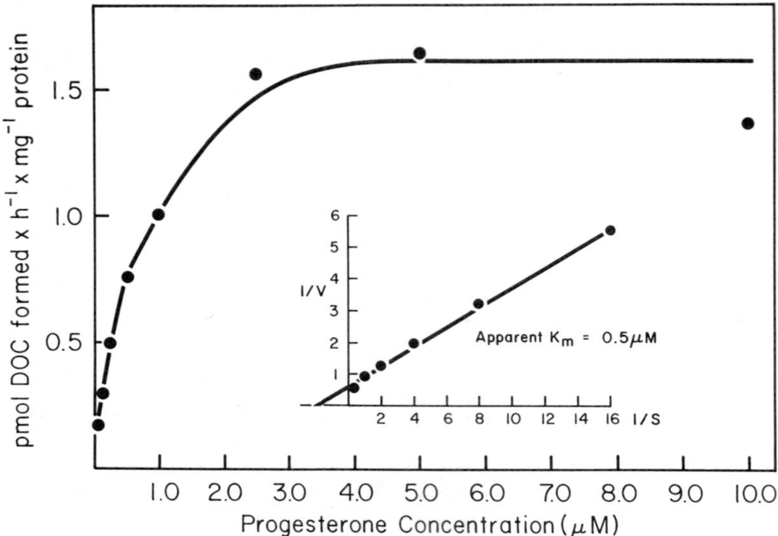

FIGURE 2. Steroid 21-hydroxylase activity in kidney of adult guinea pig. Microsome-enriched preparations were incubated for 30 min in medium that contained progesterone in concentrations ranging from 0.02 to 10 μM. The apparent K_m of the enzyme for progesterone was 0.5 μM.

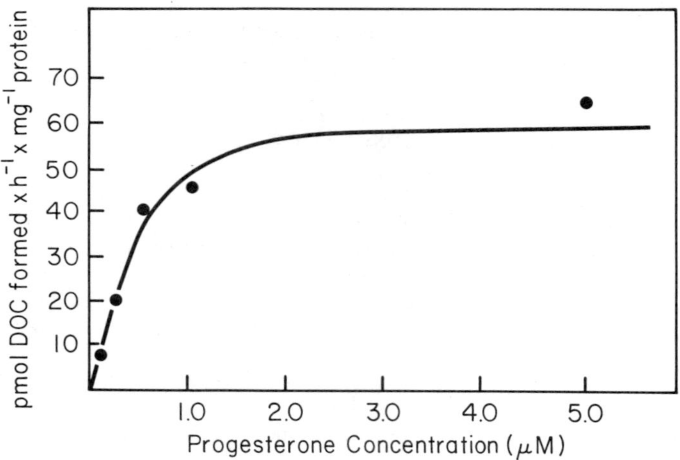

FIGURE 3. Steroid 21-hydroxylase activity in kidney of fetal rabbit. Microsome-enriched preparations were incubated for 30 min in medium that contained progesterone in concentrations ranging from 0.02 to 5.0 μM.

kidney, as well as in other extra-adrenal tissues. Interestingly, in the pregnant guinea pig, as in pregnant women, the levels of progesterone are quite elevated and do not fall prior to the onset of parturition.[15] We suggest that an evaluation of the regulation of steroid 21-hydroxylase activity in guinea pig kidney is appropriate, and that by use of this experimental animal model, an understanding of the physiologic and pathophysiologic importance of the formation in kidney of the mineralocorticosteroid, deoxycorticosterone, from circulating progesterone will be facilitated.

ACKNOWLEDGMENTS

The authors are grateful to Frank Hereford, Jesse Smith, and Steve Robinson for skilled technical assistance. We also thank Sylvia Williams for expert editorial assistance.

REFERENCES

1. WINKEL, C. A., L. MILEWICH, C. R. PARKER, JR., N. F. GANT, E. R. SIMPSON & P. C. MACDONALD. 1980. J. Clin. Invest. **66:** 803–812.
2. WINKEL, C. A., C. R. PARKER, JR., E. R. SIMPSON & P. C. MACDONALD. 1980. J. Clin. Endocrinol. Metab. **57:** 1354–1358.
3. MACDONALD, P. C., S. CUTRER, S. C. MACDONALD, M. L. CASEY & C. R. PARKER, JR. 1982. J. Clin. Invest. **69:** 469–478.
4. WINKEL, C. A., E. R. SIMPSON, L. MILEWICH & P. C. MACDONALD. 1980. Proc. Natl. Acad. Sci. USA **77:** 7069–7073.
5. WINKEL, C. A., M. L. CASEY, E. R. SIMPSON & P. C. MACDONALD. 1981. J. Clin. Endocrinol. Metab. **53:** 10–15.
6. CASEY, M. L. & P. C. MACDONALD. 1982. J. Clin. Endocrinol. Metab. **55:** 804–806.
7. CASEY, M. L., C. A. WINKEL & P. C. MACDONALD. 1983. J. Steroid Biochem. **18:** 449–452.
8. COMSA, J. 1973. *In* Thymic Hormones. T. D. Luckey, Ed.: 59–96. University Park Press. Baltimore, Maryland.
9. MEYER, W. J. & N. R. NICHOLS. 1981. J. Steroid Biochem. **14:** 1157–1160.
10. KALSNER, S. 1969. Br. J. Pharmacol. **36:** 582–593.
11. WINKEL, C. A., C. E. WADE, D. L. DANLEY, P. C. MACDONALD & M. L. CASEY. 1983. J. Steroid Biochem. **19:** 1635–1638.
12. WINKEL, C. A., M. L. CASEY, R. J. WORLEY, J. D. MADEN & P. C. MACDONALD. 1983. J. Clin. Endocrinol. Metab. **56:** 104–107.
13. ANTONIPILLAI, E., E. MOGHISSI, S. D. FRASIER & R. HORTON. 1983. J. Clin. Endocrinol. Metab. **57:** 580–584.
14. LOWRY, O. H., N. J. ROSEBROUGH, A. L. FARR & R. J. RANDALL. 1951. J. Biol. Chem. **193:** 265–275.
15. HEAP, R. B. & R. DEANSLEY. 1966. J. Endocrinol. **34:** 417–423.

An Approach to the Molecular Biology of Congenital Adrenal Hyperplasia[a]

BON-CHU CHUNG,[b] KARLA J. MATTESON,[b]
JOHN E. MORIN,[b] SYNTHIA H. MELLON,[c]
AND WALTER L. MILLER[b,c,d]

[b]*Department of Pediatrics*
and
[c]*Metabolic Research Unit*
University of California, San Francisco
San Francisco, California 94143

Steroid hormone synthesis is initiated by conversion of cholesterol to pregnenolone by the cholesterol side-chain cleavage enzyme, P-450_{scc}, which performs both 20- and 22-hydroxylations and scission of the 20–22 carbon bond on a single active site. This enzyme is rate limiting[1] and is regulated in the adrenal fasciculata by ACTH.[2,3] It may also be regulated by angiotensin II in the adrenal glomerulosa and by leutinizing hormone in the gonad. Supraphysiologic concentrations of ACTH increase accumulation of P-450_{scc} mRNA, suggesting ACTH regulates P-450_{scc} gene expression.[4,5] Thus, P-450_{scc} is a steroidogenic enzyme having great interest for regulatory biology even though clinical disorders of P-450_{scc} are rare, with only about 32 reported cases.[6]

Approximately 90% of patients having genetic disorders of steroid hormone synthesis have disordered 21-hydroxylation, which is mediated by another cytochrome P-450 called P-450_{c21}. Biochemical studies suggest there may be only a single P-450_{c21} which hydroxylates both progesterone and 17-hydroxyprogesterone, the latter more efficiently.[7] However, the variety of clinical phenotypes of 21-hydroxylase deficiency has caused many physicians to hypothesize two separate 21-hydroxylases. To study these biochemical and clinical problems, we have sought to clone cDNA's for P-450_{scc} and P-450_{c21} using bovine adrenals as an initial system.

MATERIALS AND METHODS

Preparation and Translation of RNA

Polyadenylated RNA (poly(A)$^+$ RNA) from fresh bovine adrenal cortices was prepared and translated as described.[8] Translation products were immunoprecipitated with rabbit anti-bovine antibodies (provided by M.R. Waterman, Dallas, Texas) and were analyzed on SDS/polyacrylamide gels as described.[8]

[a]This work was supported by NIH grant no. HD-16047 to WLM. The abbreviations used in this paper are as follows: ACTH, adrenocorticotrophic hormone; bp, base pairs; kb, kilobases; dC, deoxycytidine; dG, deoxyguanosine; SDS, sodium dodecyl sulfate.

[d]Address correspondence to: Walter L. Miller, Room 677-S, Department of Pediatrics, University of California, San Francisco, California 94143.

Construction of the Bovine Adrenal cDNA Library

Poly(A)$^+$ RNA was reverse transcribed twice into double-stranded cDNA[9,10] with the addition of 1 m*M* RNasin (Promega Biotec, Madison, Wisconsin) during the first reverse transcription. The double-stranded cDNA was trimmed with S_1 nuclease and the 3′ ends were extended with deoxycytidine as described.[9,10] The tailed cDNA was size-selected by electrophoresis through low-gelling temperature agarose. cDNA of $<$500 bp was discarded, and the remainder was handled in two batches of 500–1000 bp and 1000–4000 bp, representing 55% and 18% of the total, respectively. This cDNA was inserted into the *Pst* I site of pBR322 previously tailed with deoxyguanosine residues. The recombinant plasmids were used to transform *E. coli* MM294 by the Hanahan procedure.[11]

Hybrid-Selected Translation

Groups of four to nine clones were grown with an additional clone of bovine growth hormone cDNA[10] in 150 ml cultures in minimal medium with chloramphenicol amplification. Plasmid DNA was prepared by polyethylene glycol precipitation from a cleared lysate, cleaved with *Eco* RI or *Bam* HI to linearize the plasmid, denatured, and affixed to 2.5cm nitrocellulose filters as described.[12–14] Filters were hybridized at 42°C in 0.4 M NaCl, 50% formamide overnight with 40 μg of bovine adrenocortical mRNA and 3 μg of bovine pituitary mRNA. The filters were washed extensively with 1 × SSC/50% formamide (1 × SSC is 150 m*M* NaCl, 15 m*M* Na citrate), and the bound mRNA was eluted with water and precipitated with ethanol. The products from translation of this RNA were immunoprecipitated, displayed on 12.5% SDS/polyacrylamide gels, and analyzed by fluorography for two to four weeks. The bovine growth hormone plasmid in conjunction with the bovine pituitary mRNA (12% bovine growth hormone)[10] provides an internal control for each hybrid-selected translation.

Synthesis of Synthetic Oligonucleotides

Oligonucleotides were synthesized by the manual phosphoramidite procedure.[15] The first base (3′ end) is attached to silica gel (Biosearch, San Rafael, California); the next base, containing a DMT-blocked 5′ group (Biosearch), is added in acetonitrile in the presence of tetrazole, which catalyses the linkage, which in turn is oxidized with iodine to form a phosphodiester bond. The 5′ DMT group is removed with acid, the resin is washed clean with acetonitrile, and the process is repeated with the next (5′) base.

Oligonucleotide-Initiated cDNA Synthesis

Poly(A)$^+$ mRNA (10 μg) and 0.1 μg of the mixture of 32 15-mers were boiled for two minutes to melt mRNA secondary structure that might initiate cDNA synthesis and then were cooled rapidly in ice water. Buffer for reverse transcription was added and the mixture was warmed to 25°C to permit specific oligonucleotide/mRNA hybridization. Reverse transcriptase was added and the mixture was warmed to 42°C for 30 min. The mixture was then extracted with phenol and chloroform and chromatographed on G–75 sephadex.

Preparation of Genomic DNA

High molecular weight bovine genomic DNA was prepared by homogenizing 5 gm of fresh bovine thymus in 50 ml of phosphate-buffered saline for 1 min in a Waring blender, centrifuging at 2000 × *g* for 5 min, and resuspending the pellet in 10 ml of 10 m*M* Tris pH 7.6, 0.15 M NaCl, 2 m*M* $MgCl_2$ 0.5% NP40. This suspension was homogenized on ice by ten strokes of a loose-fitting Dounce pestle, and the nuclei were pelleted at 2000 × *g* for 20 min. The nuclei were resuspended in 10 m*M* Tris pH 7.4, 10 m*M* NaCl, 3 m*M* $MgCl_2$ and were digested with 0.25% SDS, 5 m*M* EDTA, and 0.5 mg/ml protease K at 37°C overnight. The digest was gently extracted three times with phenol and chloroform, dialyzed against 10 m*M* Tris pH 7.4, 1 m*M* EDTA, and then digested with 25 μg/ml RNase A for four hours at 37°C. The DNA was then extracted three times with phenol and chloroform, dialyzed against 10 m*M* Tris pH 7.4, 1 m*M* EDTA, precipitated with ethanol, and collected by spooling. The DNA was dissolved in water and used for restriction endonuclease digestions according to the manufacturer's instructions.

Blotting of DNA and RNA

Restriction endonuclease digests of genomic DNA were displayed by electrophoresis through 0.85% agarose gel and transferred to nitrocellulose sheets (Schleicher and Scheull BA85) as described by Southern.[16] Genomic blots probed with cDNA were incubated 48 hours at 42°C in 5 × Denhardt's Solution,[17] 50 m*M* Hepes pH 7.0, 3 × SSC, 50% formamide, 18 μg/ml salmon sperm DNA, and 40 μg/ml yeast tRNA, and were washed four times in 0.1 × SCC, 0.1% SDS, at 65°C for 30 min before autoradiography. Digests of cloned DNA were electrophoresed on 1.2% agarose and hybridized to synthetic oligonucleotide probes at 25°C. Hybridizations with 15-mers were washed six times in 6 × SCC at 25°C, twice at 37°C, and once at 42°C; hybridizations with 17-mers were washed six times in 6 × SCC at 25°C, twice at 42°C, and once at 47°C.

Screening cDNA Clones

Bacterial colonies were transferred to nitrocellulose sheets and lysed.[18] They were then probed with ^{32}P-labeled oligonucleotide-initiated cDNA at 42°C in 50% formamide, 1 × Denhardt's Solution,[17] 5 × SCC, 100 μg/ml salmon sperm DNA, and washed four times in 0.1 × SCC, 0.1% SDS at 50°C. Putatively positive clones identified by autoradiography were grown in 10 ml L-broth and were used to prepare plasmid DNA,[19] which was spotted onto nitrocellulose with a minifold apparatus (Schleicher and Schuell). The filter was hybridized to labeled 17-mers and washed under the same conditions used for Southern blots probed with 17-mers.

Sequencing

DNA restriction fragments were subcloned in M13 mp10 or M13 mp11 and sequenced from the universal primer site by the dideoxy method of Sanger *et al.*,[20] as described. All clones were sequenced on both strands by dideoxy sequencing. Sequences were confirmed by chemical degradation sequencing[22] along one strand.

RESULTS AND DISCUSSION

Side-Chain Cleavage

We prepared bovine adrenocortical mRNA, translated this in a cell-free reticulocyte system, and immunoprecipitated the P-450_{scc} and P-450_{c21} with rabbit anti-bovine antibodies. As estimated by densitometric scanning of the autoradiographs, P-450_{scc} and P-450_{c21} represented less than 0.5% and 0.05% of the newly synthesized protein, respectively. This can be used as a crude estimate of the relative abundance of the mRNAs for these proteins in total bovine adrenocortical polyadenylated RNA. The presence of several other protein bands of molecular weight indistinguishable from the P-450_{scc} band[23] indicated it would be impossible to screen clones by hybrid-arrested translation as the absence of the P-450_{scc} band would not be distinguishable. Thus, we chose to screen for P-450_{scc} clones by hybrid selection.[12–14] The low abundance of the P-450_{c21} mRNA and the low affinity of the antibody led us to discard immunologic strategies for identifying P-450_{c21} clones.

We constructed a bovine adrenal cDNA library of about 60,000 clones; 10,000 from cDNA of >1 kb and 50,000 from cDNA of 500–1000 bp, as described.[23] After screening about 700 clones, we identified two pools of four clones that were positive for P-450_{scc} in the hybrid-selection assay.[23] Analysis of the individual clones in one of these pools (Lane L of figure 2 in ref. 23) revealed two positive clones containing small inserts that did not cross-hybridize.[23] We, then, analyzed the individual clones in the second pool (Lane E of figure 2 in ref. 23) and identified a single positive clone as pBA644 (FIGURE 1).

The nucleotide sequence of the pBA644 insert was determined on both strands by dideoxy sequencing and confirmed by sequencing one strand chemically. Regions of apparent conflict or ambiguity were resequenced. The complete sequence of the 1073 bases (excluding the dC·dG tails added in the cloning procedure) is shown in FIGURE 2. The cloned insert was purified, nick-translated to a specific activity of 3.7×10^8 cpm/μg, and used to probe bovine genomic DNA cleaved with a variety of restriction enzymes. The autoradiograph of that Southern blot is shown in FIGURE 3. These data suggest the presence of multiple bovine P-450_{scc} gene sequences. The sequence of pBA644 and the pattern in FIGURE 3 are different from those of a bovine P-450_{scc} clone isolated in Dallas using the same rabbit anti-bovine P-450_{scc} antibodies (M.R. Waterman, personal communication). Thus, the identity of pBA644, the significance of the complex pattern in FIGURE 3, and the relation of pBA644 to the Dallas clone are uncertain at this time.

21-Hydroxylase

As the abundance of P-450_{c21} mRNA appeared to be less than 0.05% by cell-free translation criteria, no attempt was made to screen the cDNA library by hybrid-selected translation with antibody to P-450_{c21}. The subsequent publication of the amino acid sequences of eight cysteine-containing tryptic peptides of porcine adrenal P-450_{c21},[24] however, offered the possibility of screening our bovine cDNA library with synthetic oligonucleotide probes. The sequences of these eight peptides contain two regions of relatively low codon degeneracy. The amino acid region used from peptide T-29 has the sequence Gln-Glu-Phe-Cys-Glu, which could be encoded by the 32 different mRNA sequences, 5′ $CA^{A/G}$ $GA^{A/G}$ $UU^{U/C}$ $UG^{U/C}$ $GA^{A/G}$ 3′. The amino acid region used from peptide T-40 has the sequence Glu-His-Cys-Pro-Asp-Ile encoded by the 192 mRNA sequences, 5′ $CA^{A/G}$ $CA^{U/C}$ $UG^{U/C}$ CCX $GA^{U/C}$ AU(UCA) 3′.

Eliminating the final redundant nucleotide of the isoleucine codon, we synthesized 64 17-mers based on the T-40 sequence, 3′ $GT^{T/C}$ $GT^{A/G}$ $AC^{A/G}$ GGX $CT^{A/G}$ TA 5′, and we synthesized 32 15-mers based on the T-29 sequence, 3′ $GT^{T/C}$ $CT^{T/C}$ $AA^{A/G}$ $AC^{A/G}$ $CT^{T/C}$ 5′. The manual syntheses were separated into different aliquots when the mixture complexity reached 16. This permitted confirmation of the sequences by the Maxam-Gilbert technique using a modified gel system.[25] The oligonucleotides were

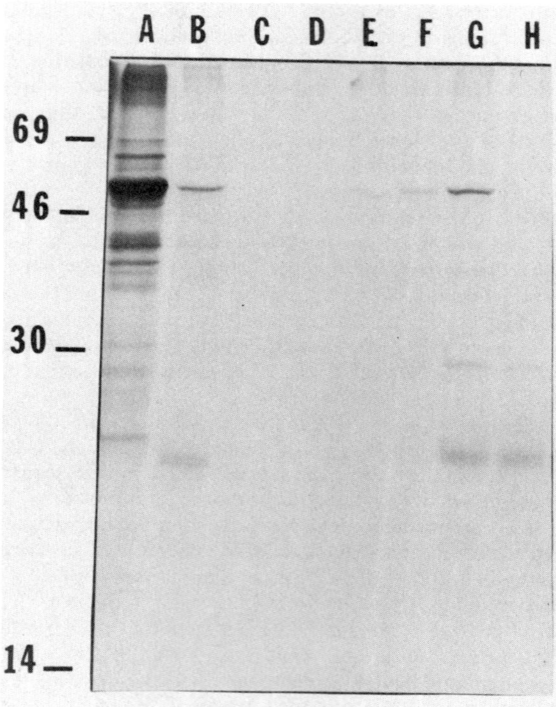

FIGURE 1. Hybrid-selected translation of individual clones from Lane E of figure 2 in Matteson *et al.*[23] Lane A: Total translation products immunoprecipitated with anti-P-450_{scc}. Lane B: Translated and immunoprecipitated products selected by pool E. Pool E included a bGH cDNA clone and the RNA included bovine pituitary mRNA (GH control). Lanes C, D, E, and F: Translated and immunoprecipitated products from each of the four clones in pool E. Lane G: Hybrid-selected translation with pBA571,[23] plus GH control. Lane H: Same as Lane G, plus exogenous excess P-450_{scc}.

complementary to the mRNA sequences so that they could hybridize to both DNA and RNA.

Since the oligonucleotides were based on a porcine amino acid sequence, we did not know if they would hybridize to bovine sequences. Silent nucleotide differences between the two species would be accounted for by the multiple oligonucleotides, but differences between the bovine and porcine amino acid sequences could prevent hybridization. Hogs and cattle are fairly close evolutionary relatives, both belonging to

TAAGTCTGAA TTTTGCAATA AGGAACTCAT GATTTGAATT ACAGTCAGCT CCCATTCCTG
ATTCAGACTT AAAACGTTAT TCCTTGAGTA CTAAACTTAA TGTCAGTCGA GGGTAAGGAC

TTTTTGCTGA CTATATAGAG CCTTCTCCAT TTTTGGCTGC AAAACATATA ATCAGTCTGA
AAAAACGACT GATATATCTC GGAAGAGGTA AAAACCGACG TTTTGTATAT TAGTCAGACT

TTTGGTATTT ATCATTTTGT GACATAATGT GTAAGAGTGT CTCGTCTGTT TGGAAAAGGT
AAACCATAAA TAGTAAAACA CTGTATTACA CATTCTCACA GAGCAGACAA ACCTTTTCCA

AGTTTCTATG ACCAGTGTGT CTCTTGGCAA ACTCTGTTAA CCTTTGTCTC ACCACTTCAT
TCAAAGATAC TGGTCACACA GAGAACCGTT TGAGACAATT GGAAACAGAG TGGTGAAGTA

TTTGTATTCC AAGGCCTTTG TTTCTCTGTT TCTCCAGGTA TCTCTTGACT TCCTACTTTT
AAACATAAGG TTCCGGAAAC AAAGAGACAA AGAGGTCCAT AGAGAACTGA AGGATGAAAA

ACCTTCCAAT CCTCTAGGAT GAAAAGGACA TCTTTTTTTT TTTTTTTGGT GTAGTTCTAG
TGGAAGGTTA GGAGATCCTA CTTTTCCTGT AGAAAAAAAA AAAAAAACCA CATCAAGATC

AAGGTCTTCA TAGAAAGGGT CAACTTCAAC TTCTTAGGCA TCAGTGGTTA GGGCATATAC
TTCCAGAAGT ATCTTTCCCA GTTGAAGTTG AAGAATCCGT AGTCACCAAT CCCGTATATG

TTGGATTACT GTAATGTTAA ATGGTTTGCT TTGGAAACTA ACCAAGATCA TTCTGTTGCT
AACCTAATGA CATTACAATT TACCAAACGA AACCTTTGAT TGGTTCTAGT AAGACAACGA

TTTGAGATTG CACCCAAATA CTGCATTTTG GACTCTTCTG TTTACTATGA GGACTACTCC
AAACTCTAAC GTGGGTTTAT GACGTAAAAC CTGAGAAGAC AAATGATACT CCTGATGAGG

ATTTAATCTA AAGGATTCTT AGGCCACAAT AGTAGATATA ATGGTCATCT GAATTATTAT
TAAATTAGAT TTCCTAAGAA TCCGGTGTTA TCATCTATAT TACCAGTAGA CTTAATAATA

AAATTTATCA ATTTTCTTCC ATTTTAGTTC ACTGAATTCT AACTTATTGA TGCTTCATTC
TTTAAATAGT TAAAAGAAGG TAAAATCAAG TGACTTAAGA TTGAATAACT ACGAAGTAAG

TTGCCATCTC CTGCTTGACC ATGTTTTTTA CCTTGATTCA TGGACCTGAC ATTCCAGGTT
AACGGTAGAG GACGAACTGG TACAAAAAAT GGAACTAAGT ACCTGGACTG TAAGGTCCAA

CCTATGCAAT ATTATTCTGT ATAGTGTCAG ACTTACTTTC ACCACCAGAC ATATCCACAA
GGATACGTTA TAATAAGACA TATCACAGTC TGAATGAAAG TGGTGGTCTG TATAGGTGTT

CTGTATATCA TTTCCGTTTT GGCCCAGCTG CTTCACTTTT TCTGGAACTA TTCATATCTG
GACATATAGT AAAGGCAAAA CCGGGTCGAC GAAGTGAAAA AGACCTTGAT AAGTATAGAC

CCCTCCACTC TTTCCCAATA GCATATTGGA CACATTCTCG AACACAGGGA GCCGGGGGAC
GGGAGGTGAG AAAGGGTTAT CGTATAACCT GTGTAAGAGC TTGTGTCCCT CGGCCCCCTG

AGGTGCTGGT TTCTTCTGGC ACACCTGGGG CAGCTGAACA CAGTGTTGAC TGGCAGACAC
TCCACGACCA AAGAAGACCG TGTGGACCCC GTCGACTTGT GTCACAACTG ACCGTCTGTG

AGCCCCACAC CAAACGCTCG CTAACACTGA CACTGTTCCC GTGATGGCCA GGGAGCCCCC
TCGGGGTGTG GTTTGCGAGC GATTGTGACT GTGACAAGGG CACTACCGGT CCCTCGGGGG

TCCCCAAAAA CCTGCTCCTG GAAGCTGGCA GGATTTGTGC CATTCATAAG GGT
AGGGGTTTTT GGACGAGGAC CTTCGACCGT CCTAAACACG GTAAGTATTC CCA

FIGURE 2. Complete nucleotide sequence of the insert in pBA644.

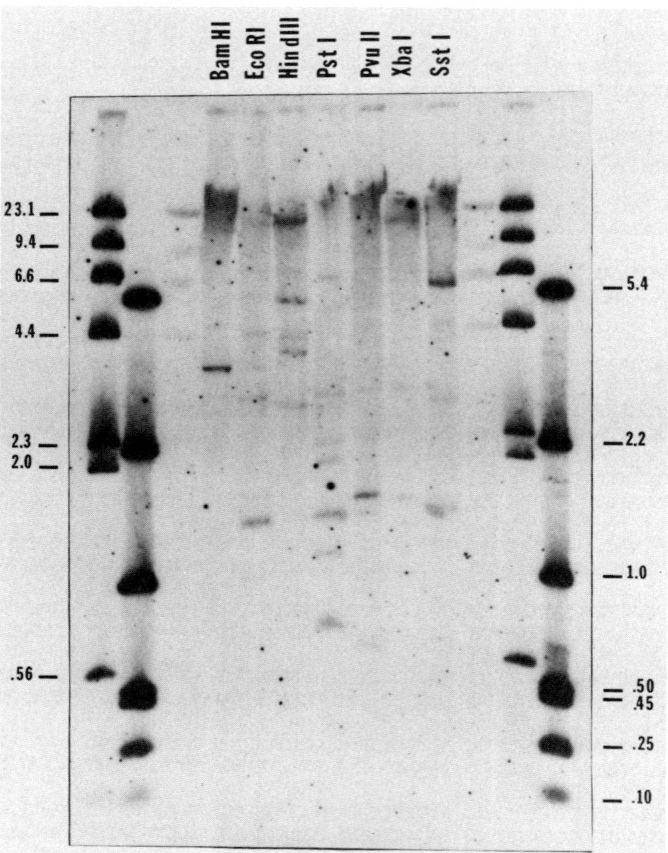

FIGURE 3. Southern blot of bovine genomic DNA cleaved with the indicated restriction endonucleases and probed with ^{32}P-labeled insert from pBA644. Molecular size standards (in kb) are from ^{32}P-labeled bacteriophage λ (indicated on the left) or bacteriophage PM2 (indicated on the right), each cleaved with *Hin*d III.

the mammalian order, *artiodactyla,* which diverged about 60 million years ago.[26] Hog and cattle amino acid homologies are generally about 85–90%, although their cytochrome c sequences are identical.[26] Therefore, we sought to determine if our oligonucleotides would hybridize to appropriate bovine sequences by probing both bovine and porcine adrenal mRNA on Northern blots.

Duplicate blots were probed with the 64 17-mers. They hybridized to porcine and bovine mRNA of about 2000 nucleotides (FIGURE 4). This is about the size we expect for P-450_{c21} mRNA; rat liver phenobarbital-induced P-450_b mRNA is reported to be 1900 bases[27] or 2150 ± 100 bases long,[28,29] while cytochrome P_1-450 induced by 3-methylcholanthrene appears in two forms of 2700 and 3500 bases.[30] Northern blots showed no hybridization of human adrenal mRNA; therefore, we used the porcine-based oligonucleotides to screen our bovine adrenal cDNA library. Direct colony

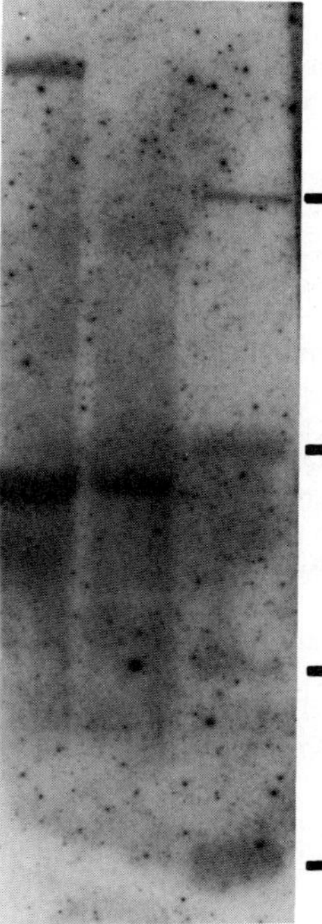

FIGURE 4. Northern blot of bovine (B) and porcine (P) poly(A)$^+$ RNA probed with the 32 15-mers. Molecular weight markers (M, *Hin*d III-cleaved, phage PM2) are 5.4, 2.2, 1.0, and 0.5 kb. The nature of the high molecular weight band in B is unknown.

FIGURE 5. Oligonucleotide-primed cDNA synthesis using bovine (B) or porcine (P) poly(A)$^+$ RNA as template. Lane C shows control synthesis with bovine mRNA and no oligonucleotides (15-mers). Molecular weight markers (left lane, *Hin*d III-cleaved phage PM2) are 5.4, 2.2, 1.0, 0.45, 0.25, and 0.1 kb.

hybridization resulted in high background. To reduce the background by increasing the washing stringency and to increase the length of our probe, we synthesized cDNA from bovine adrenal mRNA using the porcine 15-mers as specific primers. This tactic is difficult because the cDNA produced may not be specific due to cDNA synthesis initiated by mRNA self-priming or nonspecific priming. We boiled the mRNA and 15-mers, and then rapidly cooled the mixture to eliminate mRNA secondary structure that might initiate cDNA synthesis. Buffer was added and allowed to warm to 25°C and incubated for one hour to permit specific oligonucleotide hybridization. Reverse transcriptase was then added and the mixture was warmed to 42°C for 30 min. Oligonucleotide-primed reverse transcription of 10 μg of poly(A)$^+$ mRNA produced only ~1 ng of cDNA containing ~10^6 cpm. A control reaction lacking oligonucleotide primers gave virtually no incorporation of [^{32}P]dCTP (FIGURE 5). The 15-mer primed cDNA was quite heterogeneous in size, suggesting many species of cDNA were synthesized. However, the lack of self-priming indicated that this was a useful probe, so the bovine library was screened with 15-mer initiated cDNA (15-mer cDNA). The low yield of 15-mer cDNA precluded cloning this material.

Hybridization of the 15-mer cDNA to the ~10,000 clones made from cDNA larger than 1 kb yielded 54 apparent positives. These 54 clones were first screened by preparing cleared lysate from each, immobilizing the plasmid DNA on nitrocellulose using a dot-blotting apparatus (minifold), and probing with ^{32}P-labeled 17-mers (FIGURE 6). The seven clones giving the strongest signal were chosen, and plasmid

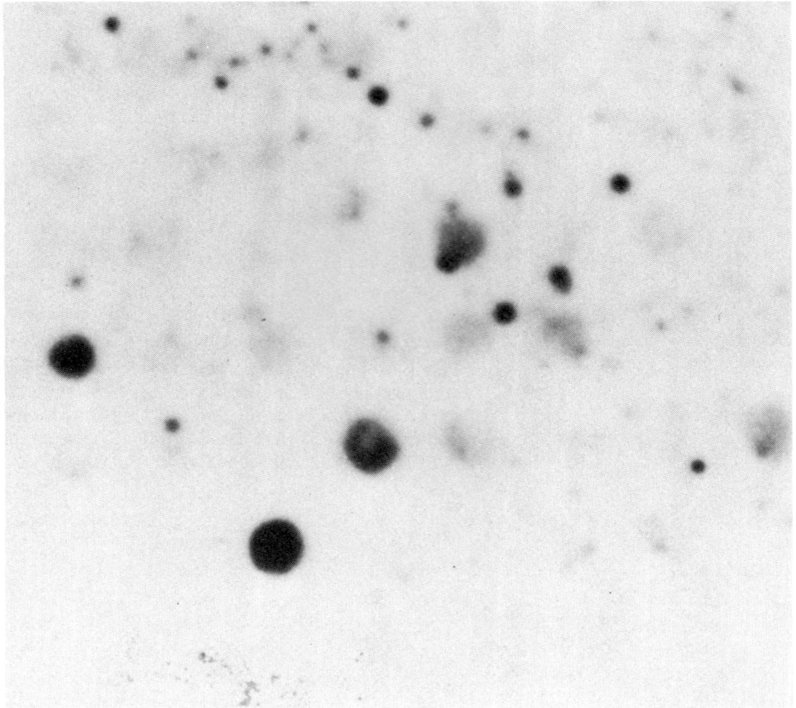

FIGURE 6. Hybridization of ^{32}P-labeled 17-mers to the 54 clones identified by hybridization to the 15-mer-initiated cDNA.

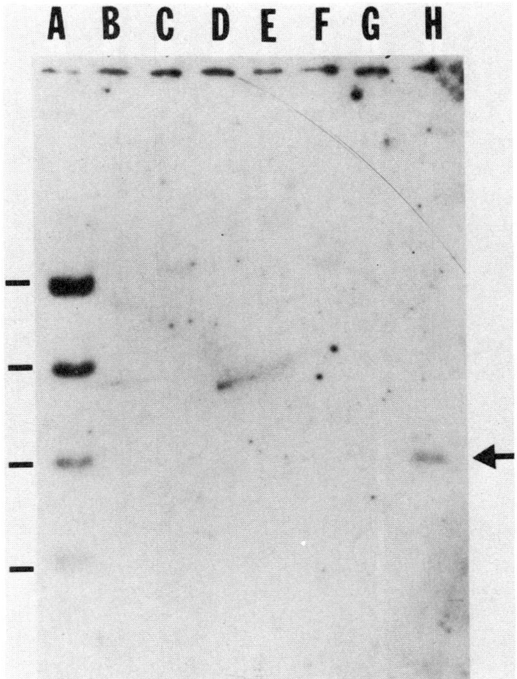

FIGURE 7. Southern blot of *Pst* I-cleaved plasmid DNA from the positive clones in FIGURE 6 hybridized to ^{32}P-labeled 17-mers. Molecular weight markers are *Hin*d III-cleaved PM2. pBA4.8 is in Lane H.

DNA was prepared from each clone, dialyzed, digested with heat-treated ribonuclease, extracted with phenol and chloroform, and precipitated with ethanol. Plasmid DNA was cut with *Pst* I, displayed by electrophoresis through 1.2% agarose, Southern-transferred, and probed with the ^{32}P-labeled 17-mers. As shown in FIGURE 7, a single clone, pBA4.8, (Lane H), was positive under highly stringent hybridization and washing conditions. When reexamined with ^{32}P-labeled 15-mers, pBA4.8 remained positive (FIGURE 8). The complete sequence of the 922 nucleotide insert (excluding dG·dC tails) in pBA4.8 is shown in FIGURE 9. This cDNA clone contains regions having a 16 out of 17 base match with our 17-mers and a 13 out of 15 base match with our 15-mers. The nucleotides adjacent to these regions encode amino acids different from those in tryptic peptides T-40 and T-29.[24] The nucleotide sequence of pBA4.8 is not significantly homologous to the sequence of the 486 coding bases of pC21a, a bovine P-450_{c21} clone simultaneously obtained by another group.[31] The origin of pBA4.8 is currently being investigated.

ADDENDUM

In the year between preparation of this manuscript describing the data we presented at the Congenital Adrenal Hyperplasia Meeting in New York (July 7–9,

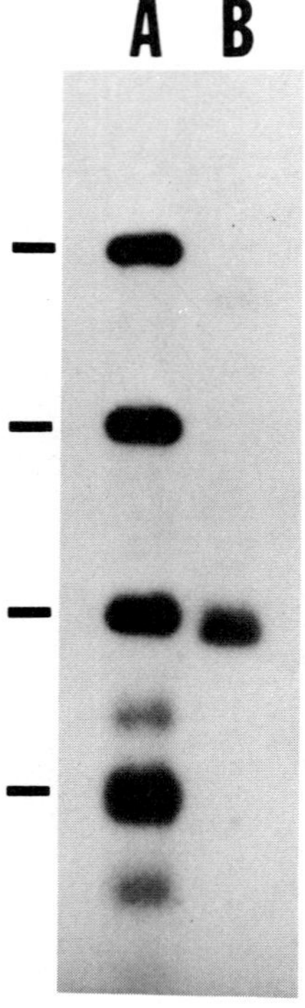

FIGURE 8. Southern blot of pBA 4.8 cleaved with *Pst* I and hybridized to ^{32}P-labeled 15-mers.

1984) and the final assembly of this volume in mid-1985, rapid developments in our laboratory have required addition of updated information.

21-Hydroxylase

Using the manual phosphoramidite procedure (Beaucage, S. L. & M. H. Caruthers. 1981. Tetrahedron Lett. **22:** 859), we synthesized 23- and 30-base oligonucleotides based on the cDNA fragment of White *et al.* (1984. Proc. Natl. Acad. Sci. USA **81:** 1986). These were used to screen a bovine genomic DNA library in λEMBL3A. We identified a phage, termed λE11, carrying the bovine P-450$_{c21}$ gene. The identity of λE11 was proven unambiguously with our new procedure for initiating dideoxy DNA sequencing directly on a double-stranded λ phage template. An 1141 bp *Eco* RI fragment of λE11 was used to probe Southern blots of bovine genomic DNA, which showed that cattle have two P-450$_{c21}$ genes. Probing of Northern blots of bovine adrenocortical RNA showed bovine P-450$_{c21}$ mRNA exists in two species of 2.2 and 2.4 kb. The 1141 bp *Eco* RI fragment was sequenced, showing three complete exons and a portion of a fourth exon. These exons encode a single open reading frame corresponding to the region in White's cDNA. The amino acid sequence predicted by the gene, however, is vastly different from that attributed to the cDNA due to four frameshift errors in the cDNA sequence. Our predicted bovine P-450$_{c21}$ is 91% homologous with the corresponding region in porcine P-450$_{c21}$ as determined by preliminary amino acid sequencing (Bienkowski *et al.*, 1985. Biochem. Biophys. Res.

Commun. **125:** 734). A detailed description of these data has been published (Chung, B., K. J. Matteson & W. L. Miller. 1985. DNA **4:** 211–220), and the whole gene has been sequenced (Chung, B., K. J. Matteson & W. L. Miller, submitted).

Side-Chain Cleavage

Morohashi *et al.* published a full-length cDNA sequence for bovine P-450_{scc} in August 1984 (1984. Proc. Natl. Acad. Sci. USA **81:** 4647). We used a new automated

```
GTTCAGATGC TGTGTCCCAT TGGGAAAGTT CAGCAGGTTA CCAGGGCCAC GGCCTCAGTC
CAAGTCTACG ACACAGGGTA ACCCTTTCAA GTCGTCCAAT GGTCCCGGTG CCGGAGTCAG

ATCCTCAGAA TCGCTGTCCC TCTTGGCAGG GACAGAGCAC CGCACCGCAG ACAGCAGCAC
TAGGAGTCTT AGCGACAGGG AGAACCGTCC CTGTCTCGTG GCGTGGCGTC TGTCGTCGTG

GTCTTCCACG GGCTTCTTGG GATTCTCCTC CAGGCTCGTC TTGATGGCTC CAGACTCAGA
CAGAAGGTGC CCGAAGAACC CTAAGAGGAG GTCCGAGCAG AACTACCGAG GTCTGAGTCT

GCAACTTCCA CTCCAACTCG TCCAAAGTCA GGTTCATGCC ACCAAACACC AGAGGTCCGG
CGTTGAAGGT GAGGTTGAGC AGGTTTCAGT CCAAGTACGG TGGTTTGTGG TCTCCAGGCC

ATAACTGAGC CTTGATGTCA CCTTCAAGGT ACACAAATAC CGTGGCAGAT TCCTATCAGG
TATTGACTCG GAACTACAGT GGAAGTTCCA TGTGTTTATG GCACCGTCTA AGGATAGTCC

GTAACTGGGT ATGCAGGTGG TTGAAATGGC TTTGATAAAC TTGACATCAG GAAACTTCCT
CATTGACCCA TACGTCCACC AACTTTACCG AAACTATTTG AACTGTAGTC CTTTGAAGGA

GGCGAGGTGC ACTCAAGTGC TGATTTATCA GGGCACAGAG GGGAATCCCT TGTTTGTAAA
CCGCTCCACG TGAGTTCACG ACTAAATAGT CCCGTGTCTC CCCTTAGGGA ACAAACATTT

GGTGCAGGAT GACCCATAAG CCCTCACCAG CTTTGGTAAC TTCTTGAACA TAATCCTTTC
CCACGTCCTA CTGGGTATTC GGGAGTGGTC GAAACCATTG AAGAACTTGT ATTAGGAAAG

CAGAGATTTC CAAAACCTCT CCAAATTTGT TCTTCAGTTG GGTCGCTTTC CATTCGGCCA
GTCTCTAAAG GTTTTGGAGA GGTTTAAACA AGAAGTCAAC CCAGCGAAAG GTAAGCCGGT

GCCTTTGCTG CCTGTACATT TCAATTGCAC GTTCGTCTTC CTCATTAAAT TCGTCTTCAT
CGGAAACGAC GGACATGTAA AGTTAACGTG CAAGCAGAAG GAGTAATTTA AGCAGAAGTA

TATCCTCCAG TTCTTCCAAA GTCATGTCTT CATATGTTTT CACAATGGAC TGCTGGAGGA
ATAGGAGGTC AAGAAGGTTT CAGTACAGAA GTATACAAAA GTGTTACCTG ACGACCTCCT

TCCGCTGCTC CTCTTCTTCT GCCTCCTTCT CCAGATCTTT CAAATCTTCC TTTGAAGGCA
AGGCGACGAG GAGAAGAAGA CGGAGGAAGA GGTCTAGAAA GTTTAGAAGG AAACTTCCGT

AGATGCCTTT TTTGCGTAAG ATGTCATTCC ACTCGGTGTC TGCGTTGGGG TCCTGCATTT
TCTACGGAAA AAACGCATTC TACAGTAAGG TGAGCCACAG ACGCAACCCC AGGACGTAAA

TCTGTCAAAT CGCTAGGGCC CTGCCGGCCA CAGCCACCCG GCCCGTGAGC TCTCTACCGC
AGACAGTTTA GCGATCCCGG GACGGCCGGT GTCGGTGGGC CGGGCACTCG AGAGATGGCG

GCACGCAGGC GCCACTCGCC TCCTCTCCCA GCCTGCCCTG AGATCTCGTC CGCCCGTTGG
CGTGCGTCCG CGGTGAGCGG AGGAGAGGGT CGGACGGGAC TCTAGAGCAG GCGGGCAACC

CCCTCCTTCT CTTGGCGCCG
GGGAGGAAGA GAACCGCGGC
```

FIGURE 9. Complete nucleotide sequence of the insert in pBA4.8.

DNA synthesizer of radical design (Warner *et al.* 1984. DNA **3:** 401) to produce 63- to 72-base bovine-sequence P-450_{scc} oligonucleotides. These were used to probe human and bovine Southern blots, indicating each species has only one P-450_{scc} gene. Probing of Northern blots of bovine and human adrenal RNA showed P-450_{scc} mRNA is ~2 kb long. Probing of Southern blots of DNA from three patients with SCC deficiency (20,22 desmolase deficiency, congenital lipoid adrenal hyperplasia) indicated this disease is not due to a deletion in the P-450_{scc} gene. The bovine-sequence oligonucleotides were also used to identify a human P-450_{scc} cDNA clone. That clone contains an 818 bp fragment of the cDNA, which contains 719 bp of coding region and 99 bp of 3′ untranslated region. This fragment has 83% nucleotide homology and 72% amino acid homology with the bovine sequence. The Southern blotting data (obtained with the oligonucleotides) in both normals and in the three patients with congenital lipoid adrenal hyperplasia were reproduced with the human cDNA probe. These data have also been submitted for publication (Matteson, K. J., B. Chung, M. S. Urdea & W. L. Miller, submitted).

REFERENCES

1. STONE, D. & O. HECHTER. 1955. Arch. Biochem. Biophys. **54:** 121–128.
2. KORITS, S. B. & A. M. KUMAR. 1970. J. Biol. Chem. **245:** 152–149.
3. JEFCOATE, C. R., E. R. SIMPSON & G. S. BOYD. 1974. Eur. J. Biochem. **42:** 539–551.
4. DUBOIS, R. N., E. R. SIMPSON, J. TUCKEY, J. D. LAMBETH & M. R. WATERMAN. 1981. Proc. Natl. Acad. Sci. USA **78:** 1028–1032.
5. DUBOIS, R. N., E. R. SIMPSON, R. E. KRAMER & M. R. WATERMAN. 1981. J. Biol. Chem. **256:** 7000–7005.
6. HAUFFA, B., W. L. MILLER, M. M. GRUMBACH, F. CONTE & S. L. KAPLAN. 1985. Clin. Endocrinol. In press.
7. KOMINAMI, S., H. OCHI, Y. KOBAYASHI & S. TAKEMORI. 1980. J. Biol. Chem. **255:** 3386–3394.
8. MILLER, W. L., S. LEISTI & L. K. JOHNSON. 1982. Endocrinology **111:** 1358–1367.
9. MILLER, W. L., D. COIT, J. D. BAXTER & J. A. MARTIAL. 1981. DNA **1:** 37–50.
10. MILLER, W. L., J. A. MARTIAL & J. D. BAXTER. 1980. J. Biol. Chem. **255:** 7521–7524.
11. HANAHAN, D. 1983. J. Mol. Biol. **166:** 557–580.
12. RICCARDI, R. P., J. S. MILLER & B. E. ROBERTS. 1979. Proc. Natl. Acad. Sci. USA **76:** 4927–4931.
13. CLEVELAND, D. W., M. A. LOPATA, R. J. MACDONALD, N. J. COWAN, W. J. RUTTER & M. W. KIRSHNER. 1981. Cell **20:** 95–105.
14. PARNES, J. R., B. VELAN, A. FELSENFELD, L. RAMANTHAN, U. FERRINI, E. APPELLA & J. G. SEIDMAN. 1981. Proc. Natl. Acad. Sci. USA **78:** 2253–2257.
15. BEUCAGE, S. L. & M. H. CARUTHERS. 1981. Tetrahedron Lett. **22:** 859–863.
16. SOUTHERN, E. M. 1975. J. Mol. Biol. **98:** 503–517.
17. DENHARDT, D. 1966. Biochem. Biophys. Res. Commun. **23:** 641–646.
18. GRUNSTEIN, M. & D. D. HOGNESS. 1975. Proc. Natl. Acad. Sci. USA **72:** 3961–3965.
19. HOLMS, D. S. & M. QUIGLEY. 1981. Anal. Biochem. **114:** 193–197.
20. SANGER, F., S. NICKLEN & A. R. COULSON. 1977. Proc. Natl. Acad. Sci. USA **74:** 5463–5467.
21. MESSING, J. & J. VIEIRA. 1982. Gene **19:** 269–276.
22. MAXAM, A. M. & W. GILBERT. 1980. Methods Enzymol. **65:** 499–559.
23. MATTESON, K. J., B. CHUNG & W. L. MILLER. 1984. Biochem. Biophys. Res. Commun. **120:** 264–270.
24. YUAN, P., S. NAKAJIN, M. HANIU, M. SHINODA, P. F. HALL & J. E. SHIVELY. 1983. Biochemistry **22:** 143–149.
25. LUTTER, L. C. 1979. Nucleic Acids Res. **6:** 41–56.
26. DAYHOFF, M. O. Atlas of Protein Sequence and Structure. National Biomedical Research Foundation, Washington, D.C.: Vol. 5 (1972), Supplement 2 (1976), Supplement 3 (1978).

27. Fujii-Kuriyama, Y., T. Taniguchi, Y. Mizukami, M. Sakai, Y. Tashiro & M. Muramatsu. 1981. J. Biochem. **89:** 1869–1879.
28. Gonzalez, F. S. & C. B. Kasper. 1982. J. Biol. Chem. **257:** 5962–5968.
29. Stupans, I., T. Ikeda, D. J. Kessler & D. W. Nebert. 1984. DNA **3:** 129–137.
30. Negishi, M. & D. W. Nebert. 1981. J. Biol. Chem. **256:** 3085–3091.
31. White, P. C., M. I. New & B. Dupont. 1984. Proc. Natl. Acad. Sci. USA **81:** 1986–1990.

Regulation of Cytochrome P-450$_{17\alpha}$ Activity and Synthesis in Bovine Adrenocortical Cells[a]

MAURICIO X. ZUBER,[b,c] EVAN R. SIMPSON,[b,c]
AND MICHAEL R. WATERMAN[b]

Departments of [b]Biochemistry and [c]Obstetrics and Gynecology
and
The Cecil H. and Ida Green Center
for Reproductive Biology Sciences
The University of Texas Health Science Center
Dallas, Texas 75235

INTRODUCTION

The rate of steroidogenesis in the adrenal cortex is regulated primarily by adrenocorticotropin (ACTH). The molecular mechanisms by which ACTH exerts control of this pathway have not been clearly elucidated. Nevertheless, it has been established that the responses of the adrenocortical cell to ACTH are of two types—rapid (acute) and long-term (chronic).[1] Both of these actions of ACTH appear to be mediated via cAMP. The acute action is characterized by an increase in the flux of cholesterol through the steroidogenic pathway resulting in rapid production of steroids. Cholesterol is made available by its release from cholesterol esters and subsequent translocation to the inner mitochondrial membrane where the cholesterol side chain cleavage enzyme (cytochrome P-450$_{scc}$) is located. The chronic response to ACTH is designed to maintain optimal levels of the steroidogenic enzymes. This regulation is manifest at the level of protein synthesis. For example, treatment of bovine adrenocortical (BAC) cells in primary monolayer culture with ACTH leads to an increase in synthesis of the steroid hydroxylase enzymes; cytochromes P-450$_{scc}$, P-450$_{11\beta}$, and P-450$_{C21}$.[2–4]

Whereas in control (untreated) BAC cells corticosterone is the predominant steroid secreted, upon ACTH treatment, the profile of steroids released into the culture medium consists primarily of 17α-hydroxylated C_{21}-corticosteroids and androgens, at the expense of corticosterone.[5] This shift in the pattern of steroidogenesis has been attributed to an increase in 17α-hydroxylase activity.[6] In addition, treatment of rabbits with ACTH results in increased production of cortisol.[7,8] This result is also thought to be due to an increase in 17α-hydroxylase activity.[9] However, the mechanism by which the increase in 17α-hydroxylase activity takes place in such systems is not known.

In order to better understand the mechanisms of action of ACTH on the steroidogenic pathway, we have used BAC cells in monolayer culture as a model

[a]This research was supported in part by USPHS grants nos. AM–28350, AM–28113, and HD–13234, and grant no. I–624 from The Robert A. Welch Foundation. MXZ was supported in part by USPHS training grant no. 5–T32HD07190.

system. With this system, we have studied how ACTH affects the activity and synthesis of 17α-hydroxylase cytochrome P-450 (P-$450_{17\alpha}$) and, by the use of a cloned DNA complementary to cytochrome P-$450_{17\alpha}$ mRNA, regulation of cytochrome P-$450_{17\alpha}$ gene expression.

17α-HYDROXYLASE ACTIVITY

In order to investigate whether the ACTH-mediated increase in 17α-hydroxylase activity is the result of a modification of existing enzyme or an increase in the total

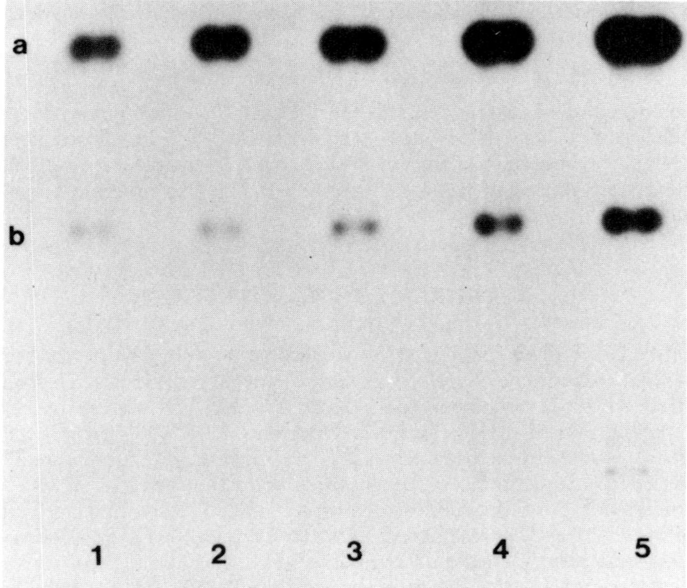

FIGURE 1. Autoradiography of TLC plate in which pregnenolone (a) is converted into 17α-hydroxypregnenolone (b) by postmitochondrial supernatant (0.3 mg/ml) prepared from freshly obtained adrenal cortex. Concentrations of substrate are respectively from lanes 1 to 5, 0.2 μM, 0.3 μM, 0.5 μM, 1.0 μM, and 2.0 μM. The bands corresponding to substrate and product were cut and the radioactivity was measured by scintillation spectrophotometry. The amounts of product formed at each substrate concentration were used to determine V_{max} and K_m values. Cyanoketone was used to inhibit 3β-hydroxy-steroid-dehydrogenase.

amount of the enzyme, or both, a kinetic study of this reaction was undertaken. FIGURE 1 shows the results of a typical experiment in which the concentration of pregnenolone is changed from 2 μM to 0.2 μM and its conversion to 17α-hydroxypregnenolone is measured. TABLE 1 shows that upon stimulation of BAC cells with ACTH, there is a large (15–20-fold) increase in 17α-hydroxylase activity, and this increase is the result of a change in the V_{max} and not due to a change in the $K_{m_{app}}$. This result suggests that the increase in activity is the result of increased amounts of enzyme. In addition, there is an increase in total cytochrome P-450 content in microsomes from ACTH-treated

TABLE 1. Effects of Treatment for 36 Hours with ACTH or Dibutyryl cAMP (Bu_2cAMP) in BAC Cell Fractions as Compared to Control Cells (−ACTH) and Fresh Gland Fractions (Gland)[a]

	V_{max} nmol/min/mg PMS Protein	$K_{m_{app}}$ μM	P-450 Content nmol/mg Microsomal Protein	Cortisol $\mu g/ml/36$ h
−ACTH	0.006	0.09	0.015	1.6
+ACTH	0.107	0.06	0.074	10.2
$+Bu_2cAMP$	—	—	0.054	9.2
Gland	0.124	0.09	0.78	—

[a] V_{max} and K_m values were obtained in the postmitochondrial supernatant fraction (PMS). P-450 content was determined in microsomes by CO difference spectra[10] and cortisol in the culture medium was measured by radioimmunoassay.[5] ACTH concentration was 1 μM, while Bu_2cAMP concentration was 1 m*M*. The changes in $K_{m_{app}}$ are not significant.

cells as compared to untreated (control) cells (TABLE 1). Although the data presented above indicate that synthesis of new enzyme rather than modification of existing enzyme might be responsible for the change in 17α-hydroxylase activity, it was necessary to study the synthesis of cytochrome P-$450_{17\alpha}$ in order to reach a formal conclusion.

CYTOCHROME P-$450_{17\alpha}$ SYNTHESIS

An antibody raised against pig testis cytochrome P-$450_{17\alpha}$[11] that cross-reacted with bovine adrenal cytochrome P-$450_{17\alpha}$ was used to study its synthesis. This was done in four different ways: (1) radiolabeling of cell proteins with [^{35}S]methionine, followed by immunoprecipitation of newly synthesized cytochrome P-$450_{17\alpha}$; (2) translation of cellular RNA *in vitro* in the presence of [^{35}S]methionine and immunoprecipitation of newly synthesized cytochrome P-$450_{17\alpha}$; (3) immunoblot analysis of the content of cytochrome P-$450_{17\alpha}$ in cells; and, (4) determination of 17α-hydroxylase activity. The results of these studies are presented in FIGURES 2 through 4. Several points should be noted from these results. In the absence of ACTH, whether at zero time or at later times, there are undetectable levels of cytochrome P-$450_{17\alpha}$. Upon stimulation with ACTH, there is a rapid increase in all the parameters studied. This indicates that as a result of increased levels of translatable mRNA specific for cytochrome P-$450_{17\alpha}$, there is an increase in the rate of synthesis of this enzyme, and as a result, cytochrome P-$450_{17\alpha}$ accumulates. Therefore, due to this, 17α-hydroxylase activity concomitantly increases.

It is of interest to note that after 48 h there is a marked decline in both translatable cytochrome P-$450_{17\alpha}$ mRNA and radiolabeled cytochrome P-$450_{17\alpha}$. However, the content of cytochrome P-$450_{17\alpha}$ and its activity remain higher than in controls. This result indicates that although synthesis declines, the levels of the enzyme are maintained, presumably because the rate of turnover is low.

MOLECULAR MECHANISMS INVOLVED IN THE REGULATION OF CYTOCHROME P-$450_{17\alpha}$

There are several questions we can ask regarding the mechanisms whereby ACTH regulates the synthesis of cytochrome P-$450_{17\alpha}$. One obvious question concerns the level at which control is exerted. In the following section, evidence is presented which indicates that, as suggested earlier, regulation occurs at the level of gene expression. We can also ask whether (as with other steroidogenic enzymes)[12] cAMP mediates the action of ACTH. TABLE 1 shows that dibutyryl cyclicAMP (Bu_2cAMP), an analog of cAMP, can mimic the actions of ACTH to increase microsomal cytochrome P-450 content and cortisol secretion. Thus, it is likely that regulation of cytochrome P-$450_{17\alpha}$ is a cAMP-mediated mechanism. Direct evidence in favor of this concept will be presented later.

An important question we considered was whether or not activation of cytochrome P-$450_{17\alpha}$ gene expression requires protein synthesis. In order to address this issue, the

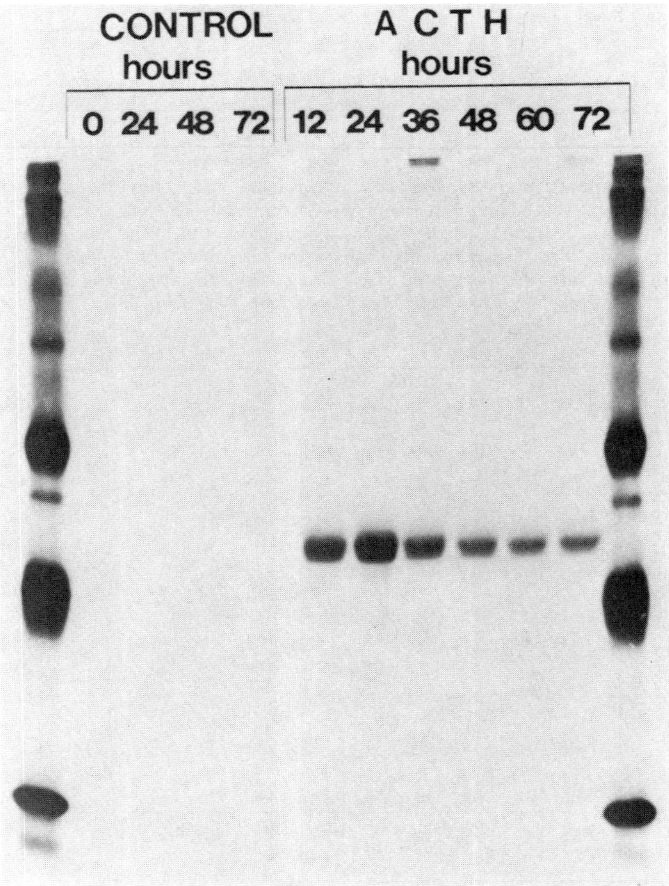

FIGURE 2. Autoradiography of polyacrylamide gel electrophoresis of newly synthesized cytochrome P-$450_{17\alpha}$ immunoisolated from *in vitro* translated RNA obtained from BAC cells treated as indicated in the figure. The optical density of each band is proportional to the amount of translatable 17α-hydroxylase mRNA present in each sample of total RNA. ACTH concentration was 1 μM. The outer lanes contain molecular weight markers and the newly synthesized cytochrome P-$450_{17\alpha}$ is found to have a molecular weight of 54 kd.

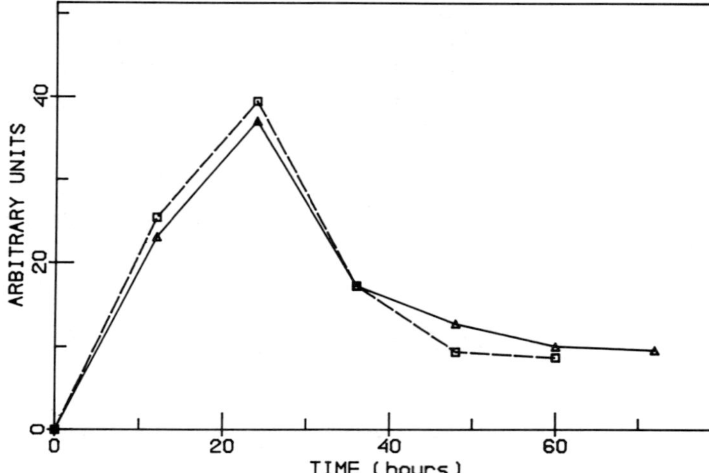

FIGURE 3. Profile of newly synthesized cytochrome P-450$_{17\alpha}$ immunoisolated from a cell labeling experiment (dashed line) and from *in vitro* translated RNA (solid line) in BAC cells treated with ACTH (1 μM) for the indicated periods of time. The data were obtained by densitometric scannings of experiments such as the one presented in FIGURE 2. This result indicates that the rate of synthesis of 17α-hydroxylase (cell labeling) is proportional to the amount of 17α-hydroxylase mRNA (*in vitro* translation).

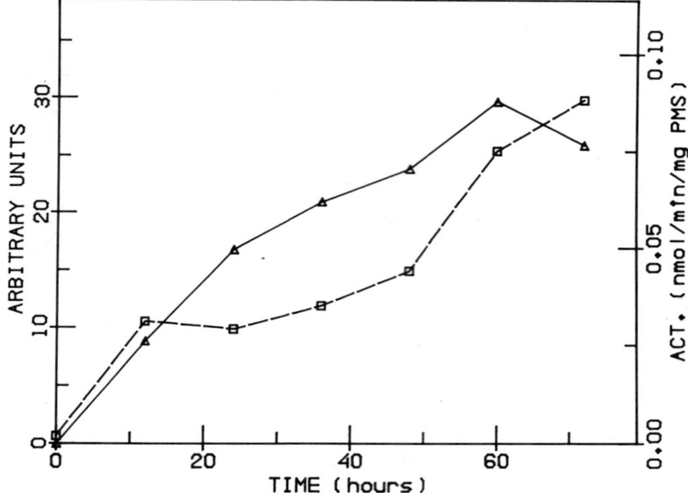

FIGURE 4. Profile of relative 17α-hydroxylase content (solid line) and activity (dashed line) in BAC cells treated with ACTH (1 μM) for the times indicated in the figure. PMS from cells treated with ACTH was used to measure the activity and levels of 17α-hydroxylase. Activity was measured as described in FIGURE 1 under saturating substrate concentrations. Relative content of cytochrome P-450$_{17\alpha}$ was obtained by densitometric analysis of the autoradiogram of an immunoblot of PMS.

following strategy was adopted. BAC cells are treated with or without ACTH and with or without cycloheximide (cyc), a translational inhibitor. RNA is then isolated and translated *in vitro,* and cytochrome P-$450_{17\alpha}$ is immunoisolated and subjected to electrophoresis and autoradiography. As shown in FIGURE 5, when cycloheximide (20 μg/ml) is present, immunoisolated cytochrome P-$450_{17\alpha}$ is undetectable in controls and minimal in the presence of ACTH. When the pattern of total translatable products is analyzed, however, no differences can be seen (data not shown), which is indicative of little or no effect of cycloheximide on total transcription during the short time course of the experiment (four hours). These results show that at least one step of translation is

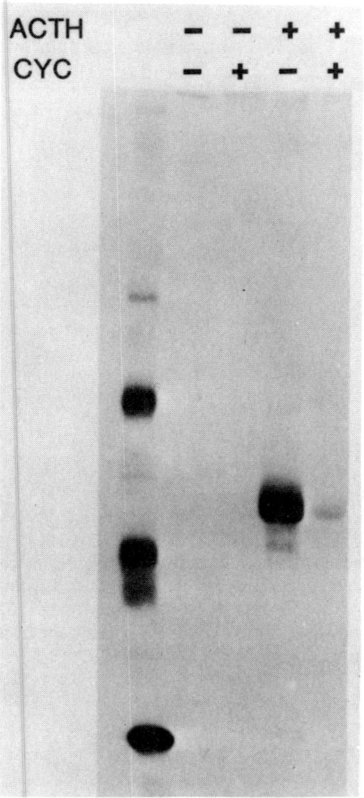

FIGURE 5. Effect of cycloheximide on levels of cytochrome P-$450_{17\alpha}$ mRNA. BAC cells were treated for four hours as indicated in the figure. In the cases where cycloheximide (cyc) was used, a 15 minute preincubation in the presence of cycloheximide (20 μg/ml) was included. RNA was extracted and translated *in vitro.* An autoradiogram was then obtained from immunoisolated cytochrome P-$450_{17\alpha}$ after gel electrophoresis. The result indicates that activation of the gene for 17α-hydroxylase requires protein synthesis.

required before activation of transcription of the gene for cytochrome P-$450_{17\alpha}$ can occur. A similar result was obtained using puromycin, which also inhibits translation, albeit by a different mechanism. Preliminary results (not shown) indicate that the same general process is true for the other steroidogenic enzymes in BAC cells.

STUDIES ON CYTOCHROME P-$450_{17\alpha}$ mRNA

By use of a cloned DNA complementary to cytochrome P-$450_{17\alpha}$ mRNA[13] (pB17α-1), we have been able to confirm the results obtained by *in vitro* translation.

We have used nick-translated pB17α-1 as a probe for Northern analysis of poly A^+ RNA from bovine adrenal gland. As shown in FIGURE 6, there is mRNA that crosshybridizes with the probe. It corresponds approximately in size to the 18S ribosomal RNA band (~2000 bp) and, thus, could contain an open reading frame large enough for a 54 kD protein such as cytochrome $P\text{-}450_{17\alpha}$. We have also conducted slot-blot analysis of RNA extracted from cells maintained in the presence or absence of ACTH and cycloheximide, as well as cells maintained in the presence or absence of Bu_2cAMP. As can be seen in FIGURE 7, this experiment confirms the action of cycloheximide to block ACTH stimulation of cytochrome $P\text{-}450_{17\alpha}$ mRNA synthesis. In addition, it is apparent that Bu_2 cAMP can mimic the action of ACTH and cause an increase in the level of cytochrome $P\text{-}450_{17\alpha}$ mRNA, indicative that the action of ACTH is mediated by cAMP.

In FIGURE 7, mRNA appears to be present in the control samples as well as in the cycloheximide treated samples. This result could be due to background usually found

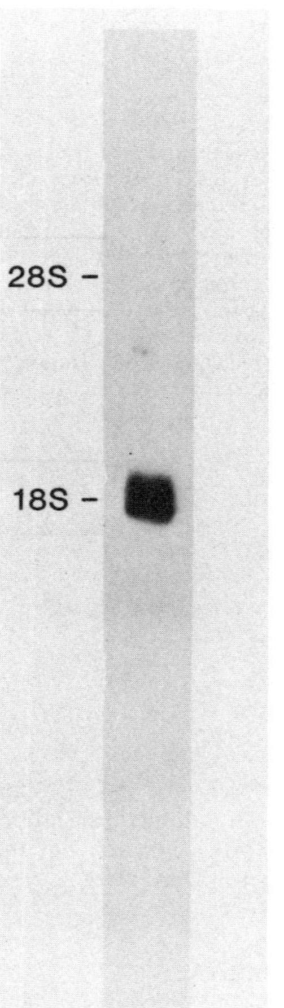

FIGURE 6. Autoradiography of a Northern blot of poly A^+ RNA obtained from bovine adrenal cortex and probed with labeled cDNA to cytochrome $P\text{-}450_{17\alpha}$. A band can be seen around 18S ribosomal RNA. These messages (~2000 bases) can easily code for a protein of 54 kilodaltons.

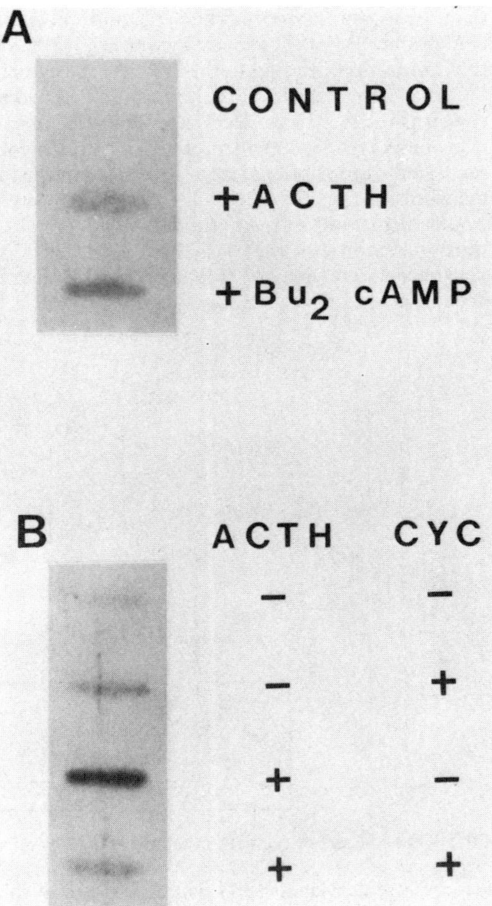

FIGURE 7. Slot blot analysis of total RNA obtained from BAC cells. In panel A, cells were treated for 36 h as indicated in the figure. In panel B, cells were treated for 4 h. In the cases when cycloheximide (cyc) is used, a 15 min preincubation period in the presence of cycloheximide (20 μg/ml) was included.

in slot blots. However, hybridization experiments of this type are very sensitive and, therefore, it may be possible to determine how much RNA, if any, is present in control cells.

CONCLUSIONS

From the data presented herein, it is possible to present a model for the mechanism of regulation of 17α-hydroxylase as shown in FIGURE 8. ACTH binds to its receptor, which activates adenylate cyclase. This results in increased cAMP levels, and by an unknown step (probably involving protein kinase), a specific mRNA coding for a protein that regulates the transcription of the gene for cytochrome P-$450_{17\alpha}$, which we have called cytochrome P-$450_{17\alpha}$ regulatory protein (17α-RP), is translated. This results in accumulation of 17α-RP in the cytosol. The newly synthesized 17α-RP is then translocated to the nucleus and presumably activates the gene coding for cytochrome P-$450_{17\alpha}$. Newly synthesized mRNA coding for cytochrome P-$450_{17\alpha}$ is then translated, leading to the synthesis of cytochrome P-$450_{17\alpha}$.

In the mechanism presented, there are two areas that need more study. The first concerns the step in which an accumulation of 17α-RP mRNA occurs. It is not known if cyclic AMP-dependent protein kinase is involved. It is also not clear if an ACTH regulated transcriptional step is required in the formation of 17α-RP. In an alternative mechanism, the 17α-RP RNA would be present in the nucleus and ACTH action would then result in the maturation and/or release of this mRNA into the cytosol. A third possibility is that 17α-RP might be a rapidly turning-over protein that is phosphorylated when ACTH is present; if the phosphorylated form was more stable, accumulation of 17α-RP would occur.

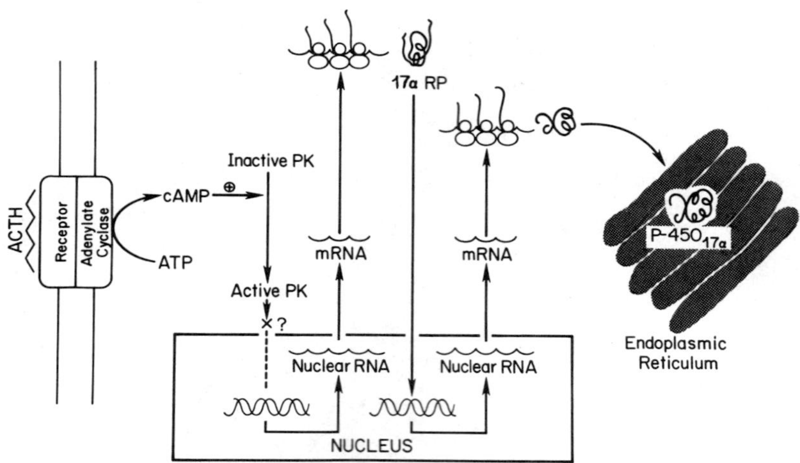

FIGURE 8. Model of ACTH action in BAC cells. After ACTH binds to its receptor on the cell membrane, adenylate cyclase is activated with a resulting increase in cAMP levels. Cyclic AMP activates a protein kinase (PK), which by a series of unknown steps results in accumulation of the messenger RNA coding for a regulatory protein (17α-RP). This mRNA is translated (this step is blocked by cycloheximide) and 17α-RP translocates into the nucleus to activate expression of the gene coding for cytochrome P-$450_{17\alpha}$.

Whichever mechanism proves to be correct, the result is an accumulation of mRNA coding for cytochrome P-$450_{17\alpha}$ that leads to increased levels of cytochrome P-$450_{17\alpha}$. Thus, 17α-hydroxylase induction mediates the switch in the steroidogenic pathway from the production of corticosterone to the formation of 17α-hydroxylated C_{21}-corticosteroids and androgens in our cell culture system. Presumably, 17α-hydroxylase also plays a key role *in vivo* to regulate the steroid profile of the adrenal cortex.

ACKNOWLEDGMENTS

We would like to thank Peter F. Hall from the Worcester Foundation for Experimental Biology, Shrewsbury, Massachusetts 01545, for kindly providing us with the cytochrome P-$450_{17\alpha}$ IgG. We also thank Alice Thomas for expert editorial assistance, and Leticia Cortez and Bill Rainey for technical assistance.

REFERENCES

1. SIMPSON, E. R. & M. R. WATERMAN. 1983. Can. J. Biochem. **61:** 692–707.
2. DUBOIS, R. N., E. R. SIMPSON, R. E. KRAMER & M. R. WATERMAN. 1981. J. Biol. Chem. **256:** 7000–7005.
3. KRAMER, R. E., E. R. SIMPSON & M. R. WATERMAN. 1983. J. Biol. Chem. **258:** 3000–3005.
4. FUNKENSTEIN, B., J. L. MCCARTHY, K. M. DUS, E. R. SIMPSON & M. R. WATERMAN. 1983. J. Biol. Chem. **258:** 9398–9405.
5. KRAMER, R. E., J. L. MCCARTHY, E. R. SIMPSON & M. R. WATERMAN. 1983. J. Steroid Biochem. **18:** 715–723.
6. MCCARTHY, J. L., R. E. KRAMER, B. FUNKENSTEIN, E. R. SIMPSON & M. R. WATERMAN. 1983. Arch. Biochem. Biophys. **222:** 590–598.
7. KASS, E. H., O. HECHTER, I. A. MACCHI & T. W. MOV. 1954. Proc. Soc. Exp. Biol. Med. **85:** 583–587.
8. FEVOLD, H. R. 1967. Science **156:** 1753–1755.
9. SALNINA, S. M. & H. R. FEVOLD. 1982. J. Steroid Biochem. **19:** 93–99.
10. OMURA, T. & R. SATO. 1964. J. Biol. Chem. **239:** 2370–2378.
11. WATANUKI, M., G. A. GRANGER & P. F. HALL. 1978. J. Biol. Chem. **253:** 2927–2931.
12. KRAMER, R. E., W. E. RAINEY, B. FUNKENSTEIN, A. DEE, E. R. SIMPSON & M. R. WATERMAN. 1984. J. Biol. Chem. **259:** 707–713.
13. ZUBER, M. X. *et al.* Unpublished results.

The Human Major Histocompatibility Complex

Genes and Proteins[a]

JACK L. STROMINGER

Department of Biochemistry and Molecular Biology
Harvard University
Cambridge, Massachusetts 02138

The Major Histocompatibility Complex (MHC) encodes three classes of proteins (FIGURE 1). The Class I proteins (HLA-A,B,C) are encoded on the telomeric side and the Class II proteins (HLA-DR,DC,SB) on the centromeric side. The Class III proteins (some of the proteins of the complement system) are encoded in the middle. Although the Class I and Class II proteins and genes are genetically related, the complement proteins (C2, BF, C4A, and C4B) are unrelated structurally to either of these. Interest in this family of genes and proteins has been catalyzed by the relationship to a number of human diseases. Some Class I and Class II proteins are related to human autoimmune diseases in that alleles of these proteins are represented to an unusually high percentage in diseased patients. For example, HLA-B27 occurs in 95% of patients with ankylosing spondylitis and in only 8% of controls. Congenital adrenal hyperplasia has been strongly linked to HLA-B47, as well as to the occurrence of a null allele at the C4B locus. Hence, this paper focused on this disease.

Class I (HLA-A,B,C) and Class II (HLA-DR,DC,SB) human histocompatibility antigens encoded on chromosome 6 in the MHC (FIGURE 1) have remarkably similar structures, despite differences in function. Each is a heterodimer composed of four extracellular domains (two of which are conserved Ig-like domains, and one of which is polymorphic) located in two chains (α or heavy, and β or light), in addition to the transmembrane and intracytoplasmic regions (FIGURE 2). From a gross structural standpoint, the antigens differ in a few ways. In the Class I antigens, three domains are in the heavy chains, while only one domain is in the light chain (β2-microglobulin). However, in the Class II antigens, each chain has two domains. Moreover, in the Class II antigens, both chains span the membrane and have small intracytoplasmic regions, while in the Class I antigens, only the heavy chain does so.

All three groups of molecules encoded in the MHC are involved in immune defense. The involvement of Class I and Class II molecules in transplantation rejection (hence, the term "histocompatibility antigens") derives from their extreme population polymorphism, which is related to their function. Transplant rejection is a by-product of essential functions of these molecules in the immune system. Class I molecules act as restricting elements in the elimination of virus infected cells, while Class II antigens are involved in cell-cell interaction in the generation of antibodies (macrophage, T helper cell, and B lymphocyte). In view of their function, it is not surprising that they are closely related structurally to immunoglobulins. An even simpler related molecule is Thy-1, an antigen found on mouse thymocytes originally, a homologue of which has

[a]This work was supported by research grants from NIH (AI–10736, AM–13230, and AM–30241).

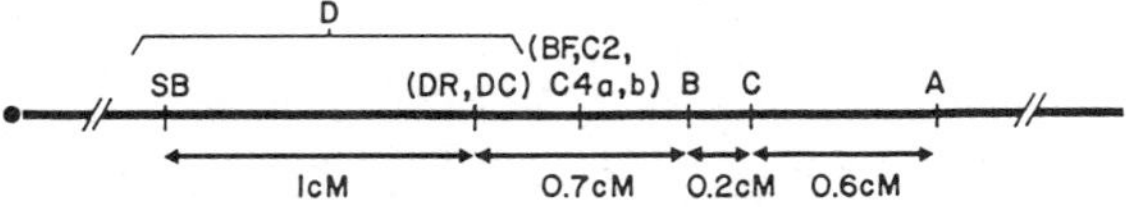

Murine Chromosome 17

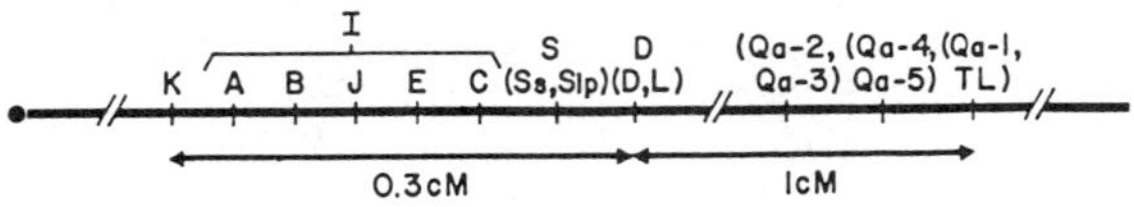

FIGURE 1. Structure of the HLA complex on chromosome 6 and its murine homologue on chromosome 17, based on genetics and serology. It should be noted that the order within each of the following groups is unknown: Ss and Slp; L and D; Qa1 and Qa2; DC and DR; C2, BF, C4A, and C4B.

apparently been found in the squid. The T cell receptor and the poly-Ig receptor of intestinal epithelium are recently discovered molecules that can be added to this list. Many of these molecules are quite old in evolution and all of them must have descended from the same ancestral gene.

The number of sets of Class II (HLA-D) antigens that are expressed in homozygous human cells is not yet entirely clear, but it is clear that the number of sets is considerably larger than is expressed in the mouse (Ia antigens). In the first place, there are three families of antigens known in man, now called DR, DC, and SB, while

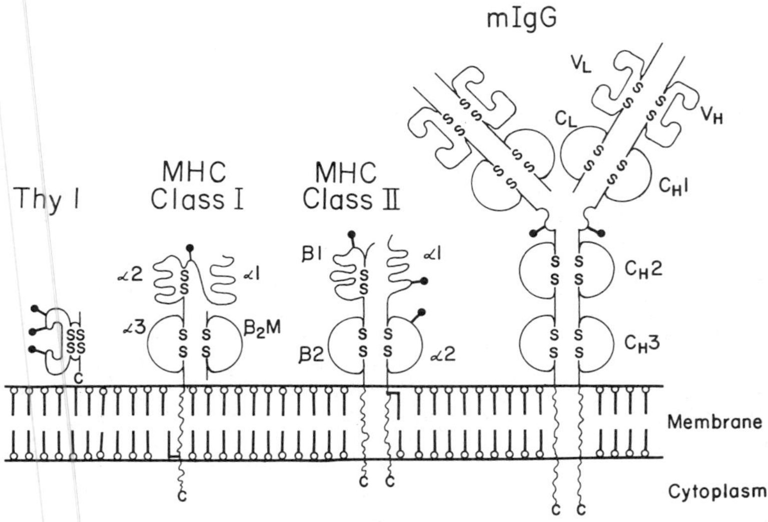

FIGURE 2. Models of membrane proteins with homology to immunoglobulin. The Class I and Class II MHC antigens each contain two immunoglobulin-like domains adjacent to the membrane.

only two families are known in the mouse, I-A and I-E (homologous to DC and DR, respectively). From a protein standpoint, DR, DC, and SB have all been separated. The DR family has been shown to be composed of at least two subsets that share a common heavy chain, but have distinct "DR-like" light chains.

From the study of the genes that encode these proteins, even more complexity is observed (FIGURE 3). In the DR family, a single heavy chain and three light chain genes have been detected; one of which encodes the classical DR serological specificity, while another encodes the MT2 and MT3 specificities. Several additional genes may occur in this family. At least two heavy chain and two light chain genes occur in the DC family. However, if all of the detected crosshybridizing sequences turn out to be complete genes, greater complexity in both the heavy and light chains of this family could occur. In the SB family, there are at least two heavy chain genes and at least two light chain genes. An additional α chain gene, called DO or DZ α, has been described.

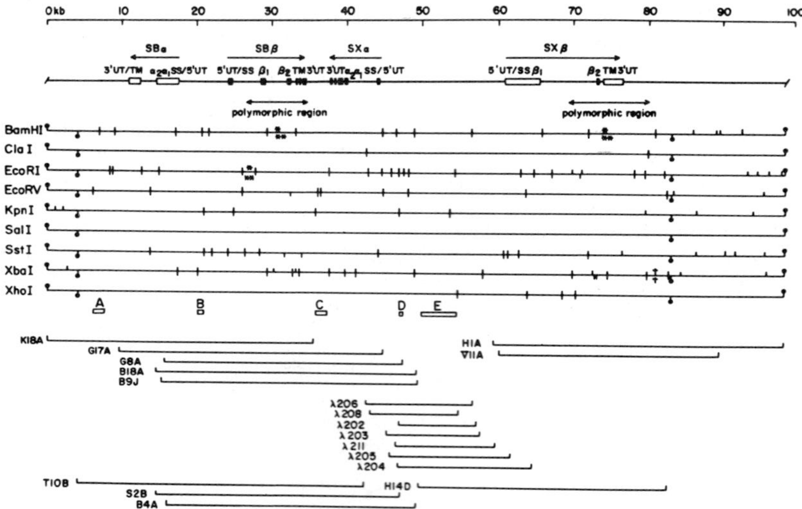

FIGURE 3A. Organization of the Class II gene region of the human MHC. Overlapping cosmid clones from the SB subregion.

It is not known presently if this corresponds to still another family or whether it has associated β chain genes. Some of the observed complexity could be due to the occurrence of nonexpressed pseudogenes. In any event, the human HLA-D region appears to be more complex than the corresponding murine Ia region (FIGURE 4), at least as so far observed, and it appears to have been created by a large gene expansion. If so, this would represent the only example presently known in which the complexity of a genetic region in man is considerably greater than that in the mouse. However, an alternate possibility, though it may seem less likely, that the mouse has undergone gene deletion over evolutionary time, is not entirely excluded. The fact that the human Class II region is much larger than that in the mouse was already known from recombination frequencies (FIGURE 1), although the data had not been so interpreted.

The Class I genes of both man and mouse are extremely complex. In the range of 20 to 30 genes have been detected in this region in both species, and only a small number

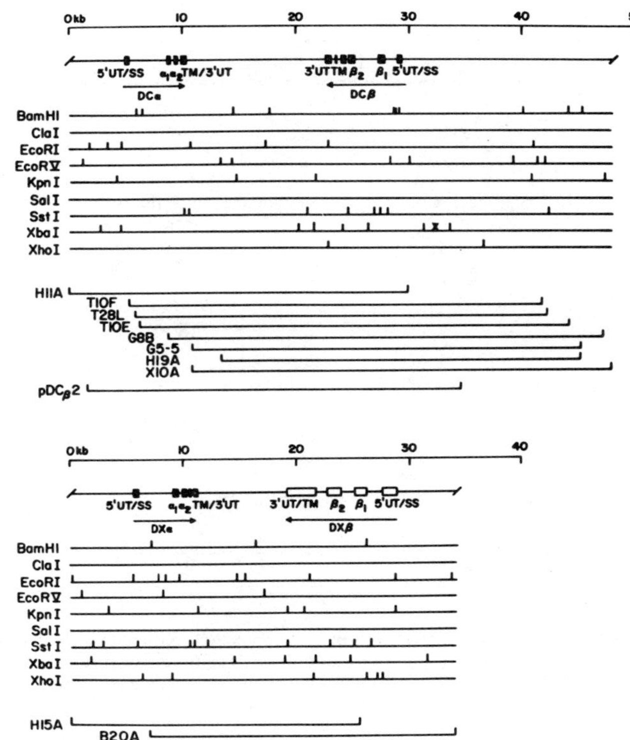

FIGURE 3B. Organization of the Class II gene region of the human MHC. Overlapping cosmid clones from the DC-DX subregion.

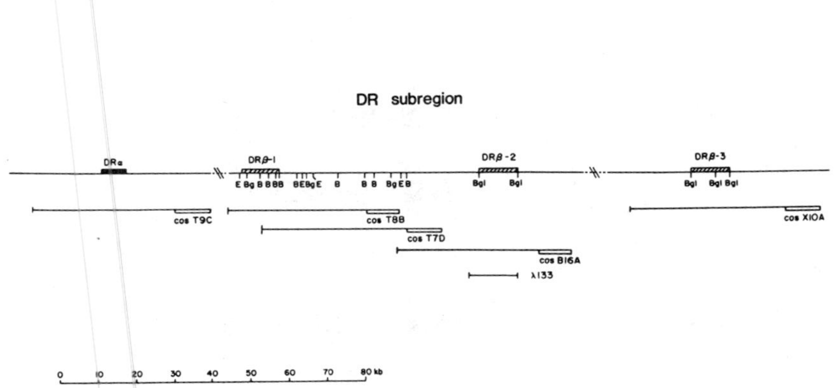

FIGURE 3C. Organization of the Class II gene region of the human MHC. Overlapping cosmid clones from the DR subregion.

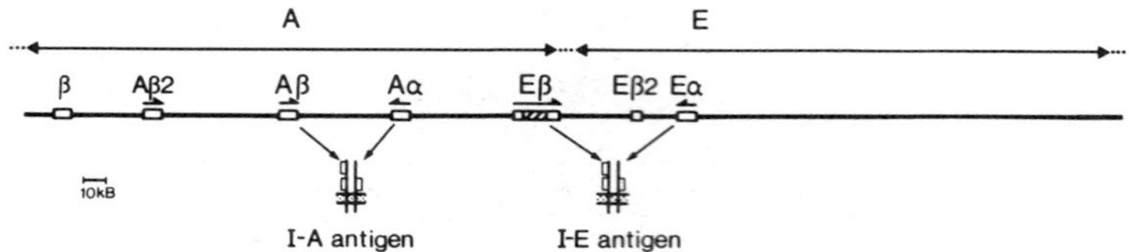

FIGURE 4. Organization of the Class II gene region of the murine MHC.

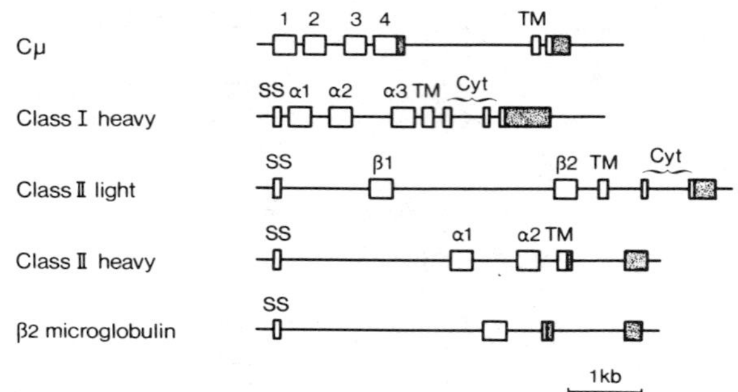

FIGURE 5. Genomic structure of five related genes (SS, signal sequence; TM, transmembrane region; CYT, cytoplasmic region; exons representing extracellular domains are so numbered). Shaded boxes denotes 3′ untranslated regions. Cμ represents the constant region of a Cμ heavy chain gene.

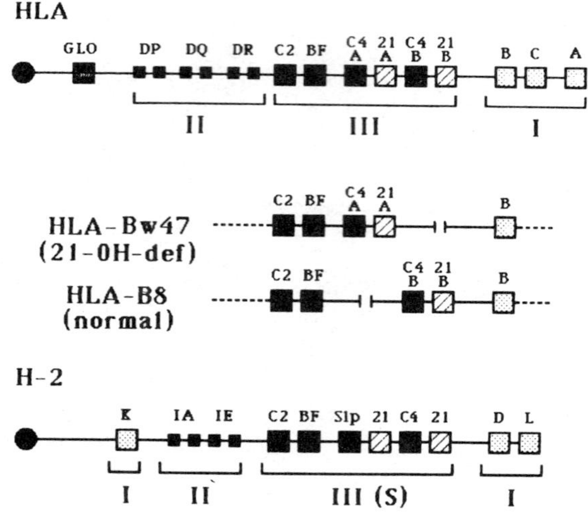

FIGURE 6. The Class III region of the human MHC. Deletions of C4A or C4B together with a 21-steroid hydroxylase gene.

of these (about five) can be ascribed to the classical HLA-A,B,C or H-2K,D,L molecules. The nature and functions of the remainder and what portion of them are pseudogenes is presently unknown. An outline of the structures of Class I and Class II genes, as well as of an immunoglobulin heavy chain constant region gene, is shown in FIGURE 5.

Several practical applications of the DNA probes generated in the course of these studies have become apparent. In the first place, polymorphisms can be detected at the DNA level (by Southern blots) just as they can be detected serologically. For example, the polymorphism of the heavy chain of the DC antigens is readily detected by this method. Other polymorphisms are harder to define with the probes available, although they can be detected. It is anticipated that as better probes become available, the typing of individuals will be more accurately accomplished at the DNA level than is presently possible serologically. In addition, the probes have been used in *in situ* hybridization experiments to localize several genes on human chromosome 6 rather precisely. Moreover, they have been used to study a chromosomal rearrangement of chromosome 6 involving reciprocal translocations of chromosome 6 with chromosome 14. The precise break point in this translocation could be defined with the use of these probes, and it is likely that much more accurate diagnosis of genetic abnormalities in man will be possible in the future. These probes are also being used to study possible polymorphisms in HLA-linked autoimmune diseases (e.g., multiple sclerosis, juvenile-onset diabetes, and rheumatoid arthritis).

An extremely important recent development has been the use of cosmids encoding the Class III genes, which are the components of the complement system in the middle of the human MHC, to detect a genetic defect in congenital adrenal hyperplasia. This disease is due to a deficiency of 21-steroid hydroxylase. A cDNA clone corresponding to this enzyme was obtained and used as a probe. Two genes for 21-steroid hydroxylase, which flanked the complement C4A and C4B genes, were detected. One of these is deleted along with the C4B gene in homozygous congenital adrenal hyperplasia, and the other is deleted in a phenotypically normal individual along with the C4A gene (FIGURE 6).[b]

REFERENCES

1. SHACKELFORD, D. A., J. F. KAUFMAN, A. J. KORMAN & J. L. STROMINGER. 1982. Immunol. Rev. **66:** 133–187.
2. AUFFRAY, C., A. BEN-NUN, M. ROUX-DOSSETO, R. N. GERMAIN, J. G. SEIDMAN & J. L. STROMINGER. 1983. EMBO J. **2:** 121–124.
3. AUFFRAY, C., J. KUO, R. DEMARS & J. L. STROMINGER. 1983. Nature **304:** 174–177.
4. ROUX-DOSSETO, M., C. AUFFRAY, J. W. LILLIE, A. J. KORMAN & J. L. STROMINGER. 1983. *In* Gene Expression, UCLA Symp. Mol. Cell. Biol., vol. 8. D. Hamer & M. Rosenberg, Eds.: 481–490. Alan R. Liss, Inc. New York.
5. ROUX-DOSSETO, M., C. AUFFRAY, J. W. LILLIE, J. BOSS, D. COHEN, R. DEMARS, C. MAWAS, J. G. SEIDMAN & J. L. STROMINGER. 1983. Proc. Natl. Acad. Sci. USA **80:** 6036–6040.
6. SCHAMBOECK, A., A. J. KORMAN, A. KAMB & J. L. STROMINGER. 1983. Nucleic Acids Res. **11:** 8663–8674.
7. KAUFMAN, J. F., C. AUFFRAY, A. J. KORMAN, D. A. SHACKELFORD & J. L. STROMINGER. 1984. Cell **36:** 1–13.

[b]This short account is based mainly on the following recent references from our laboratory. References to the work of others can be found in the papers cited.

8. Auffray, C., J. W. Lillie, D. Arnot, D. Grossberger, D. Kappes & J. L. Strominger. 1984. Nature **308:** 327–333.
9. Kirsch, I. R., C. C. Morton, W. E. Nance, G. A. Evans, A. J. Korman & J. L. Strominger. 1984. Proc. Natl. Acad. Sci. USA **81:** 2816–2820.
10. Arnot, D., C. Auffray, J. Boss, D. Grossberger, D. Kappes, A. Korman, J. Kuo, J. Lillie, K. Okada, M. Roux-Dosseto, A. Schamboeck & J. L. Strominger. 1983. *In* Progress in Immunology V, 5th International Congress of Immunology. Y. Yamamura & T. Tada, Eds.: 203–214. Academic Press. Tokyo.
11. Collins, T., A. Korman, C. Wake, J. Boss, D. Kappes, W. Fiers, K. Ault, M. Gimbrone, Jr, J. L. Strominger & J. Pober. 1984. Proc. Natl. Acad. Sci. USA **81:** 4917–4921.
12. Boss, J. M. & J. L. Strominger. 1984. Proc. Natl. Acad. Sci. USA **81:** 5199–5203.
13. Shackelford, D. A. & J. L. Strominger. 1984. Hum. Immunol. **9:** 159–174.
14. Morton, C. C., J. Brown, W. E. Nance, D. E. Woods, I. R. Kirsch, G. A. Evans, A. J. Korman & J. L. Strominger. 1984. *In* Immunogenetics—Its Application to Clinical Medicine. Proceedings of the Conference on Immunogenetics, 1983, Tokyo. T. Sasazuki & T. Tada, Ed.: 239–251. Academic Press. Tokyo.
15. Knudsen, P. J., J. McLean & J. L. Strominger. 1985. A monoclonal antibody that recognizes the α chain of HLA-DR antigens. Hum. Immunol. In press.
16. Kappes, D. J., D. Arnot, K. Okada & J. L. Strominger. 1984. Structure and polymorphism of the HLA Class II SB light chain genes. EMBO J. **3:** 2985–2993.
17. Okada, K., H. Prentice, J. Boss, D. Levy, D. Kappes, T. Spies, R. Raghupathy, R. Mengler, C. Auffray & J. L. Strominger. 1985. SB subregion of the human major histocompatibility complex: Gene organization, allelic polymorphism, and expression in transformed cells. EMBO J. **4:** 739–748.
18. White, P. C., D. Grossberger, B. J. Onufer, D. D. Chaplin, M. I. New, B. Dupont & J. L. Strominger. 1985. Two genes encoding steroid 21-hydroxylase are located near the genes encoding the fourth component of complement in man. Proc. Natl. Acad. Sci. USA **82:** 1089–1093.
19. Perlmutter, D. H., H. R. Colton, D. Grossberger, J. L. Strominger, J. G. Seidman & D. D. Chaplin. 1985. Expression of complement proteins C2 and Factor B in transfected L-cells. J. Clin. Invest. In press.
20. Sorrentino, R., J. Lillie & J. L. Strominger. 1985. Molecular characterization of MT3 antigens by two-dimensional gel electrophoresis, NH_2-terminal amino acid sequence, and Southern blot analysis. Proc. Natl. Acad. Sci USA **82:** 3794–3798.
21. Okada, K., J. Boss, H. Prentice, T. Spies, R. Mengler, C. Auffray, J. Lillie, D. Grossberger & J. L. Strominger. 1985. Gene organization of the DC and DX subregions of the human major histocompatibility complex. Proc. Natl. Acad. Sci. USA **82:** 3410–3414.

Structure Organization and Expression of the Major Histocompatibility Class III Genes

HARVEY R. COLTEN

Harvard Medical School
Boston, Massachusetts 02115

INTRODUCTION

In the past few years, complementary DNA (cDNA) clones corresponding to a subunit of the first component,[1] the second[2,3] and fourth[4-6] components of complement, factor B[7-9] of the alternative complement pathway, C3,[10] and C5[11] have been isolated. The availability of these cDNA probes has made it possible to investigate the structure of the corresponding genes, to map the complement gene loci, to provide some insight into the structural basis for genetic polymorphisms and deficiencies, and to examine the molecular mechanisms that account for tissue and species specific changes in mononuclear phagocyte maturation and response to mediators of inflammation. In this review, I will discuss the complement genes that are within the major histocompatibility complex (MHC). These include the genes that code for the second component of complement (C2), factor B of the alternative complement pathway, and the fourth component (C4). These are designated the Class III major histocompatibility complex genes to distinguish them from the polymorphic genes that code for cell surface glycoproteins of the Class I and Class II types (discussed elsewhere in this volume).

FACTOR B

Factor B is a 90,000 dalton single chain glycoprotein component of the alternative pathway of complement activation.[12] Cleavage of factor B by the serine proteinase, factor D, generates an amino-terminal fragment, Ba (~30,000 daltons), and a carboxy-terminal fragment, Bb. The latter bears the active enzymatic site. The Bb fragment, in complex with a cleavage product of the third component C3b, functions as one of the C3 cleaving enzymes. Additional functions of activated factor B include facilitation of macrophage-mediated cytotoxicity,[13] induction of macrophage spreading,[14] and activation of plasminogen.[15] Synthesis of the factor B primary translation product (M_r ~83,000)[7] is directed by a 2.6 kilobase (kb) polyadenylated messenger RNA (mRNA).[16] Core glycosylation is accomplished by a dolicholphosphate-dependent pathway and the carbohydate is further modified just prior to secretion.[17] Single loci for the human[18] and murine[19] factor B genes (~6 kb) have been identified. A 500 base pair 5′ flanking region, which presumably contains the control sequences regulating transcription, separates this gene from the 3′ end of the C2 gene. The human factor B gene is divided into 18 exons.[20] Three of these exons, coding for regions within the Ba fragment, show significant sequence homology to one another and may, therefore, have evolved by tandem duplication. Active site residues in the Bb fragment are encoded on separate exons. With the exception of a unique exon found in Bb, the organization of the 3′ two-thirds of the factor B gene is similar to that observed for other serine proteinase genes.

A comparison of the nucleotide and derived amino acid sequences for human and murine factor B reveals a high degree of homology.[9] Overall, for the 477 amino acids determined in the Bb fragment of murine factor B, 83% of the residues are identical to the corresponding region of the human protein. When conservative substitutions are considered, the homology is 90%. No deletions or insertions were detected. Even in the 3′ untranslated region, there is 59% nucleotide sequence homology, and an unusual polyadenylation signal is found in both species. This unusual signal is also found in the human C4 cDNA sequence.

Polymorphic variants of human and murine factor B have been identified by differences in electrophoretic mobility of the plasma protein,[21] but, to date, no genomic polymorphisms have been observed in human factor B DNA when sought by Southern Blot Analysis;[3] that is, no restriction fragment length polymorphisms (RFLP) were detected in a survey of DNA samples from 26 unrelated individuals digested with 23 different restriction endonucleases. On the other hand, several RFLP genomic variants of murine factor B have been recognized and have been used to provide a more detailed description of this region of the MHC.[22]

Factor B is synthesized in liver and in cells of the monocyte/macrophage series. The bulk of factor B found in plasma is produced in liver,[21] but, in extrahepatic sites, local production of factor B and modulation of the rate of synthesis in response to inflammation may serve an important role in host defense mechanisms.[23] In tissue culture, monocytes synthesize factor B only after a three day lag, whereas tissue macrophages produce large amounts of factor B within the first few hours in culture.[24] Regulation of the rate of synthesis is controlled primarily at a pretranslational level inasmuch as changes in the cellular content of factor B mRNA correlate with the changes in the amount of factor B protein synthesized.[25]

THE SECOND COMPONENT

The second (C2) component is a single chain glycoprotein (M_r 102,000) that is similar in structure and function to factor B.[26] The carboxy-terminal segment C2a (M_r 60,000), which is generated by Cl cleavage of the native protein, bears the active enzymatic site. In analogy to the alternative pathway C3 cleaving enzyme (C3bBb), the C2a fragment complexes with a fragment of the fourth component (C4b2a) to generate the classical pathway C3 convertase. The C2 gene is represented as a single copy per haploid,[18] but its precise size and fine structure are not yet known. Three primary translation products for C2 are generated by human liver mRNA in a cell-free synthesizing system.[27] Therefore, although not yet demonstrated directly, it is believed that alternative transcription or posttranscriptional processing events give rise to three forms of C2 mRNA. The predominant mRNA (2.9 kb) directs synthesis of an 84 kd C2 protein which is glycosylated and secreted within 1–2 hours. Two forms of C2 of lower molecular mass (79 kd and 70 kd) remain cell-associated for at least 12 hours after synthesis. The function of the cell-associated C2 proteins is not known.

Studies of C2 biosynthesis in mononuclear phagocytes demonstrated changes in the proportion of C2-producing cells and in the rates of synthesis as a function of cell maturation and tissue from which the cells were isolated. No C2-producing cells are detected in guinea pig marrow, but about 10% of blood monocytes produce C2 specific hemolytic plaques.[28] The proportion of tissue macrophages that produce C2 ranges from about 45% in spleen and peritoneal cavity to approximately 2–3% in lung. Bronchoalveolar macrophages synthesize C2 at a rate 5–10 times greater than the rate of synthesis in the peritoneal macrophage, partially offsetting the 20-fold difference in proportion of C2-producing cells. Hence, the absolute amount of C2 generated from comparable numbers of unfractionated peritoneal and alveolar macrophages is similar. The increase in C2 secretion associated with inflammation is mediated by at least two mechanisms: (a) an increase in macrophage cell number, and (b) an increase in rate of synthesis per cell. The latter is due to an increase in C2 mRNA per cell, but this is proportional to the increase in total cellular RNA and total protein secretion in the inflammatory cell exudate.[29] It should be noted, however, that the portion of C2-producing cells in an inflammatory exudate is similar to that in a resting cell population; i.e., there is no evidence for selective recruitment of complement producing cells.[30]

In human mononuclear phagocytes, the rate of C2 synthesis is also a function of cell maturation, tissue site, and response to stimuli that elicit an inflammatory exudate. Hemolytically active C2 is secreted by peripheral blood monocytes maintained in culture for several months.[31] During the first 3–4 days in culture, little or no C2 is secreted, but afterwards the rate of C2 synthesis increases progressively, reaching a level (after two months *in vitro*) three times that observed at one week. Synthesis of C2 is initiated without lag and rates of synthesis are several times greater in human macrophages from breast milk or bronchoalveolar lavage than in monocytes harvested from peripheral blood of the same individuals.[24] A particularly striking difference in the amount of C2 specific mRNA has been noted in comparing the bronchoalveolar and breast milk macrophage to peripheral blood monocytes;[25] that is, the amount of C2 mRNA per cell is 4- to 11-fold greater in breast milk macrophages and 14- to 95-fold greater in bronchoalveolar macrophages than in peripheral blood monocytes during the first day in culture.

Genetic polymorphism of C2 has been recognized by banding patterns following isoelectricfocusing of the plasma protein.[21] Additional variants have been detected by Southern Blot Analysis of human DNA from normal individuals.[3] Preliminary data indicate that murine C2 DNA also displays length polymorphisms when cut with appropriate restriction enzymes.[22]

Among individuals of Western European origin, C2 deficiency is the most common of the genetic deficiencies of the complement proteins.[21] The disorder is inherited as an autosomal codominant trait found in linkage disequilibrium with certain HLA haplotypes. Recent studies indicate that monocytes from homozygous C2 deficient individuals do not contain C2 mRNA and, therefore, fail to synthesize the protein (Cole *et al.,* unpublished). On the other hand, the C2 gene is present in homozygous deficient individuals and no major insertions or deletions have been detected so that it appears the defect is in transcription or posttranscriptional processing.

THE FOURTH COMPONENT

The fourth component of complement (C4) is a serum glycoprotein composed of three disulfide linked polypeptides.[32] Synthesis of the 185,000 dalton single chain precursor of C4, pro-C4,[33,34] is directed by a polyadenylated mRNA of approximately five kilobases (kb).[4,6] The beta chain is the amino-terminal subunit, the alpha chain is the central subunit, and the gamma chain is the carboxy-terminal segment of pro-C4.[36,37] These chains are separated in the precursor protein by arginine-rich intersubunit-linking peptides[4,6] similar to the linking peptides between the beta and alpha subunits of pro-C3[10] and pro-C5[11] precursors of the third and fifth complement components. Postsynthetic processing of pro-C4 involves proteolytic excision of the linking peptides, sulfation,[37] modification of residues within the region of the thiolester site, and glycosylation of alpha and beta chains.[17,38] Finally, extracellular modification of the alpha chain results in the apparent loss of a carboxy-terminal peptide of uncertain length.[39] Reviews of the literature on tissue and species specific regulation of C4 biosynthesis have been published recently,[23,40] so this aspect will not be discussed further.

Variation in subunit size, based on genetic factors, has been detected in human[41] and mouse[42-44] serum C4. For example, the alpha chain of C4 derived from one of the human C4 loci (C4A) is ~2,000 daltons heavier than the alpha chain product of the other locus (C4B). A structural variant in human beta chain has also been recognized,[45] but this variant is found in C4 protein derived from both loci.

Polymorphisms of C4 protein derived from each locus are detected by electrophoresis in nondenaturing gels.[21] Multiple variants at each locus have been recognized and the relatively high frequency of triplication and deletions at the C4A and C4B locus increases the complexity of this region. Porter has hypothesized that this variability may be of functional importance in the association of certain disorders of immune regulation and particular MHC Class III haplotypes.[46] Additional variants of the C4 genes have been detected at the DNA level.[47,48] Some of these RFLP variants in C4 permit subdivision of populations indistinguishable by C4 protein type.

In the mouse, structural polymorphisms of C4 have been detected for each of the subunits, and in strains bearing the S^{w7} haplotype, differences in C4 specific hemolytic activity have been correlated with the structural variants.[49] A C4-like protein (Slp) that lacks hemolytic activity is also found in certain strains, but its expression is under hormonal control.[50] Slp is synthesized as a single chain precursor that is processed in a manner similar to that described for C4.[51] Murine C4 RFLP DNA variants have been recognized in standard inbred strains.[22] Coupled with genomic polymorphisms in mouse factor B and C2, it is now possible to subdivide the S-region of the murine MHC in accordance with the nomenclature adopted for the balance of the H-2 region.

Genetic deficiency of the fourth component of complement in guinea pigs is inherited as an autosomal codominant trait; no C4 is found in the plasma of homozygous deficient animals.[52] Cells and tissues from C4 deficient guinea pigs do not synthesize C4,[53] but somatic cell hybrids of C4 deficient peritoneal macrophages and HeLa cells yield clones capable of synthesizing functionally active human C4.[54] Initial cell-free biosynthetic studies suggested that mRNA from homozygous C4 deficient animals directs synthesis of a heterogeneous mixture of polysome-bound C4 peptides, but not intact pro-C4.[55] In subsequent studies, we failed to detect these polysome-bound C4 peptides.[4] A human C4 cDNA clone was used to pursue this question further. This clone specifically hybridizes with a five kb mRNA in normal guinea pig liver. In the C4 deficient guinea pig liver, a low abundance seven kb mRNA species specifically hybridized with the C4 cDNA clone, but no mature C4 mRNA was detected. We believe that this 7 kb C4 mRNA species is an intermediate precursor of mature C4 mRNA; i.e., that C4 deficiency in the guinea pig is due to a post-transcriptional processing defect. Direct studies of C4 mRNA processing, a detailed structural analysis of the mutant C4 gene, and functional studies of genomic clones in transfected cells will be required to fully define the basis of the genetic deficiency. Obviously, a similar approach can now be applied to the analysis of complement deficiencies in humans and other species.

ORGANIZATION OF THE MHC CLASS III GENES

Recently, the murine complement genes within the S region of the major histocompatibility complex were mapped utilizing cDNA probes for the Class III genes and a series of overlapping cosmid clones. The initial results indicated that genes for C2 (Chaplin *et al.*, unpublished), factor B, and two "C4-like" genes designated C4x and C4y are within a region of ~250 kb and are oriented in the same direction (5′–3′).[56] Identification of the second of the two C4-like genes (C4y) as C4 was accomplished by showing that murine L cells transfected with a cosmid that contains the entire coding sequence plus 5′ flanking sequences, synthesize and secrete biologically active C4.[19] Similar experiments utilizing cosmids containing the C4x gene have not resulted in expression of this gene in the transfected cells, but by using another approach, Ogata (unpublished) has shown that this is, in fact, the Slp gene. It is possible that androgen stimulation may be required for Slp expression. Alternatively, it is possible that insufficient 5′ flanking sequences were included in the particular cosmid used, thus, excluding information critical for even constitutive expression. Based on these experiments, we concluded that the order of the genes within the S region of the murine MCH is C2-Factor B- Slp- C4. The mouse MHC Class III genes are, therefore, oriented in a sequence similar to their human counterparts.[18] The high degree of fidelity in transcriptional processing, translation, postsynthetic modification, and secretion of C4 in cells transfected with the C4 gene has considerable implications for the further analysis of the mechanisms for genetic and microenvironmental regulation of MHC Class III gene expression.

Based on family and population studies, it was not possible to determine whether the Class III genes were between HLA-B and -DR or outside of this region.[21] The use of deletion mutant cell lines and a C4 cDNA probe established that the Class III gene complex is, in fact, between HLA-B and DR.[57] Hence, a more detailed map of the Class III genes and their relationship to the balance of the MHC has been elucidated.

CONCLUSIONS

In the past few years, following the isolation and characterization of cDNA clones corresponding to the complement proteins, substantial progress has been made in several different, but related areas. Most relevant to the subject of this symposium was the generation of a more detailed map of the Class III MHC gene complex. In addition, however, fundamentals of immunopathology, host defense mechanisms, and more general cell biological questions have been addressed.

REFERENCES

1. REID, K. B. M., D. R. BENTLEY & K. J. WOOD. 1984. Cloning and characterization of the cDNA for the B chain of normal human serum Clq. Philos. Trans. R. Soc. London **306:** 345–354.
2. BENTLEY, D. R. & R. R. PORTER. 1984. Isolation of cDNA clones for human complement component C2. Proc. Natl. Acad. Sci. USA **81:** 1212–1215.
3. WOODS, D. E., M. D. EDGE & H. R. COLTEN. 1984. Isolation of a cDNA clone for the human complement protein C2 and its use in the identification of a restriction fragment length polymorphism. J. Clin. Invest. **74:** 634–638.
4. WHITEHEAD, A. S., G. GOLDBERGER, D. E. WOODS, A. F. MARKHAM & H. R. COLTEN. 1983. Use of a cDNA clone for the fourth component of human complement (C4) for analysis of a genetic deficiency of C4 in guinea pig. Proc. Natl. Acad. Sci. USA **80:** 5387–5391.
5. CARROLL, M. C. & R. R. PORTER. 1983. Cloning of a human complement component C4 gene. Proc. Natl. Acad. Sci USA **80:** 264–267.

6. OGATA, R. T., D. C. SHREFFLER, D. S. SEPICH & S. P. LILLY. 1983. cDNA clone spanning the alpha-gamma subunit junction in the precursor of the murine fourth complement component (C4). Proc. Natl. Acad. Sci. USA **80:** 5061–5065.
7. WOODS, D. E., A. F. MARKHAM, A. T. RICKER, G. GOLDBERGER & H. R. COLTEN. 1982. Isolation of cDNA clones for the human complement protein factor B, a class III major histocompatibility complex gene product. Proc. Natl. Acad. Sci. USA **79:** 5661–5665.
8. CAMPBELL, R. D. & R. R. PORTER. 1983. Molecular cloning and characterization of the gene coding for human complement protein factor B. Proc. Natl. Acad. Sci. USA **80:** 4464–4468.
9. SACKSTEIN, R., H. R. COLTEN & D. E. WOODS. 1983. Phylogenetic conservation of a class III major histocompatibility complex antigen, factor B: Isolation and nucleotide sequencing of mouse factor B cDNA clones. J. Biol. Chem. **258:** 14693–14697.
10. WIEBAUER, K., H. DOMDEY, H. DIGGELMAN & G. FEY. 1982. Isolation and analysis of genomic DNA clones encoding the third component of mouse complement. Proc. Natl. Acad. Sci. USA **79:** 7077–7081.
11. LUNDWALL, A. B., R. A. WETSEL, T. KRISTENSEN, A. S. WHITEHEAD, D. E. WOODS, R. L. OGDEN, H. R. COLTEN & B. F. TACK. 1985. Isolation of a cDNA clone encoding the fifth component of human complement. J. Biol. Chem. **260:** 2108–2112.
12. CHRISTIE, D. L., J. GAGNON & R. R. PORTER. 1980. Partial sequence of human complement component factor B: Novel type of serine esterase. Proc. Natl. Acad. Sci. USA **77:** 4923–4927.
13. HALL, R. E., R. M. BLAESE, A. E. DAVIS III, J. M. DECKER, B. F. TACK, H. R. COLTEN & A. V. MUCHMORE. 1982. Cooperative interaction of factor B and other complement components with mononuclear cells in the antibody independent lysis of xenogeneic erythrocytes. J. Exp. Med. **156:** 834–843.
14. SUNDSMO, J. S. & O. GOTZE. 1981. Human monocyte spreading induced by factor Bb of the alternative pathway of complement activation. J. Exp. Med. **154:** 763–777.
15. SUNDSMO, J. S. & L. M. WOOD. 1981. Activated factor B (Bb) of the alternative pathway of complement activation cleaves and activates plasminogen. J. Immunol. **127:** 877–880.
16. SACKSTEIN, R. & H. R. COLTEN. 1984. Molecular regulation of MHC class III (C4 and factor B) gene expression in mouse peritoneal macrophages. J. Immunol. **133:** 1618–1626.
17. MATTHEWS, W. J., G. GOLDBERGER, J. T. MARINO, L. P. EINSTEIN, D. J. GASH & H. R. COLTEN. 1982. The major histocompatibility complex (MHC) linked complement protein C2, C4, and factor B: Effect of glycosylation on their secretion and catabolism. Biochem. J. **204:** 839–846.
18. CARROLL, M. D., R. D. CAMPBELL, D. R. BENTLEY & R. R. PORTER. 1984. A molecular map of the major histocompatibility complex class III region of man linking the complement genes C4, C2, and factor B. Nature **307:** 237–241.
19. CHAPLIN, D. D., R. SACKSTEIN, D. H. PERLMUTTER, J. H. WEIS, T. A. KRUSE, J. COLIGAN, H. R. COLTEN & J. G. SEIDMAN. 1984. Expression of hemolytically active murine fourth component of complement in transfected L cells. Cell **37:** 569–576.
20. CAMPBELL, R. D., D. R. BENTLEY & B. J. MORLEY. 1984. The factor B and C2 genes. Philos. Trans. R. Soc. London **306:** 367–378.
21. ALPER, C. A. 1981. Complement and the MHC. *In* The Role of the Major Histocompatibility Complex in Immunobiology. M.E. Dorf, Ed. Garland Press. New York.
22. SACKSTEIN, R., M. H. ROOS, P. DEMANT & H. R. COLTEN. 1984. Subdivision of the S-region of the murine major histocompatibility complex by identification of genomic polymorphisms of the class III genes. Immunogenetics **20:** 321–330.
23. COLTEN, H. R., F. S. COLE, R. SACKSTEIN & H. S. AUERBACH. Tissue and species specific regulation of complement biosynthesis in mononuclear phagocytes. *In* Proceedings of the Fourth Leiden Conference on Mononuclear Phagocytes. Martinus Nijhoff Publishers. The Netherlands. In press.
24. COLE, F. S., E. E. SCHNEEBERGER, N. A. LICHTENBERG & H. R. COLTEN. 1982. Complement biosynthesis in human breast-milk macrophages and blood monocytes. Immunology **46:** 429–441.
25. COLE, F. S., H. S. AUERBACH, G. GOLDBERGER & H. R. COLTEN. 1985. Tissue specific

pretranslational regulation of complement production in human mononuclear phagocytes. J. Immunol. **134:** 2610–2616.

26. COOPER, N. R. 1975. Enzymatic activity of the second component of complement. Biochemistry **14:** 4245–4249.
27. PERLMUTTER, D. H., F. S. COLE, G. GOLDBERGER & H. R. COLTEN. 1984. Distinct primary translation products from human liver mRNA give rise to secreted and cell-associated forms of complement protein C2. J. Biol. Chem. **259:** 10380–10385.
28. ALPERT, S. E., H. S. AUERBACH, F. S. COLE & H. R. COLTEN. 1983. Macrophage maturation: Differences in complement secretion by marrow, monocyte, and tissue macrophages detected with an improved hemolytic plaque assay. J. Immunol. **130:** 102–107.
29. AUERBACH, H. S., R. D. BAKER, W. J. MATTHEWS & H. R. COLTEN. 1984. Molecular mechanism for feedback regulation of C4 biosynthesis in guinea pig peritoneal macrophage. J. Exp. Med. **159:** 1750–1761.
30. COLE, F. S., W. J. MATTHEWS, J. T. MARINO, D. J. GASH & H. R. COLTEN. 1980. Control of complement synthesis and secretion in bronchoalveolar and peritoneal macrophages. J. Immunol. **125:** 1120–1124.
31. EINSTEIN, L. P., E. E. SCHNEEBERGER & H. R. COLTEN. 1976. Synthesis of the second component of complement by long-term primary cultures of human monocytes. J. Exp. Med. **143:** 114–126.
32. SCHREIBER, R. D. & H. J. MULLER-EBERHARD. 1974. Fourth component of human complement: Description of a three polypeptide chain structure. J. Exp. Med. **140:** 1324–1335.
33. HALL, R. E. & H. R. COLTEN. 1977. Cell-free synthesis of the fourth component of guinea pig complement (C4): Identification of a precursor of serum C4 (pro-C4). Proc. Natl. Acad. Sci. USA **74:** 1707–1710.
34. ROOS, M. H., J. P. ATKINSON & D. C. SHREFFLER. 1978. Molecular characterization of the Ss and Slp (C4) proteins of the mouse H-2 complex: Subunit composition, chain size polymorphism, and an intracellular (pro-Ss) precursor. J. Immunol. **121:** 1106–1115.
35. GOLDBERGER, G., G. N. ABRAHAM, S. WILLIAMS & H. R. COLTEN. 1980. Primary structure of pro-C4 the precursor of the fourth component of guinea pig complement. J. Biol. Chem. **255:** 7070–7074.
36. KARP, D. R., K. L. PARKER, D. C. SHREFFLER, & D. CAPRA. 1981. Characterization of the murine C4 precursor (pro-C4): Evidence that the carboxy-terminal subunit is the C4 gamma chain. J. Immunol. **126:** 2060–2061.
37. KARP, D. R. 1983. Posttranslational modification of the fourth component of complement. J. Biol. Chem. **258:** 12745–12748.
38. ROOS, M. H., S. KORNFELD & D. C. SHREFFLER. 1980. Characterization of the oligosaccharide units of the fourth component of complement (SS protein) synthesized by murine macrophages. J. Immunol. **124:** 2860–2863.
39. CHAN, A. C., K. R. MITCHELL, T. W. MUNNS, D. R. KARP & J. P. ATKINSON. 1983. Identification and partial characterization of the secreted form of the fourth component of human complement: Evidence that it is different from major plasma form. Proc. Natl. Acad. Sci. USA **80:** 268–272.
40. COLTEN, H. R. 1984. Expression of the MHC class III genes. Philos. Trans. R. Soc. London **306:** 355–366.
41. ROOS, M. H., E. MOLLENHAUER, P. DEMANT & C. RITTER. 1982. A molecular basis for the two locus model of human complement component C4. Nature **298:** 854–855.
42. CARROLL, M. C. & J. D. CAPRA. 1979. Studies on the murine Ss protein. Demonstration that the S region encodes the structural gene for the fourth component of complement. Proc. Natl. Acad. Sci. USA **76:** 4641–4645.
43. FERRIERA, A., J. MICHAELSON & V. NUSSENZWEIG. 1980. Polymorphism of the chain of mouse C4 controlled by the S region of the major histocompatibility complex. J. Immunol. **125:** 1178–1182.
44. NATSUUME-SAKAI, S., J. I. HAYARAWA & M. TAKAHASHI. 1978. Genetic polymorphism of murine C3 controlled by a single codominant locus on chromosome 17. J. Immunol. **121:** 491–498.

45. MAUFF, G., M. STEUER, M. WEEK & K. BENDER. 1983. The C4 beta-chain: Evidence for a genetically determined polymorphism. Hum. Genet. **64:** 186–188.
46. PORTER, R. R. 1983. Complement polymorphism, the major histocompatibility complex and associated diseases: A speculation. Mol. Biol. Med. **1:** 161–168.
47. PALSDOTTIR, A., S. J. CROSS, J. H. EDWARDS & M. C. CARROLL. 1983. Correlation between a DNA restriction fragment length polymorphism and C4A6 protein. Nature **306:** 615–616.
48. WHITEHEAD, A. S., D. E. WOODS, E. FLEISCHNICK, J. E. CHIN, E. J. YUNIS, A. J. KATZ, P. S. GERALD, C. A. ALPER & H. R. COLTEN. 1984. DNA polymorphism of the C4 genes: A new marker for analysis of the major histocompatibility complex. N. Engl. J. Med. **310:** 88–91.
49. ATKINSON, J. P., K. MCGINNIS, L. BROWN, J. PETEREIN & D. SHREFFLER. 1980. A murine C4 molecule with reduced hemolytic efficiency. J. Exp. Med. **151:** 492–497.
50. PASSMORE, H. C. & D. C. SHREFFLER. 1970. A sex-linked serum protein variant in the mouse: Inheritance and association with the H-2 region. Biochem. Genet. **4:** 351–365.
51. PARKER, K. L., M. H. ROOS & D. C. SHREFFLER. 1979. Structural characterization of the murine fourth component of complement and sex-limited protein and their precursors: Evidence for two loci in the S region of the H-2 complex. Proc. Natl. Acad. Sci. USA **76:** 5853–5857.
52. ELLMAN, L., I. GREEN & M. M. FRANK. 1970. Genetically controlled total deficiency of the fourth component of complement in the guinea pig. Science **170:** 74–75.
53. COLTEN, H. R. & M. M. FRANK. 1972. Biosynthesis of the second and fourth components of complement *in vitro* by tissue isolated from guinea pigs with genetically determined C4 deficiency. Immunology **22:** 991–999.
54. COLTEN, H. R. & R. PARKMAN. 1972. Biosynthesis of the fourth component of complement (C4) by C4-deficient guinea pig-HeLa cell hybrids. Science **176:** 1029–1031.
55. HALL, R. E. & H. R. COLTEN. 1978. Genetic defect in biosynthesis of the precursor form of the fourth component of complement. Science **199:** 69–90.
56. CHAPLIN, D. D., D. E. WOODS, A. S. WHITEHEAD, G. GOLDBERGER, H. R. COLTEN & J. G. SEIDMAN. 1983. Molecular map of the murine S region. Proc. Natl. Acad. Sci. USA **80:** 6947–6951.
57. WHITEHEAD, A. S., H. R. COLTEN, C. C. CHANG & R. J. DEMARS. 1985. Localization of the human MHC-linked complement gene between HLA-B and HLA-DR using HLA mutant cell lines. J. Immunol. **134:** 641–643.

Molecular Cloning of Steroid 21-Hydroxylase[a]

PERRIN C. WHITE,[b,c] MARIA I. NEW,[c] AND BO DUPONT[b]

[b] *Laboratory of Human Immunogenetics*
Memorial Sloan-Kettering Cancer Center
New York, New York 10021

[c] *Division of Pediatric Endocrinology*
Department of Pediatrics
Cornell University Medical Center
New York, New York 10021

INTRODUCTION

Congenital adrenal hyperplasia due to 21-hydroxylase (21-OH) deficiency affects approximately one in 5000 births and is, thus, one of the most common inborn errors of metabolism.[1] It is inherited as a monogenic autosomal recessive trait closely linked to the HLA major histocompatibility complex.[2] In this defect of cortisol biosynthesis, there is impaired conversion of 17-hydroxyprogesterone to 11-deoxycortisol. There are several forms of 21-OH deficiency that differ in severity and in HLA associations.[3] An isolated enzyme defect in the zona fasciculata of the adrenal cortex, where cortisol is synthesized,[4] is characteristically associated with HLA-Bw51(5). This is referred to as "simple virilizing" disease because elevated levels of 17-hydroxyprogesterone result in excessive adrenal androgen synthesis. HLA-Bw47;DR7 and Bw60(40) are associated with a severe "salt-wasting" form of the disease where, in addition to the defect in the zona fasciculata, aldosterone synthesis in the zona glomerulosa (conversion of progesterone to 11-deoxycorticosterone) is impaired.[4] If untreated, this can result in shock or death in the neonatal period from inability to conserve sodium. There are two related "variant" forms associated with HLA-B14;DR1 in which patients have a relatively mild defect in cortisol synthesis, as measured by elevated baseline levels of 17-hydroxyprogesterone, and either remain asymptomatic ("cryptic" form) or develop symptoms of virilization in late childhood or at puberty ("late-onset" form).[5]

In addition to these HLA-B;DR associations, genetic linkage disequilibrium has been documented between complement allotypes and 21-OH deficiency.[6] Of greatest interest are patients carrying the Bw47 antigen, who invariably have the unique haplotype HLA-Bw47;C4A*1;C4B*Q0 (null);Bf*F; C2*C. The Bw47(w4) antigen is serologically very similar to the more common antigen B13(w4), and these two antigens form identical banding patterns on isoelectric focusing.[7] We, therefore, proposed that the HLA haplotype Cw6;Bw47; 21-OH*NULL; C4A*1; C4B*Q0;DR7 arose by a large genetic deletion or rearrangement of the original Cw6;B13;21-OH*NORMAL;C4A*3; C4B*1;DR7 haplotype.[6] This implied that genomic DNA containing the 21-OH and C4B loci might be identified by comparing

[a]This work was supported by grant nos. CA–22507 and CA–08748 from the National Cancer Institute and HD–00072 and RR–47 from the National Institutes of Health.

DNA from an individual with 21-OH deficiency who was homozygous for HLA-Bw47 and DNA from a normal individual homozygous for HLA-B13.

The underlying biochemical defect in 21-OH deficiency has never been directly determined. *In vitro* 21-OH activity can be produced with two microsomal proteins from the adrenal cortex, an NADPH-dependent cytochrome reductase and a cytochrome P-450,[8] termed P-450_{C21}. P-450_{C21} is one of a class of heme-containing monooxygenases with molecular weights of about 50,000 which vary widely in substrate specificity and organ distribution. It is one of four cytochromes P-450 which are required to convert cholesterol to cortisol[9]—the C22,27 side chain is cleaved by a P-450 to form pregnenolone; the 3β-hydroxyl is dehydrogenated, yielding progesterone, which is successively hydroxylated by three different cytochromes P-450 at the 17α, 21, and 11β positions to yield cortisol.

As P-450_{C21}, but not cytochrome reductase, is substrate-specific, we hypothesized that the HLA-linked defect involved a structural gene for this protein. Therefore, we isolated a bovine adrenal cDNA clone encoding part of P-450_{C21} in order to test this hypothesis.[10] We found that a structural gene encoding P-450_{C21} is indeed deleted when 21-OH deficiency is associated with HLA-Bw47.[11]

CONSTRUCTION OF A cDNA CLONE ENCODING P-450_{C21}

The method of producing this cDNA clone (FIGURE 1) will be briefly summarized.[10] (Reference 12 is a manual more fully describing many of these techniques.) P-450_{C21} was purified to homogeneity; serum from rabbits immunized with purified P-450_{C21} precipitated a single protein from the products of an *in vitro* translation reaction using bovine adrenal mRNA. This protein comigrated with P-450_{C21} on SDS-polyacrylamide gel electrophoresis. After sucrose gradient sedimentation, mRNA encoding P-450_{C21} was found in the 19S fraction, indicating that the mRNA was about 2200 nucleotides long. This fraction was reverse transcribed into double-stranded complementary DNA (cDNA) and inserted into the Pst I restriction endonuclease recognition site of the plasmid vector pBR322 by the dC-dG tailing procedure. *E. coli* transformed with recombinant plasmids were screened with an *in situ* immunoassay using anti-P-450_{C21} serum and ^{125}I staphylococcal protein A. The rationale of such assays is that inserts in the structural gene of a bacterial protein (in this case, β-lactamase encoding by pBR322), if in the correct orientation and reading frame, may be transcribed and translated to yield a fusion protein carrying antigenic determinants of the protein of interest. Bacteria carrying one plasmid, pC21a, bound anti-P-450_{C21} serum. This plasmid hybridized mRNA encoding P-450_{C21}, and DNA sequencing demonstrated that pC21a contained a 520 bp insert encoding approximately the middle third of the P-450_{C21} polypeptide.

DNA ANALYSIS

Initial studies were performed on DNA from individuals who were homozygous for HLA-B13 or HLA-Bw47. To assure ready availability of DNA, continuously proliferating lymphoblastoid cell lines were created by *in vitro* transformation with Epstein-Barr virus.[13] Line PLH was derived from a patient with 21-OH deficiency who was homozygous for HLA-Bw47;DR7 and line GMA was derived from a normal individual homozygous for B13.

DNA was extracted from cultured cells or peripheral blood leukocytes, as

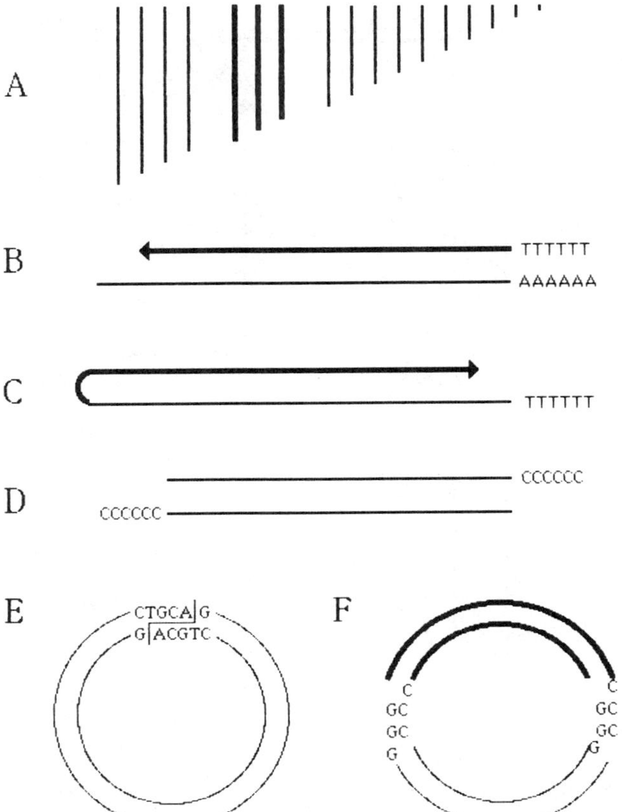

FIGURE 1. Cloning of cDNA: (A) Messenger RNA (mRNA) is fractionated on the basis of size by sucrose gradient sedimentation and the fraction containing the mRNA of interest is identified by *in vitro* translation. (B) The first strand of cDNA is synthesized by reverse transcriptase using an oligo(dT) primer that binds to the poly(A) "tail" of the mRNA. (C) The mRNA is removed by boiling and the second strand of cDNA is synthesized by DNA polymerase I. (D) The "hairpin" loop produced in step C is opened by digestion with S1 nuclease and poly(dC) tails are added to the 3′ ends of the cDNA by terminal transferase. (E) The plasmid vector pBR322 is digested with restriction endonuclease Pst I, which cuts the circular molecule at a single recognition site. Poly(dG) tails are added to the linearized plasmid. (F) The plasmid and cDNA molecules are joined by complementary base-pairing of their respective (dG) and (dC) tails. These recombinant molecules are absorbed by bacteria treated with calcium chloride and confer tetracycline resistance on the bacteria. All bacteria in each tetracycline-resistant colony carry the same plasmid.

described,[14] and analyzed by "Southern" hybridization[15] (FIGURE 2). The studies aimed to demonstrate a difference between the DNA from normal and 21-OH deficient persons, which involved a structural gene for P-450$_{C21}$.

If a change in DNA sequence forms, deletes, or moves a recognition site for a particular restriction enzyme, this is reflected in the size distribution of DNA

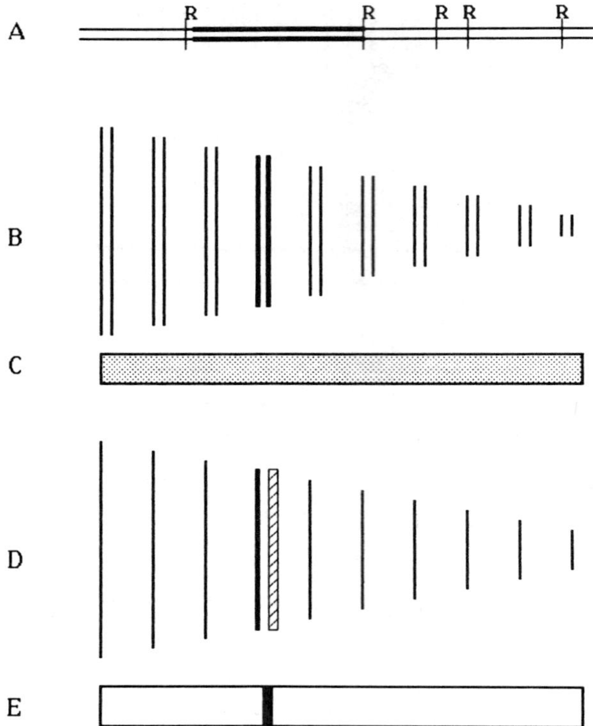

FIGURE 2. Southern hybridization: (A) Genomic DNA is digested with a restriction enzyme that has recognition sites, "R". The sequence of interest is indicated by the bold line. (B) The fragments produced by this digestion are fractionated on the basis of size by electrophoresis in an agarose gel. (C) Since about one million fragments are produced when human DNA is digested with typical enzymes, staining of the DNA in the gel produces a blur. (D) The DNA is rendered single-stranded by treatment with alkali and is blotted to a nitrocellulose membrane. The membrane is incubated with a cloned radioactive "probe" for the sequence of interest, which has also been rendered single-stranded. The probe hybridizes with the fragment containing the sequence of interest. (E) This hybridization is detected as a discrete band by autoradiography.

fragments produced by digestion with that enzyme and displayed on an agarose gel. Such differences may be detected by rendering the digested DNA single-stranded by alkali denaturation, transferring the DNA to a nitrocellulose filter, and hybridizing the immobilized DNA with cloned DNA containing a sequence of interest which has been rendered radioactive and has also been denatured. This "probe" (in this case, the insert of pC21a) forms a double-stranded structure with homologous sequences present in the DNA on the filter, which are seen as discrete bands on autoradiography. If, in different individuals, differences in size or number of bands are seen after digestion with a restriction enzyme and hybridization with a particular probe, this is referred to as a "restriction fragment length polymorphism."

Thus, DNA samples from the GMA and PLH cell lines were digested with various restriction endonucleases, subjected to agarose gel electrophoresis, and blotted to nitrocellulose. Blots were hybridized with the denatured radioactive pC21a probe and washed under conditions of moderate stringency (65°C in 0.15 M NaCl/0.015 M Na citrate). "Stringency," the degree of homology and length of homologous region required for stable hybridization, is increased by raising temperature or lowering salt concentration. Hybridizing fragments were detected by autoradiography (FIGURE 3).

After analysis in this manner, the normal DNA sample from GMA yielded two

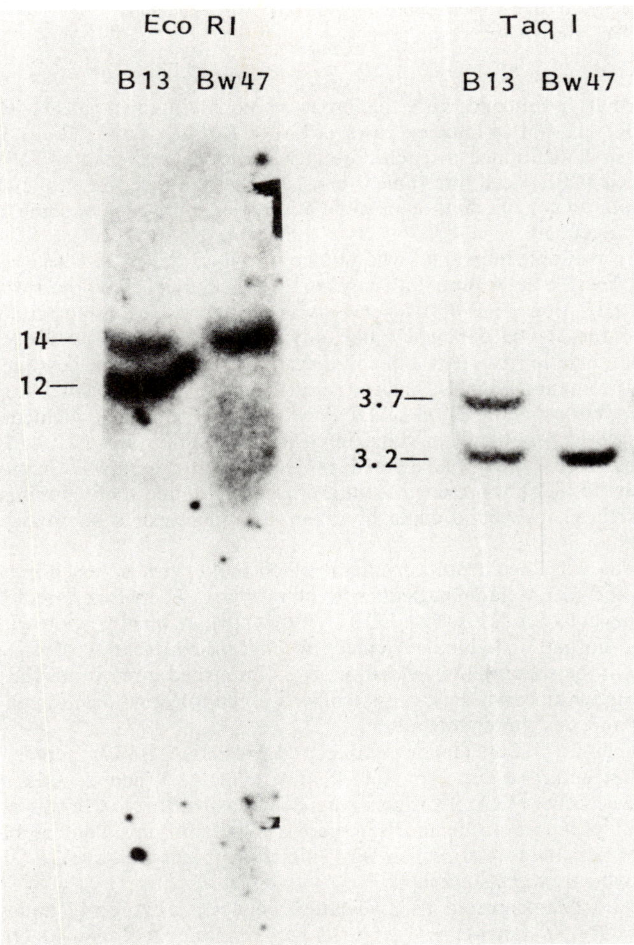

FIGURE 3. Hybridization of pC21a to DNA from normal and 21-OH deficient cell lines. DNA was digested with restriction endonucleases, Eco RI or Taq I, as indicated. Digests were subjected to electrophoresis in, respectively, 0.6% or 1.0% agarose, blotted to nitrocellulose, and hybridized with the radioactively labeled insert of pC21a. Lanes labeled "B13" contained DNA from line GMA (normal, B13 homozygous) and lanes labeled "Bw47" contained DNA from line PLH (21-OH deficient, Bw47 homozygous). Sizes of hybridizing fragments were determined by comparing electrophoretic mobilities with a Hind III digest of bacteriophage lambda DNA.

FIGURE 4. Expected hybridization pattern of DNA from individuals heterozygous for HLA-Bw47 after digestion with Taq I. The chromosome carrying Bw47 yields one hybridizing fragment of 3.2 kb, while the other chromosome yields two of 3.7 and 3.2 kb. In heterozygotes, therefore, the 3.2 kb band appears more intense than the 3.7 kb band.

fragments that hybridized with the probe at equal intensity after digestion with either Eco R1 (12 and 14 kilobase pairs, or kb) or Taq I (3.7 and 3.2 kb). One of these bands (the first mentioned in each digest) was absent in digests of DNA from the 21-OH deficient PLH cell line. This is consistent with a deletion of one of two genes, but these data do not rule out a partial deletion of a single gene or some other type of DNA rearrangement.

The only available individual who was homozygous for HLA-Bw47 was the donor of the PLH line. To determine if all cases of 21-OH deficiency on the Bw47 haplotype result from a deletion of a P-450$_{C21}$ gene, it was necessary to examine patients who were homozygous for 21-OH deficiency, but only heterozygous for Bw47. Detection of a heterozygous deletion is often difficult because it may not be possible to reliably quantitate the amount of DNA and the resulting strength of hybridization in different samples. In this case, digestion of DNA with Taq I normally yielded two readily resolvable hybridizing bands of equal intensity and similar size (3.7 and 3.2 kb). As only one band (3.7 kb) was deleted in DNA from the Bw47 homozygote, the unaffected band (3.2 kb) provided an internal standard, and so the diminished relative intensity of the 3.7 kb band could be taken as evidence of a heterozygous deletion (FIGURE 4).

DNA was extracted from peripheral blood leukocytes of ten unrelated normal individuals and six unrelated patients with homozygous 21-hydroxylase deficiency who were heterozygous for HLA-Bw47;DR7 (FIGURE 5). After digestion with Taq I, all samples for normal individuals yielded two hybridizing bands of equal intensity. Samples from the six patients yielded, in five, diminished intensity of the 3.7 kb band relative to the 3.2 kb band, thus, consistent with a heterozygous deletion; in one patient, the 3.7 kb band was completely absent.

Thus, when 21-OH deficiency is associated with HLA-Bw47, there is a deletion of a P-450$_{C21}$ structural gene ($\chi^2 = 15.9$, $p < .0001$). When 21-OH deficiency is associated with other HLA-B antigens, there is occasionally (1/6 in this experiment) a deletion of at least part of a gene. Often there is no polymorphism detectable with Taq I and it is not possible to distinguish Bw47 heterozygotes who have 21-OH deficiency from those who are merely carriers.

These data demonstrated an association between HLA-Bw47 and an apparent deletion of a P-450$_{C21}$ structural gene. As recombination between 21-OH deficiency and HLA-B is very rare, it was also important to make sure that the deletion segregated with HLA-Bw47 in family studies (i.e., that these two traits were closely linked). Linkage and association are not synonymous terms. "Association" of two traits means that in a collection of unrelated individuals, those having one trait have an increased likelihood of having the second trait. "Linkage" of two traits means that within a pedigree, an individual with one trait will also inherit the second unless recombination occurs. For example, insulin-dependent diabetes is associated with particular HLA antigens, but is not inherited as a simple trait that can be linked to HLA; 21-hydroxylase deficiency is always HLA-linked (if a person has 21-hydroxylase deficiency, an HLA-identical sib will also), but not all 21-OH deficiency genes are carried on characteristic HLA haplotypes.

DNA samples were obtained from members of the family of one of the 21-OH deficient patients and again analyzed by digestion with Taq I and Southern hybridization using the pC21a probe (FIGURE 6). In individuals who carried HLA-Bw47, the heterozygous deletion of the 3.7 kb band could be detected by comparison with the intensity of the 3.2 kb fragment. This cosegregation of HLA-Bw47 and the 3.7 kb deletion extended over three generations.

These experiments demonstrate that a polymorphism (i.e., the deletion of the 3.7 kb Taq I fragment) detectable with the pC21a probe can serve as a marker for 21-OH deficiency. In addition, it is likely that the deleted gene is normally located near the C4B locus encoding the fourth component of complement because the C4B locus carries a null allele and is presumably deleted on the same Bw47;DR7 haplotype which carries a deletion of a P-450_{C21} gene.

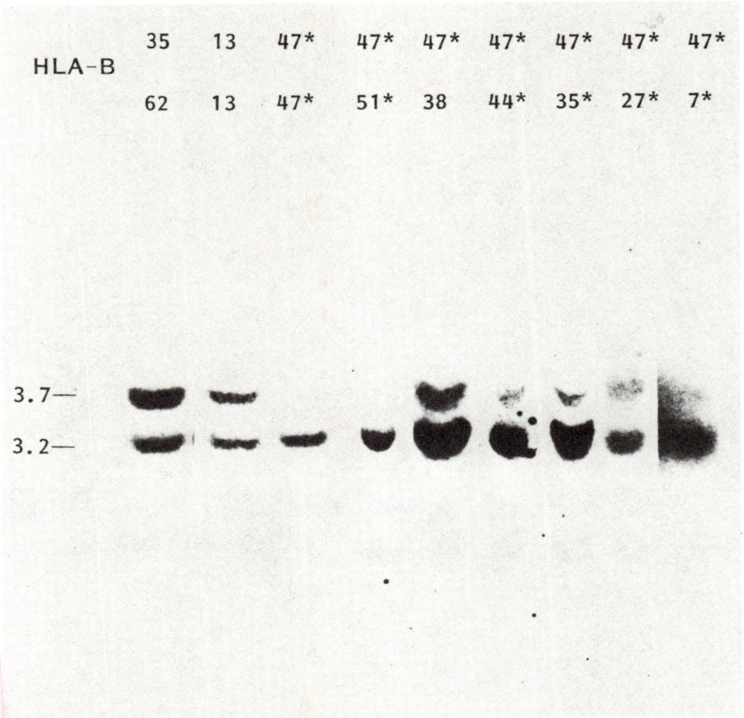

FIGURE 5. Hybridization of pC21a to DNA from additional patients with 21-OH deficiency. DNA from peripheral blood leukocytes was digested with Taq I and analyzed as described in FIGURE 3. Lanes are labeled with the HLA-B antigens carried by each individual; asterisks indicate alleles carrying 21-OH deficiency. The second and third lanes from left are a copy of the Taq I digests displayed in FIGURE 3. DNA from nine additional normal individuals yielded a pattern identical to that in the first and second lanes.

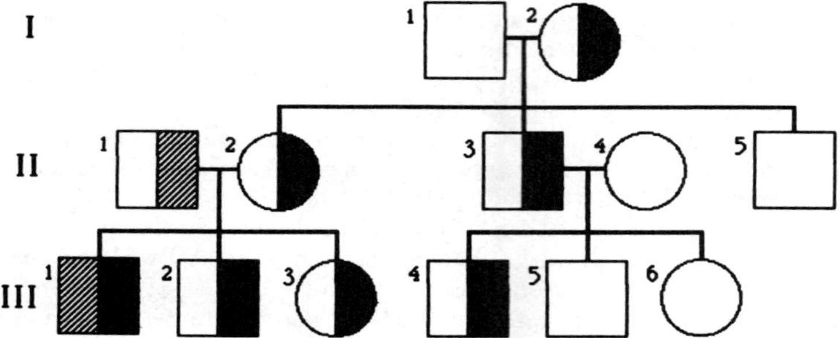

FIGURE 6A. Segregation of Taq I polymorphism in a family carrying 21-OH deficiency (pedigree of the family). The proband is III-1; III-2, III-3, II-1, and II-2 have been identified as heterozygous carriers of 21-OH deficiency by hormonal testing. Black alleles indicate 21-OH deficiency with HLA-Bw47, and crosshatched alleles, 21-OH deficiency with B35.

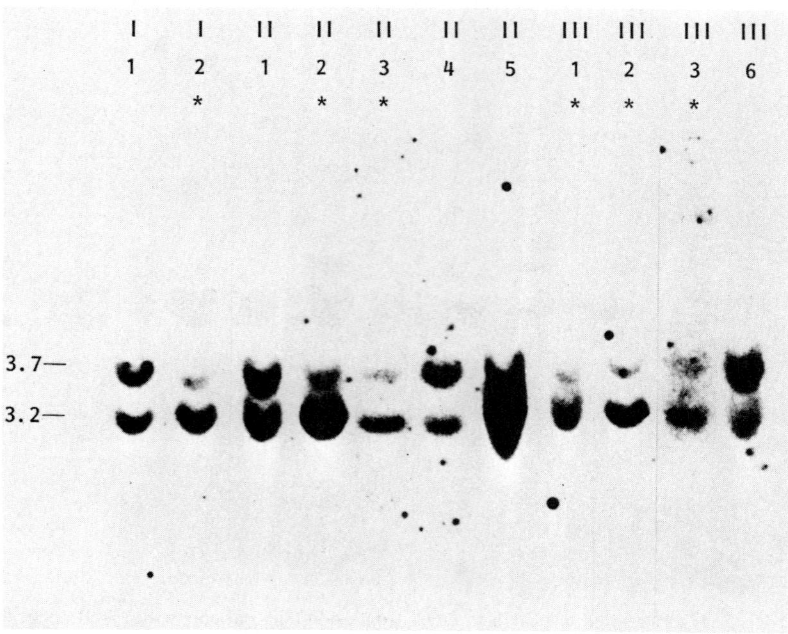

FIGURE 6B. Segregation of Taq I polymorphism in a family carrying 21-OH deficiency; analysis of Taq I digests of DNA from family members, as described in FIGURE 3. Lanes are labeled with pedigree number as defined in FIGURE 6A. Asterisks indicate carriers of the Bw47 haplotype.

FINE MAPPING OF 21-OH GENES

Several human cosmid clones (kindly provided by D. Grossberger and J. Strominger) isolated with a cDNA probe for C4 also hybridize with pC21a. Mapping of these clones shows that there are two 21-OH genes, each located within 2 kb of the 3′ end of the C4A or C4B gene, and carrying Taq I fragments of 3.2 or 3.7 kb, respectively.[16]

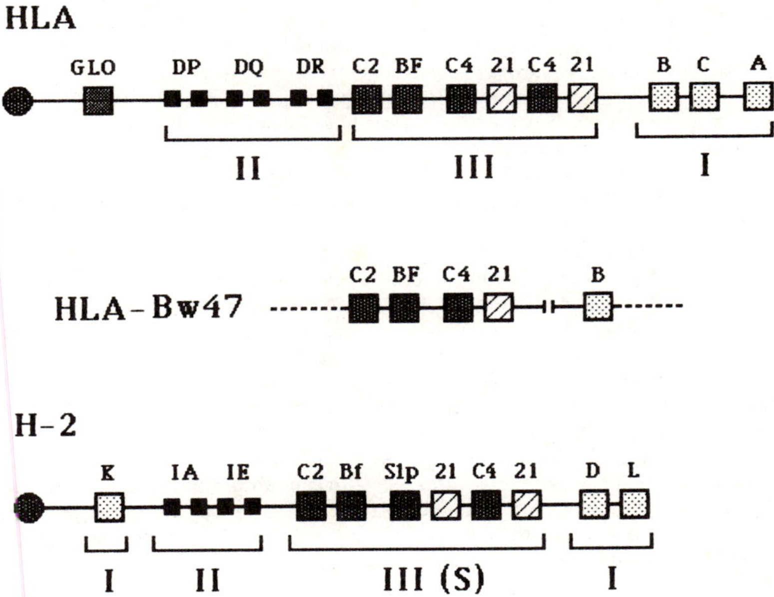

FIGURE 7. Schematic of the major histocompatibility complex in man and mouse. The HLA complex in man is on the sixth chromosome. The gene for the red cell enzyme glyoxalase I (GLO) is centromeric to the HLA complex. The Class I and Class II transplantation antigens are described elsewhere.[27] The Class III genes consist of several components of the serum complement pathway. It is proposed that the two genes encoding the fourth component of complement (C4) alternate with the two 21-hydroxylase genes ("21"). In the chromosome carrying HLA-Bw47, one 21-OH gene and the adjacent C4 gene appear to be deleted. The mouse H-2 complex is similarly arranged, but one Class I gene (H-2K) is centromeric to the Class II genes. The "S" region contains the Class III genes. "Slp" is a hemolytically inactive homolog of C4, expressed only in males of certain strains of mice. The 21-OH genes again alternate with the C4 genes.[18]

The precise arrangement of the C4 and P-450_{C21} genes has also been determined in the mouse where the major histocompatibility (H-2) complex has been extensively studied. A set of overlapping cosmid clones covering the "S" region containing the C4 genes[17] has been probed with pC21a in collaboration with D. Chaplin and J.G. Seidman. We have found that there are also two P-450_{C21} genes in the H-2 complex, one located less than 6 kb from the 3′ end of each of the two C4 genes.[18]

A schematic map of the HLA and H-2 complexes, including the 21-OH genes, is shown in FIGURE 7.

CLINICAL SIGNIFICANCE

It is notable that 21-OH deficiency associated with HLA-Bw47 results from an apparent deletion of only one of two genes. The second gene might be a pseudogene or it might be expressed only at certain times during ontogeny or in other organs. The kidney[19] and liver[20] contain 21-hydroxylase activity, although it is not known how structurally similar these enzymes are to the adrenal P-450.

It will require study of additional patients to determine the nature of the mutations resulting in different disease phenotypes. This information has potential clinical applicability to the prenatal diagnosis of 21-OH deficiency, which is currently performed by HLA typing of the fetus[21] (once the HLA type of the proband is known) and by determining 17-hydroxyprogesterone levels in the amniotic fluid.[22] Both of these tests require an amniocentesis at about 16 weeks of gestation. When a sufficient number of informative DNA polymorphisms have been found, it should be possible to obtain sufficient material by chorionic villus biopsy[23] to offer prenatal diagnosis to many families before the end of the first trimester. If a female fetus is diagnosed as being affected with 21-OH deficiency, treatment of the mother with dexamethasone would be expected to suppress the fetal adrenal gland and thus prevent virilization of the external genitalia.[24]

Since 21-OH genes are present in both the human and mouse major histocompatibility complexes, it is likely that the 21-OH gene(s) became linked to the MHC prior to mammalian speciation. It is not clear if this linkage has functional significance; it is possible that the 21-OH genes have remained in the MHC because of their close proximity to the C4 genes. Alternatively, as all other genes in this region have roles in immune regulation,[25] it is possible that 21-OH exerts an immunomodulatory role by regulating glucocorticoid and androgen biosynthesis. It is interesting that the HLA-B8;DR3 haplotype, which is negatively associated with 21-OH deficiency, has strong positive associations with many diseases with an autoimmune etiology, such as insulin-dependent diabetes mellitus and Graves' disease.[26] Determining the molecular genetic basis of one HLA-linked disease may thus make it possible to study the inherited component of other HLA-associated endocrinopathies.

ACKNOWLEDGMENTS

We thank Ms. Barbara Onufer and Mr. Daniel New for technical assistance. PCW is a Clinical Scholar of the Norman and Rosita Winston Foundation.

REFERENCES

1. New, M. I., B. Dupont, K. Grumbach & L. S. Levine. 1982. *In* The Metabolic Basis of Inherited Disease. J. B. Stanbury, J. B. Wyngaarden, D. S. Fredrickson, J. L. Goldstein & M. S. Brown, Eds. 973–1000. McGraw–Hill. New York.
2. Dupont, B., S. E. Oberfield, E. M. Smithwick, T. D. Lee & L. S. Levine. 1977. Lancet **ii:** 1309–1311.
3. Dupont, B., R. Virdis, A. J. Lerner, C. Nelson & M. S. Pollack. 1984. *In* Histocompatibility Testing 1984. E. D. Albert, M. P. Baur & W. R. Mayr, Eds.:660. Springer-Verlag. Heidelberg.
4. Kuhnle, U., D. Chow, R. Rapaport, S. Pang, L. S. Levine & M. I. New. 1981. J. Clin. Endocrinol. Metab. **52:** 534–544.
5. Pollack, M. S., L. S. Levine, G. J. O'Neill, S. Pang, F. Lorenzen, B. Kohn, G. F.

RONDANINI, G. CHIUMELLO, M. I. NEW & B. DUPONT. 1981. Am J. Hum. Genet. **33:** 540–550.

6. O'NEILL, G. J., B. DUPONT, M. S. POLLACK, L. S. LEVINE & M. I. NEW. 1982. Clin. Immunol. Immunopathol. **23:** 312–382.
7. YANG, S. Y., Y. MORISHIMA, N. H. COLLINS, T. ALTON, M. S. POLLACK, E. J. YUNIS & B. DUPONT. 1984. Immunogenetics **19:** 217–231.
8. KOMINAMI, S., H. OCHI, Y. KOBAYASHI & S. TAKEMORI. 1980. J. Biol. Chem. **255:** 3386–3394.
9. FINKELSTEIN, M. & J. M. SHAEFER. 1979. Physiol. Rev. **59:** 353–406.
10. WHITE, P. C., M. I. NEW & B. DUPONT. 1984. Proc. Natl. Acad. Sci. USA **81:** 1986–1990.
11. WHITE, P. C., M. I. NEW & B. DUPONT. 1984. Proc. Natl. Acad. Sci. USA **81:** 7505–7509.
12. MANIATIS, T., E. F. FRITSCH & J. SAMBROOK. 1982. Molecular Cloning. Cold Spring Harbor Laboratory, Cold Spring Harbor, NY.
13. HANSEN, J. A., S. N. FU, P. ANTONELLI, M. KAMOUN, J. N. HURLEY. B. DUPONT & H. G. KUNKEL. 1979. Immunogenetics **1:** 51–64.
14. WYMAN, A. R. & R. WHITE. 1980. Proc. Natl. Acad. Sci. USA **77:** 6754–6758.
15. SOUTHERN, E. M. 1975. J. Mol. Biol. **98:** 503–517.
16. WHITE, P. C., D. GROSSBERGER, B. J. ONUFER, D. D. CHAPLIN, M. I. NEW, B. DUPONT & J. L. STROMINGER. 1985. Proc. Natl. Acad. Sci. USA **82:** 1089–1093.
17. CHAPLIN, D. D., D. E. WOODS, A. S. WHITEHEAD, G. GOLDBERGER, H. R. COLTEN & J. G. SEIDMAN. 1983. Proc. Natl. Acad. Sci. USA **80:** 6947–6951.
18. WHITE, P. C., D. D. CHAPLIN, J. H. WEIS, B. DUPONT, M. I. NEW & J. G. SEIDMAN. 1984. Nature **312:** 465–467.
19. WINKEL, C. A., E. R. SIMPSON, L. MILEWICH & P. C. MACDONALD. 1980. Proc. Natl. Acad. Sci. USA **77:** 7069–7073.
20. DIETER, H. H., U. MUELLER-EBERHARDT & E. F. JOHNSON. 1982. Biochem. Biophys. Res. Commun. **105:** 515–520.
21. POLLACK, M. S., D. MAURER, L. S. LEVINE, M. I. NEW, S. PANG, H. M. NITOWSKY, G. SACHS, M. DUCHON, I. MERKATZ, R. P. OWENS & B. DUPONT. 1979. Lancet **i:** 1107–1108.
22. PANG, S., M. LOO, M. S. POLLACK, B. DUPONT, O. GREEN, G. CLAYTON & M. I. NEW. 1984. Pediatr. Res. **172A.**
23. WEATHERALL, D. J. & J. M. OLD. 1983. Mol. Biol. Med. **1:** 151–155.
24. EVANS, M. I., G. P. CHROUSOS, D. MANN, J. W. LARSEN, I. GREEN, J. MCCLUSKEY, D. L. LORIAUX, J. C. FLETCHER, G. KOONS, J. OVERPECK & J. D. SCHULMAN. 1985. J. Am. Med. Assoc. **253:** 1015–1020.
25. KLEIN, J., A. JURETIC, C. N. BAXEVANS & Z. A. NAGY. 1981. Nature **291:** 455–460.
26. DUPONT, B. 1983. *In* Recent Progress in Pediatric Endocrinology. G. Chiumello & M. Sperling, Eds.: 171–183. Raven Press. New York.
27. STEINMETZ, M. & L. HOOD. 1983. Science **222:** 727–733.

Index of Contributors

Albert, E. D., 71–75
Alper, C. A., 28–35
Awdeh, Z., 28–35

Balsamo, A., 85–89
Bernardi, F., 85–89
Bidlingmaier, F., 71–75
Bongiovanni, A. M., *ix–x*
Boué, A., 41–45
Boué, J., 41–45
Brkljačić-Šurkalović, Lj., 36–40

Cacciari, E., 85–89
Capelli, M., 85–89
Casey, M. L., 232–237
Cassio, A., 85–89
Cassorla, F., 174–181
Christiansen, F. T., 76–84
Chrousos, G. P., 156–164, 182–192
Chung, B-C., 238–251
Cicognani, A., 85–89
Clayton, G., 111–129
Cobain, T. J., 76–84
Colten, H. R., 269–276
Comite, F., 174–181
Coon, M. J., 216–224
Cordaro, C. I., 85–89
Couch, R. M., 165–173
Couillin, P., 41–45
Crigler, J. F., Jr., 28–35
Cutler, G. B., Jr., 174–181, 182–192

Dawkins, R. L., 76–84
De Felice, M., 46–51
Dumić, M., 36–40
Dupont, B., 111–129, 277–287
Dyas, J., 193–202

Evans, M. I., 156–164

Fahmy, D. R., 193–202
Feingold, J., 41–45
Finlayson, M., 225–231
Fleischnick, E., 28–35
Fletcher, J. C., 156–164
Forest, M. G., 130–147
Fukushi, M., 103–110

Garlepp, M. J., 76–84
Gedeon, A., 76–84
Gerald, P. S., 28–35
Ghizzoni, L., 225–231
Green, O., 111–129

Hall, P. F., 203–215
Höller, W., 71–75
Hors, J., 41–45
Hughes, I. A., 193–202

Inouye, K., 216–224
Irie, M., 103–110

Johnson, E. F., 225–231

Kaštelan, A., 36–40
Kaufman, H., 52–64
Knorr, D., 71–75
Kuttenn, F., 41–45

Laron, Z., 52–64
Levine, L. S., 65–70
Liem, H. H., 225–231
Loo, M., 111–129
Loriaux, D. L., 156–164, 174–181, 182–192

MacDonald, P. C., 232–237
Maio, M., 46–51
Matsuura, N., 103–110
Matteson, K. J., 238–251
Maurer, D. H., 148–155
McCluskey, J., 76–84, 156–164
Mellon, S. H., 238–251
Miller, W. L., 238–251
Morin, J. E., 238–251
Muller-Eberhard, U., 225–231

Naruse, H., 103–110
New, M. I., 1–27, 90–102, 111–129, 225–231, 277–287
Nussbaum, R., 111–129

Pang, S., 90–102, 111–129
Paolini, M., 85–89
Pertzelan, A., 52–64
Pescovitz, O. H., 174–181
Piazzi, S., 85–89
Pirazzoli, P., 85–89
Pollack, M. S., 111–129, 148–155

Rappaport, R., 41–45
Raum, D., 28–35
Robinson, J., 193–202
Roitman, A., 52–64

Salardi, S., 85–89
Scholz, S., 71–75
Schulman, J. D., 156–164
Shimozawa, K., 103–110
Simpson, E. R., 252–261
Spence, D. A., 90–102
Strominger, J. L., 262–268
Stuckey, M. S., 76–84
Suzuki, E., 103–110

Takasugi, N., 103–110
Tsuji, A., 103–110

Valentino, R., 46–51

Walker, R. F., 193–202
Waterman, M. R., 252–261
White, P. C., 277–287
Wilton, A. N., 76–84
Winkel, C. A., 232–237
Winter, J. S. D., 165–173
Winterer, J., 182–192

Yunis, E. J., 28–35

Zamir, R., 52–64
Zappacosta, S., 46–51
Zappulla, F., 85–89
Zuber, M. X., 252–261